History by Algorithms

Zvi Lotker

History by Algorithms

AI and the Future of Historical Research

Springer

Zvi Lotker
Faculty of Engineering
Bar-Ilan University
Ramat-Gan, Israel

ISBN 978-3-031-93626-5 ISBN 978-3-031-93627-2 (eBook)
https://doi.org/10.1007/978-3-031-93627-2

This Springer imprint is published by the registered company Springer Nature Switzerland AG
The registered company address is: Gewerbestrasse 11, 6330 Cham, Switzerland

Preface

Herodotus of Halicarnassus, the Father of History, first recorded history "so that the actions of people will not fade with time, including other things and especially the cause for which they went to war with one another." In our current age of big data, we record history as it happens and soon move on, overwhelmed by unprecedented modern documentation. History is a complex, evolving field that is rapidly expanding both in resources and accessibility. Our stories are buried while they are being written (Fig. 1).

Thus, the need for intelligent machines to understand our personal and cultural narratives becomes more imperative than ever. Having machines understand what our history is and why it happened would allow a dialogue that makes use of man's directing and machine's endless memory [1]. The first step to getting machines to understand narrative was taken in my book, *Analyzing Narratives in Social Networks*.

Fig. 1 Cartoon about a dialogue between man and machine about history

[2] We now propose to continue this journey into the different structure of never-ending stories: Historical narratives.

Overview

The book is divided into three main parts. In Part I, we discuss the mathematical language of history; Part II uses simple models to analyze historical law written in mathematical language; and the Part III discusses the implications of general Large Language Models (LLMs) for the study of history.

Synopsis

For machines to participate in the study of history, it is necessary to develop mathematical tools capable of describing historical narratives. In the first part of the book, we take the first steps toward achieving this goal. Tools such as big-O notation and the relative to history are discussed in Chap. 1.

We continue to develop mathematical tools needed to describe historical narrative in Chap. 2, where we discuss the implementation of Category Theory to the study of history and suggest that the definition of historical concepts should be defined using function diagrams and natural language definition simultaneously. For example, we use a function diagram to define reading against the grain. This approach reveals that reading against the grain actually compares past and present archives In our function diagram, we are forced to discuss the present as the element of comparison of the present to the past. This exact definition suggests that there is a reversed reading against the grain where the past archives try to read the future archives.

In Chap. 3, we develop the mathematical tools for describing battles, a central theme of history. We use the Lanchester differential equation in two-dimensional space-time and extend it for the case where one army surrounds another. This is the central element in the study of history and a typical narrative of great battles. Analyzing the Battle of Cannae led by Hannibal, we were able to study it in greater detail than ever before. Calculations were made using the extended Lanchester equation. The mathematical tools used in this chapter are differential equations, which are basic tools in exact science, but are rarely used in studying history.

In Chap. 4, we extend the concept of the counters into clocks. In many digital history research books, the counters are heavily used. The standard methodology is to define subsets of words and then count their appearance in the documents. From that, the researchers tend to conclude some facts about the document. However, there is no methodology for comparing these counters, and what is their meaning from a mathematical point of view? We identify those counters with clocks and develop the theory of many clocks with respect to historical narratives. We show how to find the time when the clocks deviate from each other. This method allows machines to point to the critical event in the historical narrative. This concludes Part I of the book, which develops the mathematical tools for historical narratives.

In Chap. 5, we introduce the concept of Networks and Macrohistory. We start with the definition of a network and provide several examples of graphs and their basic data structures. Then, we move to discuss some basic computational geometry with application to history. We end up this chapter with a causal graph.

In Chap. 6, we introduce Microhistory. We use social networks as a mathematical model that describes society. We develop several rules of thumb on social networks, which allow us to resurrect ancient networks only from the seizure of societies. This allows us to compare ancient social networks and compute the meaning of the destruction of old cities such as Carthage by the Romans. A social network is a mathematical model that describes the fabric of society and is therefore useful when one tries to model historical events.

Chapter 7 is the introduction to Part II of the book, where we discuss the implications of the historical narrative on mathematical language. The following chapters will provide several examples of historical processes that can be modeled using simple mathematical models and provide windows to the mathematical laws of history.

Chapter 8 discusses the historical question: Is the history of the core similar or different from the history of the periphery? This question is central to the philosophy of history since most of our data related to historical revolution/process is of the core, and we are missing the data of the periphery. By modeling the relationship between the core and periphery through social networks, we are able to categorize social revolutions and provide critical conditions on when the history of the core is relevant to the history of the periphery.

Chapter 9 provides an example of historical law that focuses on the author instead of the historical event. The law says that any tremendous historical figure needs an antipodal figure that will be compared to him and will be worth it. This law comes not from history but from the historical narrative. Therefore, the author of history tends to pick evil characters in history after the witness disappears. In the context of philosophy, this uses the fact that according to Kant, when we analyze reality, we always have to go through the perception of reality. In science, there is always reality and perception of reality. This chapter also discusses the connection between history and distributed computing. It mainly analyzes the ability of society to reach an agreement after the witnesses die through game theory and shows that it is necessary for collective memory to form the historical canonical text. Without the existence of a canonical historical text, the narrative will evaporate.

In Chap. 10, we analyze the conversion of the averaging process in the collective memory formation game and show that if the nodes have a correct categorization of their neighbors' opinions, the average process converges to the Nash equilibrium and can be used as a separate scheme that defines the different historical entities. A simple example is gun control in the USA. If you know a person's opinion about gun control, you can glean their political opinion.

In Chap. 11, we analyze the collective memory formation game in the case where some nodes have the wrong categorization of their neighbors' opinions. In such instances, the average process converges to zero opinion, and the historical event becomes unimportant. The different parties cannot disagree on the historical event using just the averaging process as the separation process. However, if we wish to use the historical events as a separation event, we need to move the averaging process away from zero. This can be done by amplifying the probabilities in each step to amount to 1. This suggests that propaganda is necessary to use such events as separation.

In Chap. 12, we model a stochastic terrorism network and explain the mechanism that generates random violence events using queuing theory.

Chapter 13 defines the general historical machine and sets the scene for generating history by machines. In the chapter, we define the relationship between historical entities, machines that write history, and the historical narrative. The central idea in this chapter is universal consciousness, which is the subject of historical narratives such as society, state, country, and more. This ends the second part of the book.

In Chap. 14, we begin the third part of the book, which explains how history transforms from text-based science to science based on information.

In Chap. 15, we provide a general introduction to machine learning and the way AI can be used in the study of digital history. The chapter is divided into two parts. The first part is a general discussion of AI and the way historians may use it. The second part discusses LLM and prompt engineering in general.

Chapter 16 deals with the new ways in which historical narrative can be represented. We move from textual-based to video-based historical narrative. The chapter provides several algorithmic methods to take any text and transform it into a video.

Chapter 17 explains how to use LLM as an expert system in historical text analyses. We provide several examples of expert systems, such as psychiatric experts, political science experts, and emotion analysis experts. We develop a method that takes any historical document, breaks it into paragraphs, transforms it into numerical values, aggregates those numbers into functions, and runs standard analysis tools on those functions. To demonstrate this method, we provide an analysis of Hitler's Mein Kampf [3] and Churchill's My Early Life [4].

In Chap. 18, we develop a methodology to analyze historical videos together with machine learning tools to study history.

Lastly, Chap. 19 discusses Fake History from the perspective of the philosophy of science. This chapter provides several rules of thumb for Fake History and its evolution.

Download Materials

Code and data from the book are available at https://github.com/zvilo/History-by-Algorithm. Any future updates to the materials will be made directly in the same GitHub repository.

Why Should You Read This Book?

Most of the books written about digital history synthesize the philosophy of history with technical details from computer science. However, digital history borrows many philosophical aspects from computer science.

According to McLuhan [5], the medium is the message. If that is the case, in digital history the message is the machine, since the machine is the medium. Therefore, to understand digital history, one needs to understand the philosophy of the machine.

This book tries to demonstrate how the philosophy of the machine, as developed by computer scientists, influences the study of digital history. Historians face technical difficulties when encountering the language of computer science for the first time. However, we believe that historians who wish to work in the area of digital history should be familiar not only with the philosophy of history but also with the philosophy of computer science, since digital history is the fusion of both disciplines.

To bridge the language barrier between historians and other readers with a humanities background, we provide several appendices filled with prompts that can be used to converse with modern AI tools. We believe that these prompts can ease the process of bridging the language gap, making the whole process possible.

Those with a background in exact sciences are much more familiar with the mathematical language used in the book. For them, reading the book should be easier. We hope that it will prompt them to develop the exact science of digital history.

One of the advantages of the book is that it exposes the reader to a variety of toolboxes from exact science that can be used during research in digital humanities. Most of the books in digital history study history using computer science programming languages. For example, many works are based on counting the frequency of specific words in historical texts. For now, let's call them "counters". However, none of them ask what the mathematical properties of these counters are and how we can compare them. The purpose of this book is not to analyze a historical text, but rather to develop the mathematical theory needed for doing digital history.

The main difference between exact science and the humanities is that while exact science tries to predict, the humanities try to interpret. This is the major obstacle when performing digital humanities. The problem is that once we transform the historical document from words to numbers, we may forget that we are still talking about interpretation, not prediction. Numbers tend to have a unique interpretation, and therefore the research tends to lose the freedom of interpretation. Researchers in digital humanities need to remember that in order to save the freedom of interpretation, we are forbidden to fully define how to transform the text into numbers. We should never forget that when we cross the bridge from text to numbers, we perform interpretation, and different bridges will generate different numbers, and therefore, we will have different interpretations. By understanding that transforming the text to numbers is an interpretation, we maintain the desired freedom of interpretation.

The book can be used as a textbook for a single-semester advanced course on digital history in the Faculty of Engineering.

Zvi Lotker
Faculty of Engineering
Bar-Ilan University
Ramat-Gan, Israel

References

1. Herodotus and Alfred Denis Godley. *Herodotus: With an English Translation by AD Godley*. W. Heinemann, 1961.
2. Lotker Zvi. *Analyzing Narratives in Social Networks Taking Turing to the Arts Authors*. Springer, 2021.
3. Adolf Hitler. *Mein kampf*, volume 1. Motilal Banarsidass, 2014.
4. Winston Churchill. *My early life*. DigiCat, 2022.
5. Marshall McLuhan. The medium is the message. In *Communication theory*, pages 390–402. Routledge, 2017.

Acknowledgements Many people contributed to the making of this book. First and foremost, I would like to thank Dr. Roman Fedorenko for the tremendous amount of time, thought, and effort he devoted to this project. His support, critical insights, patience, and hard work were invaluable—this book would not exist without him.

I am also grateful to Rotem Alter for carefully proofreading the manuscript and correcting numerous language mistakes, to Maya Sharon for proofreading, and to Gaya Golan, who helped me with the early version of the book.

For the beautiful illustrations that open each chapter, I owe special thanks to Raffael Blumenberg. Our conversations during the creative process of these figures also led to valuable insights about the book itself.

Finally, I want to thank my partner, Hamutal Tsamir, for her thoughtful discussions on history, and my beloved children, Hillel and Daniel Lotker-Hassine, for their patience and understanding throughout this journey.

Competing Interests The author has no competing interests to declare that are relevant to the content of this manuscript.

Contents

About the Author

Zvi Lotker is Full Professor in the Faculty of Engineering at Bar-Ilan University, Israel since 2018. He graduated with a double B.Sc. in Mathematics, Computer Science, and Industrial Engineering, from Ben Gurion University in 1991. In 1997, he obtained an M.Sc. in Mathematics, and in 2003, obtained his Ph.D. in Distributed Algorithms, both from Tel Aviv University. He was Postdoctoral Fellow at the CWI in Amsterdam, MPI in Saarbrücken Germany, and Mascotte in Nice France from 2003 to 2006. He was Full Professor in the Communication Systems Engineering department at Ben Gurion University in Beer Sheva in the Communication Systems Engineering department from 2010 to 2020. His main research areas are communication networks, online algorithms, sensor networks, and recently, social networks.

List of Figures

Part I
The Mathematical Language of Digital History

Chapter 1
Asymptotics in Digital Humanities

1.1 Introduction

The field of digital humanities is aptly named for its basis in both computer science and the humanities. The humanities have given us concepts of society and culture such as democracy, perspective, significance, and trade. In the meantime, computer science has contributed concepts of analysis and presentation, such as data structures, engineering, algorithms, and recursion. Digital humanities synthesize both parent fields, fusing its methods, perspectives, and vocabulary. When researching the digital humanities, it is crucial to understand both sides of the fusion.

One central concept in computer science is *big-O* notation. Big-O notations are used to describe the behavior of an algorithm. Specifically, they describe how an algorithm performs at its extremes. As an algorithm's input grows larger and larger, big-O notations describe how computational space and run-time are affected. Big-O notations classify different algorithms by asking the question: "As the input approaches infinity, what resources does the algorithm require?".

Building algorithms within the framework of big-O frees their relevance from hardware limitations. Moore's Law states that the number of transistors in a dense integrated circuit doubles every two years or so. This prediction has been maintained since 1975, and, though the pace of transistor doubling has fallen since the 2010s, Moore's Law continues to guide the production of modern technology and algorithms. As a result, the computing capabilities of machines have skyrocketed over the decades. It would be unfair to directly compare two algorithms' behaviors if one ran on a modern machine while the other lagged on an older model. With big-O notation, however, both algorithms are classified by the limiting behavior at their extremes. Two algorithms of the same order are categorically the same, whether they run on a machine from 1975 or 2025.[1]

[1] Even if Moore's Law becomes irrelevant, the notion applies to anything that amplifies machine resources. For example, the size of the machine.

Z. Lotker, *History by Algorithms*,
https://doi.org/10.1007/978-3-031-93627-2_1

Human society is experiencing changes in the same way. Historian and political scientist Benedict Anderson, best known for his influential book on nationalism *Imagined Communities*, explains how the evolution of mass communication technology enables the formation of modern nations [1]. As European societies transitioned through mass printing, newspapers, radios, and televisions, their concept of nationality evolved and shifted. If an algorithm based on the effects of nationalism were to be defined, it would be heavily affected by the technological limitations of its period.

It is possible to write laws in the humanities that are based on big-O so that they can be flexible enough to correspond with the changes of humanity's capabilities [2]. Technology changes humanity. For example, it is unfair to compare the travel capabilities of humans before and after the invention of the train. There, big-O notations are the perfect tool for the digital humanities, allowing us to work better with the "hardware" and "software" of human societies throughout historical change by ignoring the constant and putting the emphasize on the order of magnitude of the events.

1.2 Related Work

Big-O notation is a central concept in the study and analysis of algorithms. There are many textbooks that discuss that concept. We recommend the book [3]. Another great book that teaches big-O notation is [4]. One can have a great discussion with LLM on this subject, or use the Wikipedia [5] and Wolfram mathworld [6] databases as well.

1.3 Definition

1.3.1 Big-O Notation

Let the function to be estimated $f : \mathbb{N} \to \mathbb{R}$ be a real-valued function and let the comparison function $g : \mathbb{N} \to \mathbb{R}$ be a real-valued function. Let both functions be defined on some unbounded subset of the positive real numbers, and $g(x)$ be strictly positive for all large enough values of x. We will write

$$f(x) = O\left(g(x)\right) \tag{1.1}$$

if the absolute value of $f(x)$ is at most a positive constant multiple of $g(x)$ for all sufficiently large values of x. That is, the above equation is true if a positive real number M and a real number x_0 exist, such that

$$|f(x)| \leq Mg(x) \quad \text{for all } x \geq x_0. \tag{1.2}$$

This allows us to state the upper bounds of a function. In many contexts, the assumption that we are interested in the growth rate as the variable x goes to infinity is left unstated.

1.3.2 Big-Omega Notation

A variant of big-O notation, *big-Ω* notation allows us to state the lower bounds of a function. We will write

$$f(x) = \Omega\,(g(x)) \tag{1.3}$$

if the absolute value of $f(x)$ is at least a positive constant multiple of $g(x)$ for all sufficiently large values of x. That is, the above equation is true if $g(x) = O\,(f(x))$.

1.3.3 Big-Theta Notation

Finally, *big-Θ* allows us to state the upper *and* lower bounds of a function. We will write

$$f(x) = \Theta\,(g(x)) \tag{1.4}$$

if

$$f(x) = O\,(g(x)) \text{ and } g(x) = O\,(f(x))\,. \tag{1.5}$$

1.4 The Problem with Big-O

The main problem with big- O is that its definition needs to be more coherent. Our interest lies in the behavior of the functions as their variables approach infinity. As we grow theoretical variables to extremely large or extremely small values, any constants that modify the equations are overwhelmed by higher-order polynomials. In practice, there is an equally infinite intermediary area in which, given more than one changing variable, the function's behavior is less predictable. Therefore, building a bridge between theory and practice isn't easy.

One way to solve this difficulty is to construct a model that explains the phenomenon of interest. After verifying the model using the scientific method, the asymptotic of the phenomenon can be analyzed.

Later in this chapter, we use this approach to demonstrate the use of big-O notation in explaining several interesting cases in history.

1.5 Elites in Social Networks

Human societies always form in layers. Tribes have their chiefs and members, kingdoms have their nobility and commoners, and modern nations have their rich and poor. The power dynamics may be further stratified to get more complex class structures, but the simplest nontrivial form of society throughout history can be viewed as that of core and periphery (Fig. 1.1).

We tend to think about the ultra-rich and the ultra-powerful as composing the top 1% of society. But as society grows larger, does that powerful hundredth remain a constant ratio? New articles pop up every few years echoing the Matthew principle which states that the rich get richer and the poor get poorer. If the rich condense their wealth as society grows, how can their growth remain linear?

A 2018 publication, *Elites in Social Networks: An Axiomatic Approach to Power Balance and Price's Square Root Law* [2], uses the big-O notation to examine this assumption.

Within social networks, an *elite* is a small group of nodes at the network's *core* that possess a disproportionately sizable degree of influence compared to the rest of the nodes. The other, non-elite nodes are known as the periphery: the commoners that support the 'nobility' at the center. Society naturally tends to orbit around those with the most resources. Think of past royalty enforcing the laws they passed with troops they funded; think of today's celebrities, influencers, and billionaires that guide our culture, government, and economies.

Avin et al. explore the influence that an elite core has on the rest of society using a "core-periphery" social network model. They link the concept to Price's square root law, a pillar of the qualitative study of science and scientific literature [7, 8]. *Price's law*, based on Lotka's law and echoing other power law distributions such as

Fig. 1.1 A cartoon of the power relations between the core and the periphery

Zip's law and the Pareto distribution [9, 10], claims that the square root of the total number of scientific authors writes makes up half of all published scientific papers. This means that if there were 1000 scientific papers in existence written by a total of 400 authors, then 20 of those authors would be responsible for the publication of 500 of those papers. The 20 authors around whom the scientific field orbits are the core "elite", or as Price dryly calls them, the "good scientists."

Of course, this is a rule of thumb. Power law distributions are proportional relationships that, especially when dealing with the humanities, serve best as approximations. Like big-O notation, it is the behavior of the function as it tends to the extremes that we care about; an order of magnitude can provide more relevant information than an exact amount.

Aside from the leading members of the scientific writing community, we can look to Price's law for finding the core of any group and any social network.

Rather than calculating the square root of the network's nodes V, Avin et al. found that the equivalent of Price's law within social networks was the square root of the network's edges m. In their paper, they use Price's relationship to look at Marvel comic superheroes and social media sites such as Buzznet and Flixster. It confirms their theory.

Later in this chapter, we will use the axiomatic core-periphery model to look at elite-troubled historical societies. Did Ancient Rome or Revolutionary France have an issue with the 1%? Or was it a different ratio that started the evolution of their collapse?

In the example of Price's law, half of the total papers—either the half written by the small group of elites or by the rest of the peripheral authors—represents a point of symmetry that Price called a 'balance point' [11]. This balance point is a metaphorical boundary that optimizes the separation of the elite from the periphery. It is the point at which the elite says "No thank you, we have enough team members, the rest of you are released." A social network with a balance point partition maximizes the elite-periphery influence, though the hard partitioning is an algorithmic need and not a reflection of any historical barriers of entry.

Instead of scientific papers, the currency of society is a construct of influence. Avin et al. consider influence to be a "measurable quantity between any two groups of the population." It could be money, social pressure, military power, or any other persuasive force. Influence is a directional stream that contains information, and information can be measured. The paper denotes it as I, with external influence from group X on group Y represented as $I(X, Y)$, and internal influence within group X represented as $I(X, X)$. We will get a glimpse into those streams in later chapters of the book.

Influence has additive, symmetric, and bounded properties, as well as the constant self-influence property in which each individual x exerts influence $I(x, x) = 1$ on itself. The total amount of influence inside a network is denoted by

$$I_{tot} \equiv I(V) = I(V, V) .$$

We can say that a social network is "elite-centered" if it allows for the three axiomatic properties that are defined in the paper: elite dominance, elite robustness, and elite density. We will now discuss how each property is defined and how to quantify and measure it in regards to an elite $\hat{E}$ and a periphery $\hat{P}$ in a core-periphery partition $\hat{E}, \hat{P}$.

Dominance: Elites control the population they are in, and their influence on the periphery is stronger than that of the periphery on itself. Elite dominance may be calculated as

$$\mathrm{dom}(\hat{E}) = \frac{I(\hat{E}, \hat{P})}{I(\hat{P}, \hat{P})}, \tag{1.6}$$

with an elite-centered network obeying the axiom:

(A1) Elite-dominance: $\mathrm{dom}(\hat{E}) \geq c_d$, for a fixed constant $c_d > 0$

Robustness: Elites are harder to influence from the periphery. Their internal influence holds stronger than that of the periphery of their group, keeping the elite class' cohesion. Elite robustness may be calculated as

$$\mathrm{rob}(\hat{E}) = \frac{I(\hat{E}, \hat{E})}{I(\hat{P}, \hat{E})}, \tag{1.7}$$

with an elite-centered network obeying the axiom:

(A2) Elite-robustness: $\mathrm{rob}(\hat{E}) \geq c_r$, for a fixed constant $c_r > 0$

Density: The average elite node has more influence than any given periphery member. Given that

$$\delta(X) = \frac{\log I(X)}{\log|X|}$$

denotes the log-density of a set $X \subseteq V$, density for an $(\hat{E}, \hat{P})$ partition may be calculated as

$$\mathrm{dens}(\hat{E}) = \frac{\delta(\hat{E})}{\delta(V)}. \tag{1.8}$$

An elite-centered network would obey the axiom:

(A3) Elite-density: $\mathrm{dens}(\hat{E}) \geq 1 + c_c$, for a fixed constant $c_c > 0$

So, what is the growth behavior of elites in a social network in which they are influentially dominant, resistant to influence from outside their class, and composed of individuals with higher per capita influence than the periphery? In a bounded, influence-symmetric network, their modeled growth is not a constant 1% of the population. It is sublinear.

Avin et al. formally describe it as

Theorem 1.1 (Elite Size) *Let $(\hat{E}, \hat{P})$ be a core-periphery partition that satisfies the dominance, robustness, and density axioms (A1), (A2), and (A3). In that case, the elite size is* sublinear *in the size of society, namely,*

$$c \cdot n^{\frac{\delta(V)}{\delta(\hat{E})}} \leq |\hat{E}| \leq n^{\frac{1}{1+c_c}}.$$

They go on to provide proof of how common this type of growth is among social networks. As we shall see later, history also provides proof of the sublinear growth rate of elites in elite-centered societies.

Do we ever see stable linear growth? Yes, but only in networks where the density property does not exist.

There is a possibility for such a network; The axioms are independent. According to the Axiom Independence theorem in the paper, assuming any two of them does not imply the third. Following that, there does exist an influence-symmetric network partitioned in a way that fulfills the dominance and robustness axioms, but not the density axiom. This makes for a consistently dense network, where most nodes know each other. In such a network, the core elite would grow at the same rate as the population, potentially reaching larger core sizes than would be possible in core-periphery networks.

Formally,

Lemma 1.1 (Large Cores) *Consider an influence-symmetric network. Let $(\hat{E}, \hat{P})$ be a core-periphery partition that satisfies the dominance (A1) and robustness (A2) axioms. If there exists some constant c, s.t.*

$$I(\hat{E})/|\hat{E}| \leq c \cdot I(V)/|V|,$$

then $|\hat{E}| \geq c_3 \cdot n$ for some constant c_3.

Alternatively, in unbounded networks without influence-symmetry, the elite could maintain their dominance over the periphery while staying a consistent size. The society could grow indefinitely, and the number of elites would remain the same, as their relationship with the masses became more and more lopsided to maintain the power balance. In such a network, the core elite could be much smaller than would be possible in core-periphery networks of the same size.

Formally,

Lemma 1.2 (Constant Elite Size) *In an asymmetric or unbounded network, there may exist core-periphery partitions $(\hat{E}, \hat{P})$ that satisfy the dominance, robustness, and density axioms (A1), (A2) and (A3), where the elite $\hat{E}$ has a constant number of members.*

How come elite numbers in core-periphery networks do not balloon up when more resources are made available, or shrink down for unimaginable personal wealth? Avin et al.'s paper proposes a fourth, supplementary, property of elites.

The fourth property of elites is compactness. Elites seek to maximize their own personal power and status, as well as that of their entire class. These two interests

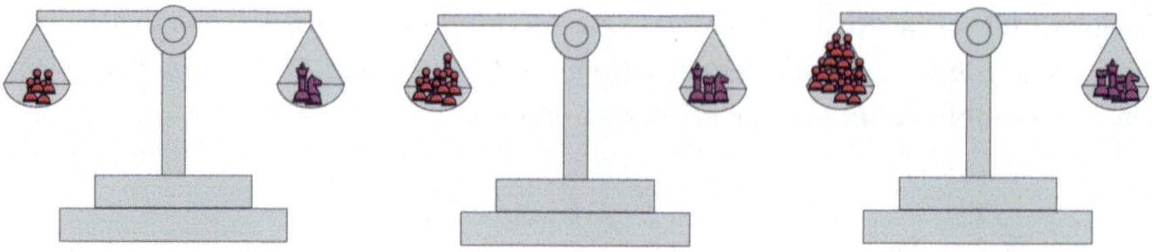

Fig. 1.2 This figure demonstrates the relationship between the core and periphery populations, showing that the relationship is not linear but rather sub-linear. The scales show that to maintain the balance between the two populations, the core must grow at a slower rate than the periphery

conflict; a larger elite group would increase the dominance and robustness of the class but would decrease each individual's resources and status. The supplementary property of compactness states that there is a minimal group size for the elites to maintain their control of the periphery while maximizing their own individual status.

Formally,

(P5) Elite-compactness: The elite is a *minimal* set of individuals that satisfy the dominance and robustness axioms (A1) and (A2).

So, when a society grows larger and more nodes can be influenced, elites will increase their portion of the population, though at a slower rate than society as a whole. The larger the society, the larger the elites' individual influence pools (Fig. 1.2).

But what happens when the personal interests of the elites win out against their class' need for continued dominance? When the elite are extremely powerful but too small to dominate the rest? Or when the elite dilute their personal wealth as their class status spreads across society? What happens when the core size is no longer the $\sqrt{m}$ ratio?

As history has shown us: social upset. Later in this chapter, we will continue to work with big-O notation based on the Avin et al. paper, and discuss specific examples of when societies lost their balance points and reestablished them through revolution.

1.6 Machine Learning and Asymptotics

As mentioned above, big O notation exists in an asymptotic concept. However, when it comes to history, we have a set of finite numbers. To overcome this problem, we propose models that allow us to generate an infinite sequence of numbers and calculate their asymptotic. In this section, we provide a different approach to tackling this problem through machine learning.

Machine learning technology has several general-purpose algorithms that can solve common tasks such as prediction, partition, and clustering.

In the prediction problem, one is given a finite sequence or a single number, and the machine learning algorithm tries to predict the single number from the input. In Wolfram Mathematica, this machine learning process is implemented by the function "Predict." We can therefore construct a network that predicts the asymptotic of a finite sequence.

Clearly, this process does not make sense in general, since any finite sequence of numbers can be extended to many different functions with different asymptotics. However, since life in general and history in particular enjoys locality and continuity, machine learning technology tends to perform very well on actual data.

1.7 Exponential Growth in Non-Linear History

In many cases, we are interested in analyzing parts of society. For example, what percentage of the exponentially growing population amounts to the elderly? Or what portion of a society in which the economy is growing exponentially is wealthy.

A mistake that even professionals make is to work in linear percentages and ignore big-O notations when dealing with exponential growth. A percentile comparison between groups of exponentially differing sizes is not an accurate analysis of their differences. Such errors can even be found in well-recognized works such as Thomas Piketty's *Capital in the Twenty-First Century* [12].

Given the set of all the people

$$pop(t) = \{1, 2, \ldots, 2^t\} \tag{1.9}$$

in an exponentially growing society, i.e.,

$$|pop(t)| = 2^t \tag{1.10}$$

we can define the birthday of a person i to be

$$birthday(i) = \log_2(i). \tag{1.11}$$

Now we can define the age of a person i at time t to be

$$Age_i = t - birthday(i). \tag{1.12}$$

The elderly population in society can be defined as

$$old = \{i \in pop : Age_i > t/2\}, \tag{1.13}$$

leading to the amount of elderly individuals to be

$$|old| = 2^{t/2}. \tag{1.14}$$

Therefore, the elderly population is never a percentage of the overall populace, but a root, i.e., $O(size(t)^{0.5})$ or $O(size(t)^{(\alpha)}$. The following theorem summarizes this behavior quite simply: The rich get richer and the old get older.

Theorem 1.2 *For all r there exists t_0 such that the number of elderly people is less than $r size(t)$ for all $t > t_0$.*

Proof By the definition of elderly, we know that the percentage of elderly at time t is

$$r(t) = \frac{1}{2^{t/2}} \tag{1.15}$$

Therefore at time the theorem follows if we select t_0 to be:

$$t_0 = \frac{2\log\left(\frac{1}{r}\right)}{\log(2)} \tag{1.16}$$

This logic can be applied to any exponential matter in our historical studies.

1.8 Examples Throughout History

Up until now, we have discussed the theory of core-periphery. Now we would like to move from theory to practice. As mentioned earlier, caution should be exercised when trying to apply the idea of asymptotics to history, especially since asymptotics work only on models and not on specific moments in time. However, it is interesting to see how simple calculations can shed some light on famous revolutions.

In the next few paragraphs, we will discuss three famous historical revolutions: Rome's transformation from republic to empire, the French Revolution, and the Russian Revolution.

The older the society, the more difficult it is to estimate its core. To overcome this obstacle, we use two methods. One is a crude estimation and the other is sampling the elite using Wikipedia. More on that technique can be found in the next section.

We use the following core-periphery rule of thumb: In a balanced society with a population size of n, the elite/core is the size of $\sqrt{n}$. When this rule is drastically broken, it is a cause for revolution. In the following examples, we analyze the history in regards to big-O theory using this simple rule of thumb.

1.8.1 Rome: Republic to Empire

In the Roman Republic, the Senate was the governing body. The toppled monarchy had 300 senators; the republic started out with a similar number, though with vastly different characteristics. Sulla expanded the senate to 600 members, Caesar raised the cap to 900, and Augustus reined their numbers back down to 600. We will give a historical background and explain how these numbers tie into our core-periphery model.

The Senate was the ruling body of the Roman Empire, but it did not care for the empire as much as it did for its base in Rome.

Rome's problems started with the end of the Second Punic War in 201 BC. It was at that time that Rome became an unrivaled Empire; there was no external force threatening it, no foreign enemy to unite the disparate classes against it. Italians who supported Rome's enemies during the war had their lands seized by the senate. The fields lay untended, restricting agricultural production and creating friction between the large pleb population and the wealthier patricians.

This conflict occurred both in the times of Sulla and Caesar. While Sulla stood with the patricians and Caesar with the plebeians, both took the same political action of expanding the size of the Senate. Both felt that Rome's core needed to expand.

During Sulla's rule, the population of Rome was estimated at about $pop = 400,000$ men. Note that $\sqrt{pop} = 632.4$. During Caesar's rule, the population of Rome was estimated at about $pop = 4,000,000$ men. In this case $\sqrt{pop} = 2,000$. In both cases, we see that no matter the side of the plebeian-patrician struggle, the politically stabilizing move was to increase the core size to be closer to $\sqrt{n}$. [13, 14]. Figure 1.3 demonstrates the shift in the balance of power in Roman society in the transition from Republic to Empire in the years 100BC-33BC.

Census numbers have been the topic of academic discussion for generations. Historians debate over which census included women and children, whether slaves were inventoried in the records, and to which geographical border the census spanned. What we can take from this information are its trends: The population grows, and the elite core must grow to match their influence.

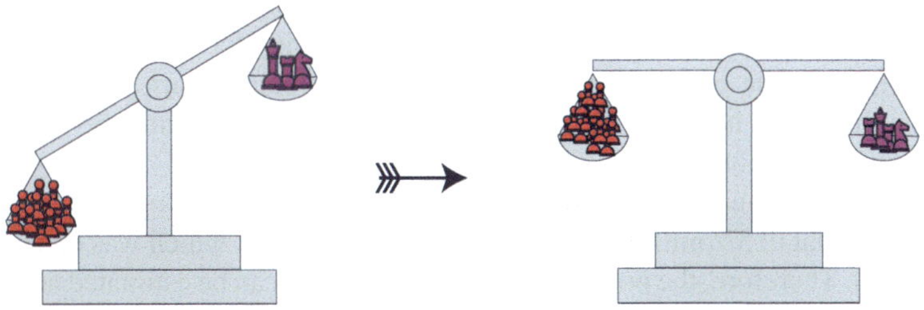

Fig. 1.3 The relationship between the Roman core and periphery populations when the elite (senate) was too small to control the periphery, and the resulting increases in the senate's size in attempts to remedy the situation

1.8.2 The American Revolution

During the 18th century, England controlled the American colony without giving them any place in the English Parliament. The Americans could not vote in the English Parliament; in fact, they were under the protection of the English monarch.

Instead, all the colonies had their own institutions which were subservient to English law. English law had three political institutions which together controlled England and its judicial system: the Monarch, the House of Lords, and the House of Commons.

The British House of Commons totalled 558 Seats [15] The House of Lords constituted of approximately 300, see the book [16].

At the time of the American Revolution, the population of England was roughly 8 million, and the population of the American colony was about 2.5 million, [17–20].

For simplicity's sake, we will ignore the rest of the British colonies at that time. Now, let us use our model of society with respect to the core and the periphery and try to evaluate the opinion of the core and the mixed society that includes the British and Americans, a total of 10.5 million individuals. Therefore, the core size is therefore 3,240. Now, let's assume that all people belonging to the American core oppose additional taxation of the colonies. The size of the American core is roughly 1,581. Let's compute the percentage of those in the core that object taxation:

$$1581/3240 = 0.48795 < 0.5 \tag{1.17}$$

Note that the percentage of Americans in the core is almost 50%, while the total population is almost a quarter. This follows the fact that the core is sub-linear. People in the core can represent several minorities or groups. The fact that Americans, even represented in the core with 48.795%, were too few to stop the taxation process.

However, looking at the speed at which the American colony grew, we notice that its population doubled between 1770 and 1775 [20]. Therefore, assuming the same speed of growth, the estimated population of the American colony should be about 5 million. In reality, however, the war slowed down growth rates and the population only reached 5 million in 1800.

Estimating England's population growth, we can compare the population in 1750, which was equal to 5.7 million, with the population in 1775, which was equal to 6.7 million [20]. Therefore, the population growth per year r can be estimated according to

$$r^{25} = \frac{6.7}{5.7} \tag{1.18}$$

Solving the equation above gives us the growth rate of the English population equal to $r = 1.0064866$, or roughly 1% per year. Now we can estimate England's population in 1780:

$$6.7r^5 = 6.92014 \tag{1.19}$$

Therefore, the total population in 1780 is estimated by 12.0709 and the core size is $12.0709^{0.5}$.

We can see that by continuing the trend in 1780, the American colony would have had enough power to influence the decision in favor of their interests. We did all the calculations for the core, i.e.,

$$5^{0.5}/(12.0709)^{0.5} = 0.643599 > 0.5 \quad (1.20)$$

These numbers suggest that 1776 was the last time Britain could have tried to maintain the American colony in the core opinion.

1.8.3 France: Viva La Revolución

Next, we move to the Enlightenment period, where population data is more accessible. In 1789, when the French Revolution kicked off, the number of citizens in France was $pop = 28,000,000$ [21]. According to our core-periphery rule of thumb, the size of a stable elite in such a population amounts to $\sqrt{pop} = 5,291.5$ individuals. According to French historian François Bluche, the number of nobles sat around 140,000 [22]. Others, such as Gordon Wright, cite larger numbers. We have chosen the minimal estimation to show that even with the strictest definition of the elite, the core was simply too large to maintain.

The move towards stabilization required a reduction in the size of the elite.

We will take this moment to make a remark on the utilization of linear relationships instead of polynomial relationships when viewing history. Looking at the elite population of revolutionary France through a linear lens, one can see that only about 0.5% of the population belonged to the elite. That seems like a reasonable amount—less than one percent. Linearly, France had one of the smallest elite classes in Europe at the time.

The issue is that compared to its contemporaries, France was larger and more populated. Linear approximations for polynomial relationships can be confusing, especially as their margin of error grows alongside the amounts.

Calculating the top 1% of a town of 100 will return the 1 individual who has amassed their wealth. Taking the $\sqrt{100}$ will yield the 10 elite that control the territory. The 9-individual difference between the two methods can be negligible, making the 1% an acceptable approximation.

Now, calculating the top 1% in a country with $100,000,000$ citizens will return an elite class size of $1,000,000$. Taking the root of $100,000,000$, however, leaves us with a stable elite class size of $10,000$. That is a staggering difference of $990,000$ individuals. With these larger amounts, the linear percentile approach that we are so comfortable with in historical settings is greatly overestimating what the core size should be. Figure 1.4 demonstrates the balance of power during the French Revolution. The same figure demonstrates the balance of power in the Russian Revolution.

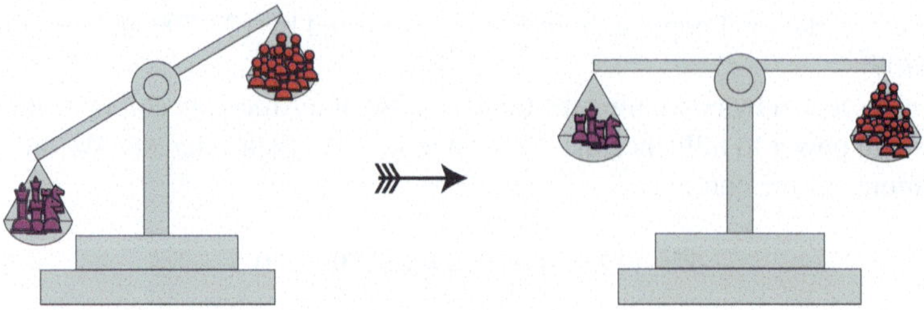

Fig. 1.4 The relationship between the French or Russian core and periphery populations when the elite (the noble) was too big to be financed by the periphery, resulting in the changing of the core

In both revolutions, the core started out too big and by the end of the revolution, many nobles lost their status—and their heads.

1.8.4 Russia: The 1917 Revolution

Our next example is the Russian Revolution of 1917. Again, we examine the relationship between the core and the periphery. Here, population estimations are much more accessible. The population of the Russian Empire around 1914 was estimated at 162,000,000, see [23], while the nobility numbered about 1,900,000, see [24], or check Wikipedia.

At first glance, the core-periphery ratio seems reasonable. There is a top 1%. However, moving from a linear perspective to a polynomial perspective can show what motivated the socioeconomic unrest: The square root of the overall population, $pop = 162,000,000$ is $\sqrt{pop} = 12,727$—more than ten times smaller than the number of existing nobility. Note that the ratio between the square root of the size of the society and the actual number of nobles is 150.

Now, let us compare these ratios to the United States during that same time period. According to the U.S. Census Bureau, the population of the U.S. in 1914 was about $pop = 99,110,000$, with 7,000 of those being millionaires (source). The American economy was much more stable at the time. $\sqrt{pop} = 9,955$.

It is common to think that the main problem that led to the Russian Revolution was a surplus of poor citizens from a polynomial perspective, we can see a different main problem: there were far too many wealthy people.

We are not the first to point out that different relationships in society, such as those between core and the periphery, are not always linear [10]. This warrants repeating. Immanuel Kant wrote about a specific type of human error, an error born of the difference between the way the world is and how we perceive it. The two perspectives are indeed hard to separate. Linear thinking—and calculation in percentages and multitudes—come naturally to us. It may be due to our education or human nature, but it seems a quixotic task to turn our attention towards polynomial thinking. That is why these mistakes keep happening in our academic efforts, and why we must keep pushing for a shift in mindset, especially in population-facing calculations such as those found in history.

1.9 Conclusion

In this chapter, we began our journey by suggesting the methodology of combining humanities with computer science. When applying algorithmic tools to the humanities, one needs to study the validity of the algorithm. This is done using tools from computer science, such as Big-O notation, axioms, and simple models. It is well known that algorithms can discriminate; they are not as objective as we would like them to be.

We briefly explained Big-O notation, discussed some of its uses, and showed that the growth of a society's core is sub-linear. Using percentages as if the core's growth is linear is a mistake that leads to misunderstandings and leaves us incapable of making accurate conclusions for societies of different sizes. The phrase "the rich get richer, and the poor get poorer" exemplifies this non-linear behavior, the effects of which are cyclically felt in our society.

The fact that society is non-linear can explain the fact that history will never come to an end. As long as there are people, there will be different stories to tell. The story is the ratio between the size of the classes in society; different numbers and proportions generate different historical behaviors.

Another important ingredient in the never-ending story of history is that there is a huge difference between the society that grows and the society that shrinks exponentially. For the most part, humans tend to live in the growing exponential society, especially over the last 300 years. However, from time to time, some part of the human population starts to shrink. China in the near future, Europe, or Japan are just a few examples. In those cases, we can see the narratives of the shrinking society.

1.10 Exercise

1. Compute the asymptotic of $f(n)$ in the following cases:
   a. $f(n) = n^2$
   b. $f(1) = f(2) = 1$, $f(n) = f(n-1) + f(n-2)$
2. Let d be an integer and a_i be a positive real number, where $i = 0, 1, \ldots, d$. Let the function $p(n)$ be defined by

$$p(n) = \sum_{i=0}^{d} a_i n^i \tag{1.21}$$

a. Write a function in Wolfram Mathematica that predicts the asymptotic of polynomial p.
b. Write a machine learning algorithm using the function "Predict" that computes the asymptotic of the function $p(n)$

3. Use the assumption that the elite size is the square root of the population to compute the size of the elite in the following countries. In each of these cases can you identify the elite by simple terms such as billionaires, party members, and so on, and compare your assumption with the actual number:

   a. China
   b. India
   c. United States

4. Assume that we have a society of size n, with a minority size of αn. The representation of the minority in the core is $\sqrt{\alpha n}$.

   a. Compute an α that is as small as possible, to prevent the majority in the core from dominating the minority.
   b. Compute an α that is as big as possible to allow the majority in the core to rule in favor of the minority.

5. As mentioned, one of the drawbacks of Big-O notation is the difficulty applying asymptotic historical data. In this exercise, we are going to partition the set of all functions into three subsets: sub-linear, linear, and superlinear. This partition is very rough and, therefore, easier to apply to history analyses. In math, a partition of a set S is a collection of non-empty, pairwise disjoint subsets of S whose union is S itself. Formally, a partition P of S satisfies:

References

1. B. Anderson, *Imagined Communities: Reflections on the Origin and Spread of Nationalism* (Verso Nooks, 2006)
2. C. Avin, Z. Lotker, D. Peleg, Y.-A. Pignolet, I. Turkel, Elites in social networks: an axiomatic approach to power balance and price's square root law. PLoS ONE **13**(10), e0205820 (2018)
3. T.H. Cormen, *Introduction to Algorithms* (MIT Press, 2009)
4. S.S. Skiena, *The Algorithm Design Manual*, vol. 2 (Springer, Berlin, 1998)
5. wikibigo (2023). https://en.wikipedia.org/wiki/Big_O_notation
6. wikibigo (2023). https://mathworld.wolfram.com/Big-ONotation.html
7. Paul Travis Nicholls, Price's square root law: empirical validity and relation to Lotka's law. Inf. Process. Manag. **24**(4), 469–477 (1988)
8. W. Glänzel, A. Schubert. Price distribution: an exact formulation of Price's "square root law". Scientometrics **7**(3–6), 211–219 (1985)

9. L. Egghe, R. Rousseau, A characterization of distributions which satisfy Price's law and consequences for the laws of Zipf and Mandelbrot. J. Inf. Sci. **12**(4), 193–197 (1986)
10. Derek de Solla Price, A general theory of bibliometric and other cumulative advantage processes. J. Am. Soc. Inf. Sci. **27**(5), 292–306 (1976)
11. D.J. de Solla Price, *Little Science, Big Science* (Columbia University Press, 1963)
12. T. Piketty, Capital in the twenty-first century, in *Capital in the Twenty-First Century* (Harvard University Press, 2014)
13. J. Beloch, *Die Bevölkerung der griechisch-römischen Welt*, vol. 60 (Duncker & Humblot, 1886)
14. T. Frank, Roman census statistics from 225 to 28 BC. Class. Philol. **19**(4), 329–341 (1924)
15. Wiki. wikisource-collective-memory (2024). https://en.wikipedia.org/wiki/1774_British_general_election
16. G.M. Ditchfield, M.W. McCahill, A Register of House of Lords Lists, 1761–1800. Parl Hist, **10**(1), 194 (1991)
17. Wiki. wikisource-collective-memory-populations-great-britain (2024). https://www.encyclopedia.com/history/encyclopedias-almanacs-transcripts-and-maps/populations-great-britain-and-america
18. Wiki. wikisource-collective-memory-populations-great-america (2024). https://www.encyclopedia.com/history/encyclopedias-almanacs-transcripts-and-maps/populations-great-britain-and-america
19. gov. wikisource-collective-memory-populations-great-britain (2024). https://www.commerce.gov/news/blog/2022/07/us-census-bureau-releases-key-stats-celebration-fourth-july
20. R.V. Wells, The population of England's colonies in America: old English or new Americans? Popul. Stud. **46**(1), 85–102 (1992)
21. M.L. Levy, La population de la France en, et 1789. Population et sociétés **233** (1989)
22. F. Bluche, J.-F. Solnon, *La véritable hiérarchie sociale de l'ancienne France: le tarif de la première capitation (1695)* (Librairie Droz, 1995)
23. A. Markevich, M. Harrison, Great war, civil war, and recovery: Russia's national income, 1913–1928. J. Econ. Hist. **71**(3), 672–703 (2011)
24. D. Smith, *Former People: The Last Days of the Russian Aristocracy* (Pan Macmillan, 2012)

Chapter 2
Function Diagrams in Digital Humanities

We ended Chap. 1 by analyzing the relationship between the core and the periphery in three major revolutions and showing that the relationship between the size of the core and the periphery was unbalanced in all three events. This fact can explain the necessity of revolution. The goal of this chapter is to develop a language that allows us to formalize and categorize different kinds of history. We will do so by using the mathematical tools borrowed from the category theorem. Figure 2.1 shows how necessary it is to develop a common language between humans and machines if we want to enable dialogue between them.

The tools we will learn in this chapter are known as *function diagrams*, or simply *diagrams*. They act as guides for developing historical data analyses. Diagrams can help us define the concepts that we are trying to study, communicate the definition to others, and inform us of the structures we build before we dive into the details. One needs to understand the relationship between the different types of data and how they generate a narrative that can be conveyed to humans while writing a machine to do so.

These diagrams are mathematical linguistic tools that allow historians and programmers to communicate the definitions they are working with and the processes that make up their historical analysis. It turns an individual's work into a potentially group endeavor.

2.1 Related Work

Diagrams are common mathematical tools used by researchers in different fields of science. More in-depth reading can be found in papers devoted to Persistence Diagrams [1], Binary Decision Diagrams [2], or Functional Decision Diagrams [3].

Z. Lotker, *History by Algorithms*,
https://doi.org/10.1007/978-3-031-93627-2_2

Fig. 2.1 A cartoon about the importance of a common language

2.2 What Is History?

Herodotus wrote the first historical book to ensure that human actions would not be forgotten over time [4]. From this we can infer that a historical event must comprise three essential components: time $\mathbb{T}$, place $\mathbb{G}$ or $\mathbb{M}_n$, and meaning $\mathbb{N}$.

The book *What is History?* [5] and its successor, *What is History, Now?* [6], define history as a narrative about the past. As a narrative, history accommodates multiple perspectives of the same events and encourages the search for contradictions without outright dismissing any version.

We distinguish between two kinds of narratives: fictive and historical. Fictive narratives originate in the author's imagination and have no relation to a historical event. In a historical narrative, the events represent what happened at some time and at some point in space in the past. From here on out, our focus will be on the historical narrative

Historical narratives are grounded in reality and constrained by geographical locations, dated time periods, and the fundamental laws that govern existence. The setting is beyond the author's control, anchored in the real world rather than an imagined one.

In the following three subsections, we will examine each component of a historical event in detail.

2.2.1 *Place*

Space is a fundamental element of history. In geopolitical history, place is primarily represented through geography—nations, cities, and locations that shape the way historical events unfold and are recounted. However, in other branches of history, such as the history of science, place is not always confined to physical geography and can be substituted with a more complex term: space. Later in this chapter, we will explore alternative conceptions of space.

Formally, we define the place in which a historical narrative occurs as the set $\mathbb{G}$. This set can be interpreted in various ways: It may represent a Euclidean space, the unit sphere, the set of all general graphs, or even the space of all matrices $\mathbb{M}_{n,n}$.

Place is also the domain of conflict, whether physical or ideological. Because conflict is central to historical analysis, the study of place inherently involves examining the structures in which conflicts occur. Conflict analysis often relies on static social networks, derived either from snapshots of dynamic social systems or from aggregated representations of entire narratives.

Another key aspect of place is the *Background*, a challenge shared with artificial intelligence research. In history, background is addressed through geographic settings and static social networks, which provide context for events and interactions. The next essential component of any historical event is time.

2.2.2 *Time*

Time is indivisible from history and stories in general. It is the main axis along which conflict develops and resolves. We study how conflict develops over time and the critical points of interest within the conflict. We denote the set of relevant times by $\mathbb{T}$. Most of the time, we use the set $\mathbb{T}$ as a subset of natural numbers $\mathbb{T} \subseteq \mathbb{N}$, or the rational number $\mathbb{T} \subseteq \mathbb{Q}$ or e $\mathbb{T} \subseteq \mathbb{R}$. The main property we demand from $\mathbb{T}$ is that it will be a well-ordered set $\leq$ by the preceding relationship. This means that any two times $t_1, t_2 \in \mathbb{T}$ could be compared, and one of them has to have happened before or at the same time as the other.

$$t_1 \leq t_2 \text{ and } t_2 \leq t_1 \Rightarrow t_1 = t_2 \tag{2.1}$$

Next, we discuss the third component of a historical event: Meaning.

2.2.3 *Meaning of Historical Event*

Meaning can take many forms and interpretations. For example, we can describe the meaning of a historical event through text, sound, painting, etc. To overcome the polyform of meaning, we represent it by a natural number.

In general, a natural number, together with a decoder and encoder, can be used as a universal tool to describe meaning. For example, all files on the computer can be decoded by binary numbers, which are natural numbers. Since most of us have videos, music, photos, etc., all those media (i.e., meaning) can be encoded using natural numbers together with a suitable decoder or encoder.

For most of this book, we will use a trivial decoder/encoder of the natural number, which will give the meaning of the number to its size, where 1 is more significant than 0. We use this very simple decoder/encoder to make this book more accessible.

We are now ready to discuss a sequence of historical events.

2.2.4 *Historical Events*

From the above, it follows that a *historical event* is

$$h \subset \mathbb{H} = \mathbb{T} \times \mathbb{G} \times \mathbb{N}. \tag{2.2}$$

Since a historical event can spread over an interval of time or subsets of times, it can happen at several points at the same time or in general subsets of the points in our space. It can also have a general meaning that we encode through natural numbers.

By the same logic, a sequence of historical events is

$$\mathscr{H} = (h_1, h_2 \ldots, h_\tau), \tag{2.3}$$

where each historical event is $h_i \subset \mathbb{T} \times \mathbb{G} \times \mathbb{N}$.

Now, an archive $\mathbb{A}rc$ is a set of all historical events relevant to our narrative.

$$\mathbb{A}rc = \{h_1, \ldots, h_\tau\} \tag{2.4}$$

In the next section, we provide a simple example that summarizes key military events in WWII according to Wikipedia [7].

2.2.5 Example World War II

To demonstrate the idea of a sequence of historical events, we summarize the list of main timelines in WWII according to Wikipedia.

It is possible to break the historical event of WWII into 7 main arenas, as seen in Fig. 2.2. Each of these historical events has a time interval, position in space, and meaning, which can be described in a single sentence.

Now, we will explain how to build a function diagram, which as we mentioned is one of the main tools in defining historical concepts. It helpfully emphasizes how the archive is processed to achieve historical analyses.

2.3 Description of Function Diagrams

Let us illustrate how to construct diagram functions through a series of examples, focusing on transforming catalogs (denoted by $\mathbb{C}at$) from Wikipedia or other digital archives into historical narratives. Wikipedia serves as a particularly useful source because of its accessibility and extensive coverage. Beyond Wikipedia, other catalog types, such as archives at time t (denoted by $\mathbb{A}rc(t)$) and lists (denoted by $List \subseteq \mathbb{C}at$), can also be employed in this process.

Our formalism is simple but important because it encapsulates the framework for historian machines. It is an example of creating a rich common language that humans and machines can understand and use to converse.

2.3.1 Category Theory

Just as a writer might organize their thoughts using paper outlines and networks, those operating in the digital sciences might organize their thoughts using the tools of category theory. Like writing exercises, category theory forms connections between elements and builds structures as a basis for writing.

Humanist researchers can take advantage of category theory as a field-bridging language. Historians who seek to employ programming would greatly benefit from category theory as a method to prime their research for algorithmic and mechanical processing. Category theory is also a communication technique that allows for easier collaboration with mathematicians, programmers, and other historians.

Wikipedia defines a *category* C as consisting of the following three mathematical entities:

1. A *class* $\text{ob}(C)$ whose elements are called *objects*.
2. A *class* $\text{ob}(C)$ whose elements are called *objects*.
3. A class $\text{hom}(C)$ whose elements are called *morphisms*, maps, or arrows. Each morphism f has a *source object* a and *target object* b. The expression f :

Start	End	Place	Meaning
Fri 1 Sep 1939	Fri 6 Oct 1939		Invasion of Poland
Fri 10 May 1940	Tue 25 Jun 1940		Battle of France
Sun 3 Sep 1939	Tue 8 May 1945		Battle of the Atlantic
Thu 30 Nov 1939	Wed 13 Mar 1940		Winter War
Tue 9 Apr 1940	Mon 10 Jun 1940		Norwegian Campaign
Mon 10 Jun 1940	Thu 13 May 1943		North African Campaign
Sun 22 Jun 1941	Wed 9 May 1945		Eastern Front of World War II

Fig. 2.2 The seven main steps of WWII as seven historical events. Each has its starting and ending time, place, and meaning

$a \to b$, would be verbally stated as "f is a morphism from a to b". The expression $hom(a, b)$—alternatively expressed as $\text{hom}C(a, b)$, $mor(a, b)$, or $C(a, b)$—denotes the hom-class of all morphisms from a to b.

4. A *binary operation* $\circ$ called *composition* of morphisms, such that for any three objects a, b, and c, we have

$$\circ : \text{hom}(b, c) \times \text{hom}(a, b) \to \text{hom}(a, c). \tag{2.5}$$

The composition of $f : a \to b$ and $g : b \to c$ is written as $g \circ f$ or gf, and is governed by two axioms:

a. *Associativity*: If $f : a \to b$, $g : b \to c$ and $h : c \to d$ then

$$h \circ (g \circ f) = (h \circ g) \circ f. \tag{2.6}$$

b. *Identity*: For every object x, there exists a morphism $1_x : x \to x$ called the *identity morphism*, 1_x for x, such that for every morphism $f : a \to b$, we have $1_b \circ f = f = f \circ 1_a$.

Building diagrams using category theory is a flexible and direct way to structure blueprints for different historical narratives. It is highly beneficial for organizing data sourced from both the humanities and the sciences.

2.3.2 *Analyzing Narratives in Social Networks*

The core of the preliminary results is found in my book, *Analyzing Narratives in Social Networks* [8]. It illustrates how to define literary conflict in the language of social networks, depicting the conflict as a partition of the nodes using functions.

Since history is a narrative, we first introduce notation form *Analyzing Narratives in Social Networks* [8]. The primary data structure in a narrative is *evolving social network* γ, which is a function

$$\gamma : \mathbb{T} \to \mathbb{M}_{n,n}[\mathbb{R}] \tag{2.7}$$

where $\mathbb{T} \subset \mathbb{R}$ represents time and $\mathbb{M}_{n,n}[\mathscr{R}]$ represents the space of all matrices over the *ring* $\mathscr{R}$ or *module*.[1] Usually, we will use $\mathscr{R}$ as the real number $\mathbb{R}$ or the integer number $\mathbb{Z}$. Using $\mathscr{R}$ allows us to utilize more complex information such as *vector space* and *matrices*. This is useful when we need more than one type of information to describe a relationship, such as between two people or countries. For example,

[1] We use $\mathscr{R}$ as a module when we need to solve general systems of linear equations.

we might need to describe two people as best friends from different nations. This flexible specificity is especially important when we model historical narratives.

2.3.3 *From Narratives to Historical Narratives*

It is important to mention how the relationship between space and time changes when moving from narratives to historical narratives.

In geopolitical history, conflict is inseparable from geography. Therefore, we will insist that ring $\mathscr{R}$ contain some geographical information in the historical narrative. Narratives, in general, are a function of time. We are able to use concepts such as objective and subjective clocks, which we will return to later in this chapter and in Chap. 4, to build dynamic networks that measure different times within a narrative. *Dynamic network* history calls for further developing such concepts as collective clocks and clocks affiliated with geography (see Chap. 4).

Additionally, we term history as a "never-ending story," wherein the world provides a mechanism for infinite plot generation. This is significantly different from the shaping of traditional narratives, where the plot is kept within stricter structures. Moreover, historians cannot make up events; they can only interpret or ignore some parts of the story. Clearly, it would be impossible to tell the complete story, and we all ignore some parts of it all the time.

2.4 One-Dimensional Historical Narrative

$$\begin{array}{ccccc} \mathbb{A}rc & \xrightarrow{f} & \mathbb{R}^2 \times \mathbb{Z}_k & \xrightarrow{\pi_4'} & \mathbb{Z}_k \\ & \searrow^{x} & \uparrow{\scriptstyle \gamma_1} & \nearrow^{\gamma_2} & \\ & & \mathbb{T} & & \end{array} \tag{2.8}$$

A one-dimensional historical narrative is a list of events with individual time stamps and a single number describing their magnitude. For example, in the Chronicle of Fatalities (also referred to as Mortality History), the events are years that are identified by their dates and the number of people who died in those years. Another example of a one-dimensional historical narrative could be historical economics analyses, where magnitudes measure numeric economic activities.

Life and death are two of humanity's most essential concepts, personally and culturally. Therefore, death rates can be used to evaluate and pinpoint critical events in human history.

It is always useful to start by describing the archive. In the case of the *Chronicle of Fatalities*, our archive contains a list of years from 100 AD to 1600 AD, and for each year $y \in \{100, 101, \ldots, 1600\}$ the number of significant deaths in the year y, denoted by $nd(y)$ that was recorded that year. For noise-canceling reasons, let L_p be the set of years with more than 10 prominent deaths. We remove the years with less than 10 prominent deaths from the set.

Clearly, the set L_p is well-ordered and we can therefore define the predecessor $y --$ of the year $y \in L_p$ to be the year that appears in L_p just before y. Formally,

$$y -- = max\{t : t < y, t \in L_p\} \tag{2.9}$$

The function $c(y)$ gives the main reason for most deaths if it exists. Now we can define our archive formally:

$$\mathbb{A}rc = \{(y, nd(y), nd(y --), c(y)) : 100 \leq y \leq 1600\}. \tag{2.10}$$

Note that the year 100 in the archive is missing the $nd(100 --)$ according to our definition. This information can be filled in manually by looking at the number of important people who died in the year 99.

2.4.1 *The Process of Building an Archive*

The next step is to explain how we construct the archives from our database. Our archives are composed of four fields: The first field is the year y, followed by the number of people who died $nd(y)$ that at year y, the third field is the number of people who died in the preceding year $nd(y --)$. The last field is the leading cause of death for most people. In our example, we have three main reasons: religious war, war (without religious cause), and epidemics. Religious war is denoted by 1, non-religious war is denoted by 2, and epidemics by 3, see Fig. 2.3.

Clearly, we have no way of knowing the number of people who died each year in the past; we cannot even do so for modern deaths. This can be overcome by using estimations of a number instead of specific numbers. But even estimations are difficult, so instead we use the number of important people who died as an

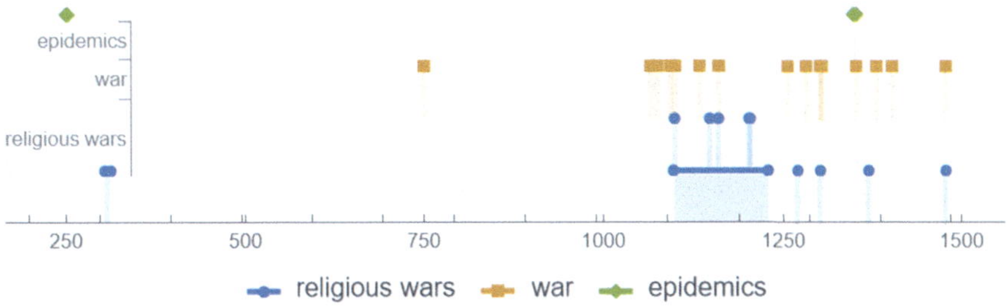

Fig. 2.3 A timeline of significant catastrophes from 100 AD to 1600 AD as computed by a simple algorithm counting the number of deaths occurring each year

estimation of the total number of people who died around the same time. There are several approaches to defining "important people," and no single way is unequivocally correct. We use a very technical definition: the person is important if they appear in English Wikipedia - and anyone can argue with that. We simply use this approach to demonstrate the method.

Formally, to evaluate the function nd we need to construct a *Simple indicator function* $\mathbb{1} : L_p \times \mathbb{N} \to \{0, 1\}$, which states that person $x \in L_p$ died in year y.

Formally, the definitions of the function are as follows:

$$\mathbb{1}(x, y) = \begin{cases} 1 & x \text{ died in year } y \\ 0 & x \text{ did not died in year } y. \end{cases} \tag{2.11}$$

Now we can define the function nd as

$$nd(y) = \sum_{x \in L_p} \mathbb{1}(x, y) \tag{2.12}$$

There are several ways to calculate the leading causes of death in a specific year, and we will show the main two. Bear in mind that we are only interested in the reason for years in which the death rate is significantly greater than standard years.

Computing the leading cause of death can be done manually, but would take a lot of work. For example, one can read the Wikipedia entry for the year 541 and look up the number of instances that the word *plague* appears. Another way is to look at the list of people who died in year y:

$$L_p(y) = \{x \in L_p : \mathbb{1}(x, y) = 1\} \tag{2.13}$$

and extract the cause and location of death. After doing this for each person, we can calculate the leading cause of death for that year.

After calculating our archives, it would be convenient to use the four projections functions $(\pi_1, \pi_2, \pi_3, \pi_4)$, $\pi_i : \mathbb{A}rc \to \mathbb{A}rc_i$ which allow us to get the i-th number in the vector of our archive, i.e.,

$$\pi_i(y) = \mathbb{A}rc(y)_i \tag{2.14}$$

i.e., $\pi_1(y) = y$ gives us the year y, and $\pi_2(y) = nd(y)$ which is the number of people who died in year i.

Now we will explain the function diagram (2.8). The first function is f and can be defined as follows:

$$f(y) = (\pi_2(y), \pi_3(y), \pi_4(y) = c(y)) \tag{2.15}$$

The next function in function diagram χ is the projection of the archive $\mathbb{A}rc$ to its first coordinate, which is the year. This function was defined by Eq. 2.14.

The next function in function diagram π_4') is the projection of archive $\mathbb{A}rc$ to its fourth coordinate, which is the main cause of death, formally:

$$\pi_4'(n_1, n_2, z) = z \tag{2.16}$$

What remains is to define γ_1 and γ_2, which we can do by composing functions. Since function χ is invertible, γ_1 is defined by

$$\gamma_1(t) = f\left(\chi^{-1}(t)\right) \tag{2.17}$$

We use the same idea to define γ_2, like so:

$$\gamma_2(t) = \pi_4'\left(f\left(\chi^{-1}(t)\right)\right) \tag{2.18}$$

Since diagram (2.8) is commutative, it follows that

$$\gamma_2(t) = \pi_4(\left(\chi^{-1}(t)\right) \tag{2.19}$$

The summary of the diagram (2.8) can be seen in Fig. 2.3, where the number 0 indicates the lack of a known catastrophe and is therefore ignored. The number 1 represents a known catastrophe caused by religious conflict, indicated in the figure by a blue circle. The number 2 indicates war and is displayed as an orange square. The number 3 represents epidemics and is shown in green. The timeline highlights catastrophic years from 100 AD to 1600 AD.

2.4.2 *Machine Learning and Anomaly Detection*

One of the classic tasks in machine learning is anomaly detection. In this problem, we are given a set of numerical vectors and ask the machine to find the anomalous vectors. For more on anomaly detection, see [9].

In Wolfram Mathematica, this is implemented with the command AnomalyDetection[data]. Once we apply this command to the data, we receive a function that can tell us whether or not an element that belongs is anomalous. Clearly, this can be applied to historical events, especially when we analyze the list *data* of dead people. We use the following data set, which contains a list of two-dimensional vectors. The first number y_i in each element represents the year, and the second $dnum_1$ represents the number of dead people. Formally, our list is

$$data = ((y_1, dnum_1), \ldots, (y_n, dnum_n)) \tag{2.20}$$

After harvesting the data set from Wikipedia, we compute the anomaly detection using the command

$$ad = AnomalyDetection[data]; \tag{2.21}$$

We then compute the machine learning function *ad* for each of the years to find if they are anomalous or not:

$$data[[Flatten[Position[ad[\#]\&/@data, True]]]] \tag{2.22}$$

The outcome of these experiments is summarized in the table below. The first column shows the beginning of the time interval, the second column shows the end of the time interval, and the third column shows the anomalous years. It is easy to understand why each date is important from a historical perspective. The nice thing about this table is that even without any historical understanding, the machine could identify the main events in Western culture. The main reason for most of our results belonging to Western culture is that we are using English Wiki pages.

Another point we need to explain is how to ensure that all morphisms exist. When working from a list of elements L_p to a final narrative, it is important that all elements p utilized for the narrative have complete data sets. If we start with a list of people, each missing morphism—a missing date of birth, for example—must be added. Otherwise, the entire element, in this case the person, must be removed from the list. We must do one or the other, as Category Theory requires that we provide all the relevant information.

Now, let us explore the theoretical background for comparative history.

2.5 Comparative History or D-Dimensional Historical Narrative

$$\begin{array}{ccc} \mathbb{A}rc & \xrightarrow{f} & \mathbb{N}^d \\ & \searrow^{\pi} & \uparrow{\scriptstyle\gamma} \\ & & \mathbb{T} \end{array} \tag{2.23}$$

In the previous section we learned how to embed a historical narrative into a one-dimensional space, allowing us to categorize historical events by encoding each category with a number.

This section will show us how to embed a historical narrative into a d-dimensional space. Embedding the historical narrative into the d-dimensional space allows us to generate a comparative historical narrative.

Many historical fields use this concept. For example, any economic analysis of historical events will apply this scheme to compare countries through the compression of numbers, reprint production capacity, or financial capacity.

We will explain the idea by analyzing the feminist movement across the world (Fig. 2.4). Let C be the set of all countries in the world. For each country $c \in C$, we will have its index id

$$id : C \to \{1, \ldots, |C|\} \tag{2.24}$$

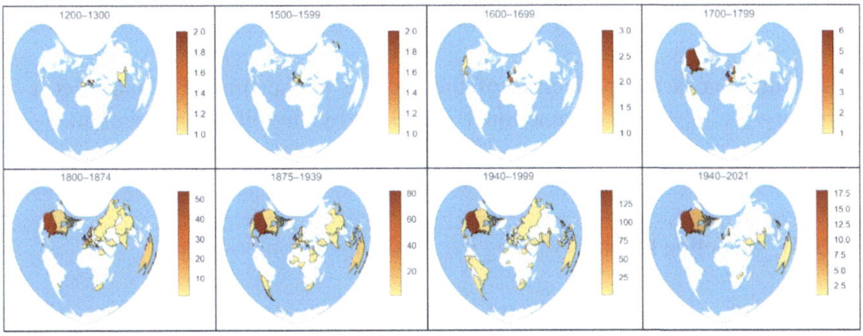

Fig. 2.4 The geographic density of active feminists throughout history, showing the shift of feminist influences from Europe to the Americas in recent centuries

which will represent its dimension. Note that for some countries, the number of feminists born there is 0. Therefore, we can ignore those countries and reduce the dimension. To simplify the representation, we assume that $|C| = d$, i.e., we remove irrelevant countries.

2.5.1 *The Process of Building Archives in Comparative History*

After giving each country its own number or index, we construct the archive as follows:

$$\mathbb{A}rc = \{(y, (n_1(y), \ldots, n_d(y)) : 1200 \leq y \leq 2021\}. \tag{2.25}$$

In this format, y represents the year and $n_i(y)$ represents the number of feminists born in or before year y in the country indexed as i. Note that the function n_i can be constructed in a similar way as done in Sect. 2.4.1. This is left as an exercise for the reader.

2.5.2 *Explaining the Comparative Function Diagram*

Once we have constructed the archive $\mathbb{A}rc$, we need to explain the comparative function diagram (see (2.23)). The diagram contains three functions: f, π, γ. We will explain all three of them.

Function f is the projection of the archive, disregarding year y. Formally, let $x = (y, (n_1, n_2, \ldots, n_d) \in \mathbb{A}rc$. Function f is defined by the following equation:

$$f(x) = (n_1, n_2, \ldots, n_d) \tag{2.26}$$

The function is the projection of the archive to the time. Formally, let $x = (y, (n_1, n_2, \ldots, n_d) \in \mathbb{A}rc$. Function π is defined by the following equation:

$$\pi(x) = y \tag{2.27}$$

Lastly, we define function γ using the following decomposition:

$$\gamma(t) = f(\pi^{-1}(t)) \tag{2.28}$$

Note that the function pi^{-1} exists by definition of $\mathbb{T}$.

In the next section, we will discuss the general technique according to which the historical narrative is embedded in a general graph or the space of all matrices $\mathbb{M}_{n,n}[\mathbb{R}]$ instead of $\mathbb{R}^d$.

2.6 History with Social Networks

Let us now extend the function diagram (2.23) to the function diagram (2.32). This section requires some basic knowledge of graph theory, which can be found in any one of the many excellent introduction books, such as [10–12]. For notation in graph theory, see Appendix A.

To explain history with a general graph, we will study the social network structure of the feminist movement. This will allow us to point out the central figures in the feminist movement throughout history (Fig. 2.5). We begin by describing archive $\mathbb{A}rc$.

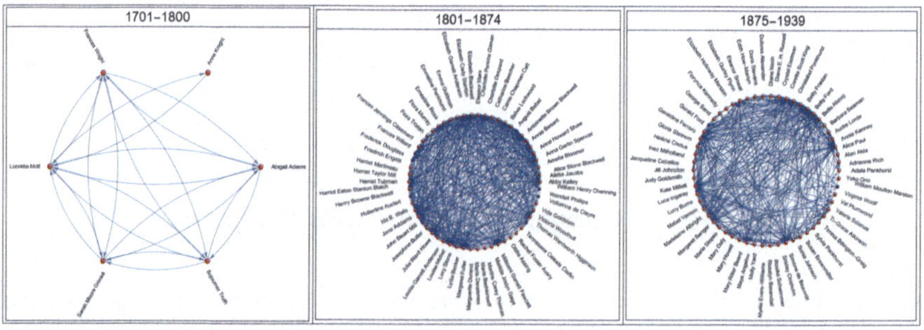

Fig. 2.5 Representation of the most important feminists from each wave of feminism, with the blue dots indicating the anchors

2.6.1 Constructing Archives in Historical Social Network

Before constructing archive $\mathbb{A}rc$, we must set some notations.

Let L_p be our list of people of interest. Since we are analyzing relationships between pairs of historical individuals, the archive in this case is constructed from pairs of individuals, accompanied by timestamps. For each $p_v \in L_p$ we denote its Wiki page by wik_v and its *hyperlink* by h_v.

Each Wiki page contains many links. Denote the set of all hyperlinks that appear in the Wiki page wik_v by $hypl_v$. For each pair of feminists $p_v, p_u \in L_p$, we can say that feminist p_v is affiliated with feminist p_u if Wiki page wik_v contains a link to Wiki page wik_u i.e.,

$$h_u \in hypl_v \tag{2.29}$$

We assume that for each feminist $p_v \in L_p$ there is a timestamp $t(p_v)$. In our example, this will be the time that feminist p_v was born, i.e.,

$$t(p_v) = b(p_v). \tag{2.30}$$

Now we are ready to construct our archive.

$$\mathbb{A}rc = \left\{(max\,(b(p_v), b(p_u))\,, (v, u)) \in \mathbb{Z} \times L_f^2 : h_u \in hypl_v\right\} \tag{2.31}$$

Our archive is constructed of two components: time, and the relationship between the two feminists. When considering the time component, we require both feminists to have been born at the time when the relationship between them is established. To determine this, we take the maximum time of their birthdays, which is equal to the latest date of birth. The second component determines whether or not feminist p_u is mentioned on the Wiki page of $hypl_v$. This means that feminist p_u influenced or was a central figure in the narrative of feminist p_v.

In the next subsection, we explain how to construct the historical narrative of a social network.

2.6.2 Explaining the History Social Network Diagram

$$\begin{array}{ccc} \mathbb{T} & & \\ {\scriptstyle \chi}\uparrow & \searrow^{\gamma} & \\ \mathbb{A}rc & \overset{\rho}{\hookrightarrow} & \mathbb{M}_{n,n}[\mathbb{Z}_2] \end{array} \tag{2.32}$$

After constructing the archive, we describe the process of generating the historical social network narrative γ using Diagram (2.32). The diagram has three functions: χ, ρ, γ. We explain all three of them.

Function χ is just a projection of the time stamp to the set $\mathbb{T}$. Formally, it is defined as follows:

$$\chi((t, (v, u)) = t \tag{2.33}$$

Next, we define function ρ. To do so, we need the following family of sets: for each $t \in \mathbb{T}$ we denote the set of all edges that appear before or at time t by

$$E_t = \{(v, u) : \text{ there exist } t' \leq t \text{ s.t.}, (t', (v, u)) \in \mathbb{A}rc\} \tag{2.34}$$

Now we can define function ρ for all $(t, (u, v)) \in \mathbb{A}rc$. Denote

$$\rho((t, (v, u)) := \left[\rho(t)_{i,j}\right] \tag{2.35}$$

and define the matrix elements of ρ as

$$\rho(t)_{i,j} := \begin{cases} 1, \text{ if } i = v, j = u, (i, j) \in E_t \\ 0, \text{ Otherwise} \end{cases} \tag{2.36}$$

The last function we need to define in Diagram (2.32) is γ:

$$\gamma(t) = \sum_{(i,j) \in E_t} \rho(t, (i, j)) \tag{2.37}$$

Note that the arrow in the diagram is hooked, teaching us that this diagram is not commutative but accumulative. It is commutative only for sets $\sum_{(v,u)\in E_t} \rho(t, ((v, u)))$, which are maximal elements, i.e.,

It is helpful to identify key players in the historical narrative using the method described in this section, since we can use the centrality function from social network theory to identify key figures in the narrative. Moreover, the fact that we can compute the function γ means that an algorithm can generate the narrative mechanically. For more on this, see [8].

In the next section, we use a function diagram to define the process of reading against the grain. The function diagram reveals a surprising relationship between the past and the future when reading against the grain.

2.7 Reading Against the Grain

As a final example, we will translate the humanities' concept of "reading against the grain" into a mathematical formulation of a function diagram. This concept refers to elucidating what the data does *not* say and pushing against any inconsistencies or ambiguities [13–15]. The basic idea is to identify the differences between past and present, that is, to compare the society of the present, for which we have better data, with the society of the past. We read against the grain by using the present's archive $\mathbb{A}rc(t+\Delta)$ as a frame and transforming it—with the help of a theoretical time machine or translating machine Tm—from the present archive to the past archive and vice versa. This process can be accomplished with large volumes of data by using translation machines combined with deep learning and machine learning [16, 17].

We start with some notation. Let $0 < \Delta \in \mathbb{R}$ and $t \in \mathbb{R}$ be real numbers. We assume that we are given two archives: $\mathbb{A}rc(t+\Delta)$, which represents the archive at time $t+\Delta$, and $\mathbb{A}rc(t)$, which represents the archive at time t. Since we assume that $\Delta > 0$, it follows that archive $\mathbb{A}rc(t)$ is more ancient than archive $\mathbb{A}rc(t+\Delta)$.

The next step to reading against the grain is to classify both archives much like we did in Sect. 2.4. We denote the method to do that by $\rho_0/(\rho_1)$ for the archive $\mathbb{A}rc(t+\Delta)/(\mathbb{A}rc(t))$, respectively. This can be done either manually or algorithmically. Denote the outcome of this process by $\mathbb{A}rc(t+\Delta)\times\mathbb{N}/(\mathbb{A}rc(t)\times\mathbb{N})$, respectively.

Since there is a difference in the archive size, our archive tends to shrink the further back we go in time. We construct another machine to generate the spirit of the archive. We denote these two machines ex_0 and ex_1. This is especially important when we try to work with machine learning, since the learning algorithm is extremely sensitive to the size of the archive. The outcome of this process is denoted by $EX(\mathscr{A}_{t+\Delta})/(EX(\mathscr{A}_t))$, respectively.

Now we would like to translate the past synthetic archive $EX(\mathscr{A}_t)$ to the later synthetic archive $EX(\mathscr{A}_{t+\Delta})$ and vice versa. To do that, we construct a translation algorithm Tm_Δ which translates the past synthetic archive $EX(\mathscr{A}_t)$ to the present synthetic archive $EX(\mathscr{A}_{t+\Delta})$, joined by the translation algorithm $Tm_{-\Delta}$ which translates the present synthetic archive $EX(\mathscr{A}_{t+\Delta})$ to the past synthetic archive $EX(\mathscr{A}_t)$. The complete process is summarized in Diagram (2.40).

Now we can compare these archives using the formula

$$ex_0\left(\rho_0\left(\mathscr{A}(t+\Delta)\right)\right) \setminus Tm_\Delta\left(ex_1\left(\rho_1(\mathscr{A}(t))\right)\right) \tag{2.38}$$

This will allow us to identify the missing grain in the past.

And just like that, we can define the missing grain in the future by performing the reverse process:

$$ex_1\left(\rho_1\left(\mathscr{A}(t)\right)\right) \setminus Tm_{-\Delta}\left(ex_0\left(\rho_0(\mathscr{A}(t+\Delta))\right)\right) \tag{2.39}$$

$$
\begin{array}{ccc}
\mathscr{A}(t+\Delta) & & \mathscr{A}(t) \\
\big\downarrow{\scriptstyle \rho_0} & & \big\downarrow{\scriptstyle \rho_1} \\
\mathscr{A}(t+\Delta)\times\mathbb{N} & & \mathscr{A}(t)\times\mathbb{N} \\
\big\downarrow{\scriptstyle ex_0} & & \big\downarrow{\scriptstyle ex_1} \\
EX(\mathscr{A}_{t+\Delta})\times\mathbb{N} & \underset{Tm_{\Delta}}{\overset{Tm_{-\Delta}}{\rightleftarrows}} & EX(\mathscr{A}(t))\times\mathbb{N}
\end{array}
\tag{2.40}
$$

2.8 Conclusion

In this chapter, we introduced the language of function diagrams to describe different historical narratives. By doing so we are able to categorize historical narratives into different categories such as one-dimensional historical narrative, comparative history, historical social network, and reading against the grain. Clearly, these examples by no means encompass every category of historical narratives. Another flaw is that all our function diagrams are limited.

All of our examples followed these procedures: First, we construct the archive; then we describe the function diagram. The construction of the archive can be described via function diagrams as well. However, we find this process complicated and recommend that you separate the stages.

Our first example was the one-dimensional historical narrative. This is the simplest historical narrative, since the archive contains events, dates, and the magnitude of the events—but it is also a powerful narrative. In our examples, we were able to pinpoint the most important events in human history in the last 2,000 years. When studying general history, such events appear over and over again. However, there is a major difference between the general approach, where some historical expert tells us which event is the most important, and ours, where the machine pinpoints the most important events just by using the numbers and disregarding any general historical knowledge. Moreover, the machine proves why these events are extremely important: the main reason is the number of people who died around that date.

The category of compared history allowed us to define a historical narrative where a single number is not sufficient, but a vector of numbers can tell the complete narrative. Many historical narratives fit into these categories, such as national comparisons between countries, wars, and battle descriptions composed of different units with their own location, as we will see in the next chapters.

In Sect. 2.6 we introduced the historical social network category. This category allows us to examine the connection between the pairs of historical entities and how they develop. We explained how to use the historical social network category in order to identify the main feminist influences throughout history. In general, this category allows us to identify the key players in the historical narrative. The archive in this category is more complicated than in the previous category.

We ended the chapter by reading against the grain. Here, we used the function diagram to provide an exact definition of what reading against the grain is. We showed that it uses the present as a reference point to compare the present to the past, allowing us to transform the concept of reading against the grain from a one-dimensional concept into a two-dimensional one where instead of comparing past and present, we can compare two different pasts. This section is more philosophical since we do not provide any examples; it is also more futuristic since we use machines that are currently unavailable to us, such as extension archive machine ex_0, ex_1 and translation archive machines. However, there is no reason to believe that such machines will not exist in the future. Moreover, this example shows the interplay between modern history and computer science; the idea that our historical analysis uses machines is crucial to the future of humanities. Lastly, in some sense machines are exchangeable with humans. Therefore, the definition of reading against the grain shows what historians must do: They need to categorize the archive, translate it from the past to the present, and take into account that the archive of the past is smaller than the archive of the present.

In the next chapter, we will discuss the history of war.

2.9 Exercise

One-Dimensional Historical Narrative

1. Figure 2.6 represents the derivative of the Russian GDP on the left and the ADP itself on the right. The figure was constructed with Wolfram CountryData on Russia: data = CountryData["Russia", "GDP", 1988, 2021]; Use the Anomaly-Detection function in Wolfram Mathematica to compare which graph is better for detecting an anomaly, the derivative of the function, or the function itself. Note that the numbers are in order of $10^1 1$. For purposes of anomaly detection, it is

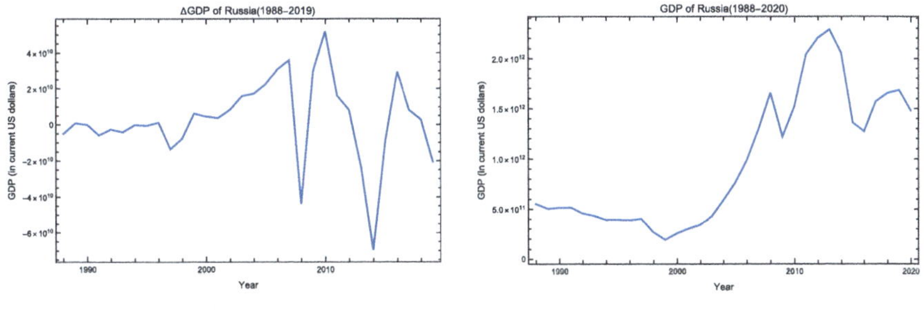

Fig. 2.6 This figure shows the derivative of the Russian GDP on the left and the Russian GDP from 1988 to 2020 on the right

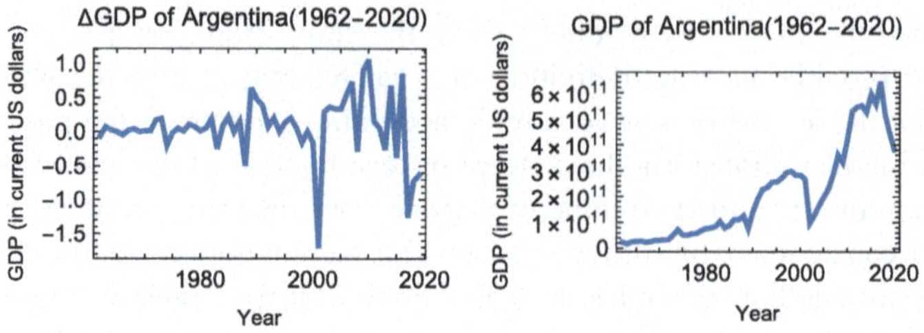

Fig. 2.7 This figure shows the derivative of Argentina's GDP on the left and Argentina's GDP from 1962 to 2020 on the right

recommended to use smaller numbers, e.g., between −10 to 10. Can you explain the result of anomaly detection?

2. Repeat Question 1, but this time apply it to Argentina. See Fig. 2.7

D-Dimensional Historical Narrative

3. Instead of the list of feminists, take the list of physicists in Wikipedia (see [18]) and use it to describe the migration of physicists from Europe to America. Repeat the process from Sect. 2.5 to generate a map similar to the maps in the Fig. 2.4.
4. The list of all battles in the Second World War can be found in [19]. Explain why the method of description in Sect. 2.5 will not represent the evolution of the war. Provide the method for generating maps that describe evolution at war. Explain how you construct the archive, and draw the relevant function diagram

Social Network Historical Narrative

5. Compute the social network of the physicists from the list of physicists in Wikipedia [18], like we did in Sect. 2.6. Use the degree centrality function to identify the key physicists.

Reading Against the Grain

6. Let $t_1 < t_2 < t_3 \in \mathbb{R}$ be real numbers representing time. Assume that you have three archives—$(\mathbb{A}rc(t_i))_{i=1}^{3}$. Explain what we mean when we say that reading against the grain becomes easier over time. For example, consider the feminist movement in the West: More and more materials about the equality of sexes appear in the archives; data about women is progressively increasing in the archives.

7. Define what it means for the reading against the grain to oscillate. For example, consider the relationships between left and right political opinions: For some periods and in different places, the left is in power, making documents and texts about it easily accessible. For instance, Marx's ideas were very popular in the 19th and 20th centuries.
8. Constitutional Originalism is a legal theory that interprets constitutional, judicial, and statutory texts based on their original meaning at the time they were adopted. For more, see [20]

   a. Describe Constitutional Originalism in the language of reading against the grain and time machine.
   b. Let's consider the following legal theory, which we will call Constitutional Futurism. According to this theory, the authors of the Constitution wrote it in such a way that the Constitution will observe the shock of changing society over time. By doing so, we forego the need to imagine what these authors thought at the time of composing the Constitution. Which theory seems to you more reasonable: the one that says that the Founding Fathers could write a document that forecast the shock of the future, or that Supreme Court judges are able to think like the Founding Fathers?

References

1. D. Cohen-Steiner, H. Edelsbrunner, J. Harer, Stability of persistence diagrams, in *Proceedings of the Twenty-First Annual Symposium on Computational Geometry* (2005), pp. 263–271
2. Akers, Binary decision diagrams. IEEE Trans. Comput. **100**(6), 509–516 (1978)
3. R. Drechsler, A. Sarabi, M. Theobald, B. Becker, M.A. Perkowski, Efficient representation and manipulation of switching functions based on ordered Kronecker functional decision diagrams, in *Proceedings of the 31st Annual Design Automation Conference* (1994), pp. 415–419
4. Herodotus, A.D. Godley, *Herodotus: With an English Translation by AD Godley* (W. Heinemann, 1961)
5. E.H. Carr, *What is History?* (Penguin UK, 2018)
6. S. Lipscomb, H. Carr, *What Is History, Now?* (Orion, 2021)
7. Wikipedia contributors. List of timelines of world war ii (2024). https://en.wikipedia.org/wiki/List_of_timelines_of_World_War_II
8. Z. Lotker, *Analyzing Narratives in Social Networks Taking Turing to the Arts Authors* (Springer, Berlin, 2021)
9. Anomaly detection (2023). https://en.wikipedia.org/wiki/Anomaly_detection
10. R. Diestel, *Graph Theory* (Springer, Berlin, 2006)
11. B. Bollobas, *Graph Theory: An Introductory Course*, vol. 63 (Springer Science Business Media, 2012)
12. C. Godsil, G.F. Royle, *Algebraic Graph Theory*, vol. 207 (Springer Science & Business Media, 2013)
13. J. Fetterley, *The Resisting Reader: A Feminist Approach to American Fiction*, vol. 247 (Indiana University Press, 1978)
14. T. Bewes, Reading with the grain: a new world in literary criticism. Differences **21**(3), 1–33 (2010)

15. C.A. McKenzie, S. Jarvie, The limits of resistant reading in critical literacy practices. Engl. Teach.: Pract. Crit. **17**(4), 298–309 (2018)
16. S.P. Singh, A. Kumar, H. Darbari, L. Singh, A. Rastogi, S. Jain, Machine translation using deep learning: an overview, in *2017 International Conference on Computer, Communications and Electronics (Comptelix)* (IEEE, 2017), pp. 162–167
17. S. Yang, Y. Wang, X. Chu, A survey of deep learning techniques for neural machine translation (2020). arXiv:2002.07526
18. wikisource (2023). https://en.wikipedia.org/wiki/List_of_physicists
19. List of world war ii battles (2023). https://en.wikipedia.org/wiki/List_of_World_War_II_battles
20. J. Gienapp, *Against Constitutional Originalism: A Historical Critique* (Yale University Press, 2024)

Chapter 3
History of War

3.1 Introduction

In the previous chapter, we developed the language of function diagrams to describe different historical processes using category theory to achieve a categorization of history studies.

In this chapter, we will use function diagrams to develop a mathematical language that allows us to study and analyze past battles (Fig. 3.1). We begin with Lanchester's differential equation. Lanchester's equation needs to be enriched to describe all battles. It has several main disadvantages: it is homogeneous in time, independent of geography, and maintains only two armies. Figure 3.2 demonstrates the necessity of extending Lanchester's equation to a nonhomogeneous differential equation. Analyzing historical battles demands a nonhomogeneous differential system of equations. The purpose of this chapter is to explain how to do so. We extend the language of Lanchester's equation to evolving differential equations, which are nonhomogeneous differential equations in time. This allows us to describe most historical battles. To complete the description of a historical battle, we analyze a common but crucial element of military strategies, such as one army encircling another. In this case, the differential equations are different from the standard Lanchester's equations.

We end up with a few historical examples of great battles.

3.2 Related Work

Mathematical modeling was used by different scientists to study historical facts, especially wars and battles. Among the classic works are Lewis F. Richardson's mathematical theory of war, analyzed in Rapoport's paper [1]. As an example of a wider approach to mathematical analyses of war and other conflicts, consider Jones's book, *Game theory: Mathematical models of conflict* [2]. In this chapter, our main

Z. Lotker, *History by Algorithms*,
https://doi.org/10.1007/978-3-031-93627-2_3

Fig. 3.1 A cartoon about how historians can use mathematical models of war

Fig. 3.2 The importance of speed in modern warfare

attention is given to the use of Lanchester's equations in historical research to analyze different historical conflicts. Deitchman used this approach in modeling guerrilla warfare [3]; Tam used it in researching the Ardennes campaign [4] and applied the model to one of the most recent conflicts - the Russian-Ukranian war [5]. We will be using the example of the Battle of Cannae, a major battle in the Second Punic War.

3.3 Lanchester's Equations

Lanchester's equations are a set of mathematical laws developed by Frederick Lanchester, a British engineer and mathematician, to model the dynamics of combat between two forces. The laws are based on the assumption that each force will try to maximize its advantages and minimize its disadvantages to gain an advantage over the other force. Lanchester developed these laws at the beginning of the 20th century, when the nature of warfare was changing rapidly due to the advent of new technologies such as aircraft and tanks. These changes made it increasingly important for military planners to understand combat dynamics, and Lanchester's laws provided a valuable tool for this purpose. These laws are still commonly used in military planning and operations research to evaluate the relative strengths and weaknesses of different military forces and strategies. They are also the basis for many military simulations and video games.

Lanchester's laws provide a simple and intuitive way to model combat dynamics between two forces: blue and red (B, R). The equations are relatively easy to understand and can be applied to a wide range of situations. Economists may use them to model competition dynamics between firms or industries; biologists may even use them to study the interactions between different species in an ecosystem.

The armies in Lanchester's model start combat at time 0. The size of the armies at the beginning of the combat is $B(0)$, $R(0)$ soldiers. In general, the number of soldiers at time t for the blue army is denoted by $B(t)$, and the red army is $R(t)$.

The change in the number of soldiers for each army at time t is proportional to the number of its enemy around that time. Since the armies are firing at each other, the number of soldiers is always decreasing. Formally, the equations are:

$$\begin{aligned} \frac{\partial R(t)}{\partial t} &= -\alpha B(t) \\ \frac{\partial B(t)}{\partial t} &= -\beta R(t) \end{aligned} \tag{3.1}$$

For each army, we have a technological parameter that allows us to determine the number of soldiers that the army kills during combat. For the blue army, the parameter in Eq. 3.1 is denoted by α, and for the red army it is denoted by β.

It is possible to solve those two systems using any computing aid such as Wolfram Mathematica, Python, or Matlab. The solution for the number of blue soldiers is given below:

$$B(t) = \frac{e^{\sqrt{\alpha}\sqrt{\beta}(-t)}\left(\sqrt{\beta}\mathrm{B0} + \sqrt{\beta}\mathrm{B0}e^{2\sqrt{\alpha}\sqrt{\beta}t} + \sqrt{\alpha}\mathrm{R0} - \sqrt{\alpha}\mathrm{R0}e^{2\sqrt{\alpha}\sqrt{\beta}t}\right)}{2\sqrt{\beta}} \tag{3.2}$$

The solution for the number of red soldiers is given below:

$$R(t) = \frac{e^{\sqrt{\alpha}\sqrt{\beta}(-t)}\left(\sqrt{\beta}\mathrm{B0} - \sqrt{\beta}\mathrm{B0}e^{2\sqrt{\alpha}\sqrt{\beta}t} + \sqrt{\alpha}\mathrm{R0} + \sqrt{\alpha}\mathrm{R0}e^{2\sqrt{\alpha}\sqrt{\beta}t}\right)}{2\sqrt{\alpha}} \tag{3.3}$$

Several papers study famous battles using Lanchester's equations: Herbert K. Weiss [6] examined combat data on the U.S. Civil War and determined the extent to which it can be explained by using Lanchester's equations; Judy H. Tam [4] found a certain concurrence between the predicted and actual force strengths of the latter stages of the battle; Jerome Bracken [7] formulated four Lanchester models of the Ardennes campaign and estimated their parameters for the data; Hung et al. [8] also devoted their work to the Ardennes campaign and adopted the concepts of the tactical factor variable and the shift time variable to improve the original Lanchester model.

In fact, the extended Lanchester model is

$$\begin{aligned} \frac{\partial R(t)}{\partial t} &= -\alpha R(t)^p B(t)^q \\ \frac{\partial B(t)}{\partial t} &= -\beta B(t)^p R(t)^q \end{aligned} \tag{3.4}$$

In the extended Lanchester model, the number of casualties is proportional to some polynomial of additional parameters p and q, where p is the exponent parameter of the shooting force, and q is the exponent parameter of the target force. However, this model offers symmetry between the two forces. For more on the extended Lanchester model, see [7, 9].

However, when analyzing flanking, the situation for the two armies could not be symmetrical. When one army flanks the other, as is especially common in historic battles, the symmetry of the armies is broken, as shown in the following sections.

3.4 Adapting Lanchester's Laws for Flanking

Many battles end when one army flanks the other. In this case, the battle often ends with the destruction of the flanked army. In this section, we develop and solve the equivalent of Lanchester's equation for flanking. This is a crucial ingredient when trying to construct a battle simulation (Fig. 3.3).

We consider the case where the red army surrounds the blue army. When the red army flanks the blue army, all red army units are able to fight, while only the units in the boundaries of the blue army are able to fight back. We use the isoparametric theorem, which relates the area $A(S)$ of a fat convex set S to the perimeter, which is the length $l(S)$ of its boundary convex set in two dimensions. Formally, the isoparametric theorem for a fat convex set states that:

$$l(S) = \Theta(\sqrt{A(S)}) \tag{3.5}$$

Therefore, Lanchester's equations become

$$\begin{aligned} \frac{\partial R(t)}{\partial t} &= -\alpha \sqrt{B(t)} \\ \frac{\partial B(t)}{\partial t} &= -\beta R(t) \end{aligned} \tag{3.6}$$

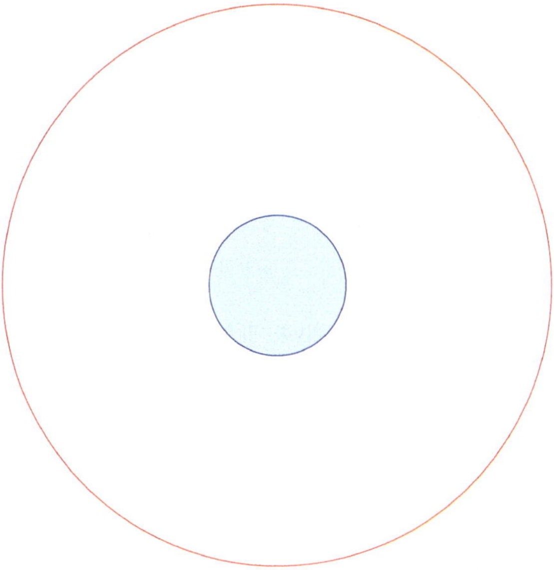

Fig. 3.3 An example of a typical flanking

where α represents the blue units' ability to strike red units and β represents the red units' ability to strike blue units at a given moment. The flanking red army R can fire using all its units and, therefore, the number of blue army units that were hit by the red army at time t is $-\alpha\sqrt{B(t)}$. Similarly, the blue army can use only the unit on its boundary and, therefore the casualties of the red army at time t is $-\alpha\sqrt{B(t)}$. These equations are meant to be used with quantities greater than 1, since for all $x > 1$ we have $x > \sqrt{x}$. The opposite happens when $x < 1$, i.e., $x < \sqrt{x}$. This makes sense as armies are made up of more than one soldier.

In order to solve Eq. 3.6, we substitute the function $B_2(t) = \sqrt{B(t)}$ inside (3.6). Once we do that, Eq. 3.6 become:

$$\begin{aligned} \frac{\partial R(t)}{\partial t} &= -\alpha B_2(t) \\ \frac{\partial B_2^2(t)}{\partial t} &= -\beta R(t) \end{aligned} \tag{3.7}$$

To solve Eq. 3.7, we first write the functions $R(t)$, $B_2(t)$ as a power series i.e.,

$$R(t) = \sum_{i=0}^{\infty} r_i t^i \tag{3.8}$$

and for the function $B_2(t)$

$$B_2(t) = \sum_{i=0}^{\infty} b_i t^i. \tag{3.9}$$

Now we can plug the number of soldiers present at the beginning of the battle by setting the constant r_0 and b_0. For example, if at the beginning of the battle the number of soldiers in the red army is 100 and the number of soldiers in the blue army is 144, it follows that

$$r_0 = 100; b_0 = 12. \tag{3.10}$$

Now, the first equation tells us that for all n:

$$(n+1)r_{n+1}t^n = -\alpha b_n t^n. \tag{3.11}$$

We can use the above equation to compute r_{n+1} as a function of b_n:

$$r_{n+1} = -\frac{\alpha b_n}{(n+1)} \tag{3.12}$$

Now, to compute the value of b_{n+1}, we use the second equation:

$$\frac{\partial B_2^2(t)}{\partial t} = -\beta R(t) \tag{3.13}$$

When we plug the power series $B_2(t) = \sum_{i=0}^{\infty} b_i t^i$ into the Eq. 3.13 we get that

$$\sum_{n=0}^{\infty}(n+1)\left(\sum_{i=0}^{n+1} b_i b_{n+1-i}\right)t^n = -\beta \sum_{n=0}^{\infty} r_n t^n. \tag{3.14}$$

Using the above equation, it follows that

$$(n+1)\left(\sum_{i=0}^{n+1} b_i b_{n+1-i}\right)t^n = -\beta r_n t^n \tag{3.15}$$

or

$$b_{n+1} = \frac{-\beta r_n - (n+1)\left(\sum_{i=1}^{n} b_i b_{n+1-i}\right)}{2(n+1)b_0}. \tag{3.16}$$

assuming that each army is standing and firing at the other army.

While the described Lanchester model is the simplest method for modeling combat dynamics, there are some extensions.

The main drawback of all these models is that they ignore geography. It is well known that the key to winning the campaign is to incorporate geography-related tactics. In general, all previous works did not use the context of geography.

$$\begin{array}{ccc} \mathbb{A}rc & \xrightarrow{f_1} & \mathbb{M}_{2,d}[\mathbb{R}] \\ & \searrow^{\pi_1} & \uparrow{\scriptstyle \gamma_1} \\ & & \mathbb{T} \end{array} \tag{3.17}$$

3.5 Incorporate Geography for the Lanchester's Model

In this section, we are going to explain how to extend the Lanchester model in such a way that it can incorporate the movement of armies. The main idea is to add a function that allows us to describe the location of each army at each time of the battle. Once we have the function, we can break the time into homogeneous time intervals where in each time interval the battle is described by a specific differential equation system. Before we explain how to do that, let us describe a more general model that can be applied to different types of armies in historical contexts.

Lanchester's laws do not take into account factors such as terrain, environmental conditions, or different unit types within an army. By adding geographical points, we can describe where specific military units are located as well as what type of unit they are. We will then be able to write nonhomogeneous differential equations in time and solve them/calculate, in the same way as Lachester's laws, the entire process of a battle in every stage. We consider each step of the battle to be homogeneous during that time interval. We can now break down the development of the battle into several time steps. Per each time step, the set of differential equations will be set at homogeneous intervals.

We assume that there are two sides in the conflict and d different types of forces on each side. For example, in first century wars, we have troops and cavalry. In this case, one might think that $d = 2$, but it is possible that $d = 4$ or even more. It depends on the way the battle is fought, as different factors may contribute to units' different abilities.

We will start with the simple version using Lanchester's laws. In this case, we assume that there is only one homogeneous army for each side, i.e., $d = 2$. We denote the red army by R, and the blue army by B. The number of soldiers in the red army at time t is denoted by $R(t)$ and, correspondingly, the number of blue soldiers in the blue army at time t by $B(t)$.

3.6 Battle of Cannae

Let us consider the Battle of Cannae, fought in August 216 BC. The Battle of Cannae was a major battle of the Second Punic War between the armies of the Roman Republic and the Carthaginian Empire and resulted in a decisive victory for the Carthaginians, led by General Hannibal. The Romans led an army with 80, 000 infantry and 6, 400 cavalry, while Hannibal led an army with 40, 000 infantry and 10, 000 cavalry. The Carthaginians won in spite of their smaller army, showcasing the tactical power of flanking.

The Romans followed their usual strategy, relying on their overwhelming numbers and their ever-pushing march forward to break enemy lines and mow down the other army. Hannibal, aware of their tactics, arranged his army in a crescent. Initial contact was made only with his center troops, who then fell back, drawing the Romans forward as the crescent inverted and collapsed around them. The Romans, emboldened by the retreating troops, pushed forward. Meanwhile, Hannibal's cavalry won against the Roman cavalry on either side of the infantry and collapsed on the backs of the Roman forces. The Romans found themselves surrounded, unable to use their traditional formation to overwhelm their opponents. They were cut off from the back and sides and suffered major losses. Some estimate that as many as 50, 000 Roman soldiers were killed, whereas Hannibal only lost 6, 000 of his front lines.

3.6.1 The Forces Before the Battle

In general, the battle of Cannae can be deconstructed into 4 smaller battles. The first battle began with the attack of the main Roman infantry forces against the Carthaginians Fig. 3.4. Hannibal drew up his troops in the shape of a crescent, and the Romans,

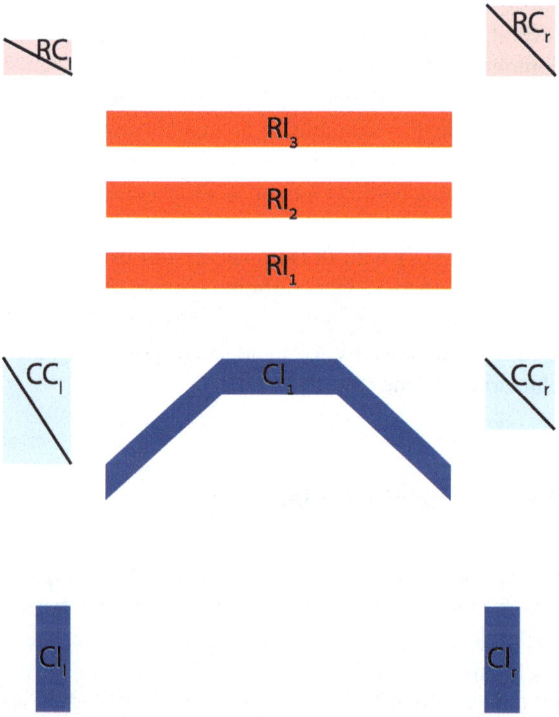

Fig. 3.4 This is the first stage of the battles. Red represents Roman infantry, light red represents Roman cavalry, blue is Carthaginian infantry, and light blue is Carthaginian cavalry

attacking in the center, advanced in pursuit the slowly retreating Carthaginians, leaving their flanks uncovered. The second and third battles were cavalry battles that made the flanking strategy possible.

3.6.2 Modeling the Battle of Cannae

We will now construct a mathematical model of the battle of Cannae. We have two armies, each consisting of five units: 3 infantry and 2 cavalry. The Roman army can be written in mathematical terms as follows:

$$R = (RI_1(t), RI_2(t), RI_3(t), RC_l(t), RC_r(t)) \tag{3.18}$$

The infantry of the Roman army $RI_1(t)$, $RI_2(t)$, RI_3 is organized in three rows. The first unit $RI_1(t)$ is the closest to the Carthaginians, and the third one $RI_3(t)$ is the furthest. The Roman army has two cavalry units. The left unit $RC_r(t)$ was led by Paulus and had 2200 cavalry. The right unit $RC_r(t)$ was led by Varos and had 4000 cavalry.

The Carthaginian army is written in mathematical terms as follows:

$$C = (CI_1(t), CI_l(t), CI_r(t), CC_l(t), CC_r(t))) \tag{3.19}$$

The infantry of the Carthaginian army $CI_1(t), CI_l(t), CI_R(t)$ was divided into one main line $CI_1(t)$ and two reserve squads $CI_l(t), CI_r(t)$ that lay hidden from the Roman forces to their left and right. They were used later in the fourth battle to flank the Roman army. The Carthaginian cavalry consisted of two units. The left unit $CC_l(t)$ was led by Hasdrubal and had 4000 Numidians and 2000 Iberians, 6000 horses in total. The right unit $CC_r(t)$ was led by Hasdrubal and Hanno and had 4000 horses.

We note that in the general model for each of these forces, we need to have several functions n, geo. For example, one may need to know the number of infantry and cavalry n at each time, as well as the position geo of each of these forces. Since the number of cavalry commanded by Hasdrubal at the beginning of the battle was 6000, we can write it as follows:

$$n(CC_r(0)) = 6000 \tag{3.20}$$

However, to simplify the model, we assume that our variable contains only the number of soldiers in each unit of the army. Instead of using the position, we encode the position into the differential equation and, by doing that, reduce the complexity dramatically. Since we encode the position into the differential equation, the system stops being homogeneous in time, and we have several systems of equations.

Now we will describe the starting position using the differential equation, as well as the evolution of the battle and the results for each of the four battles of Cannae.

3.6.3 The First Cavalry Battle

We now write the differential system that describes the first cavalry battle.

The first cavalry battle was between Hasdrubal and Paullus. As previously stated, Hasdrubal commanded 4000 Numidians and 2000 Iberians, 6000 in total, while Paullus had 2200 Roman cavalry Fig. 3.5. This battle ended with the destruction and escaping of Paullus's cavalry.

$$\begin{cases} \frac{\partial CC_l}{\partial t} = -RC_l(t) \\ \frac{\partial RC_l}{\partial t} = -CC_l(t(t) \\ CC_l(0) = 6000 \\ RC_l(0) = 2200 \end{cases} \tag{3.21}$$

From a technological point of view, we assume that the Roman and Carthaginian cavalry were equal, and subsequently choose the constant to be 1. The solution to the equations is

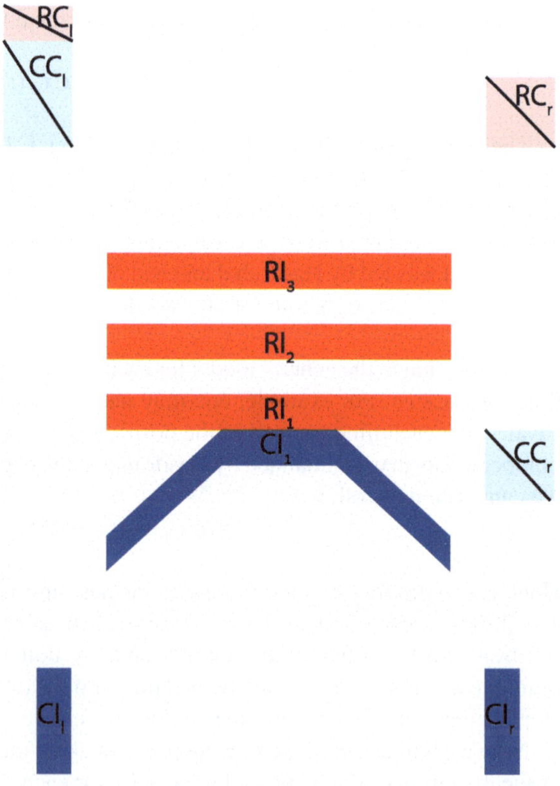

Fig. 3.5 The second stage of the battles and the first cavalry battle. The Carthaginians had 6000 cavalry, and the Romans had 2200 cavalry

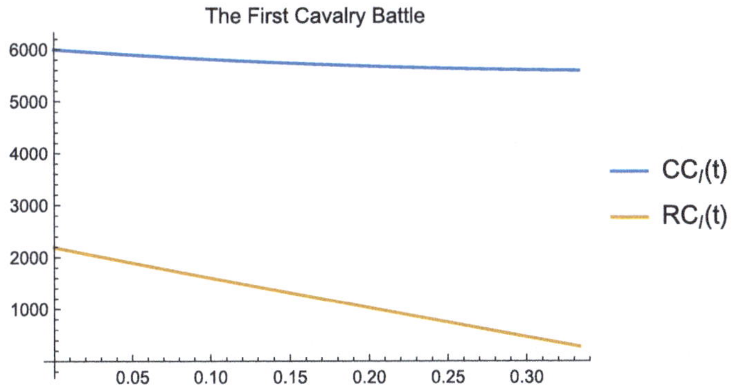

Fig. 3.6 Evolution of the battle as a function of time, where the Romans lost 90% of their forces

$$\begin{cases} CC_l(t) = 100e^{-t}\left(19e^{2t} + 41\right) \\ RC_l(t) = -100e^{-t}\left(19e^{2t} - 41\right) \end{cases} \tag{3.22}$$

According to Livy, the first cavalry battle was fierce, so we assume that the Roman cavalry fought until only 10% of their forces survived. This allows us to compute the duration of the first cavalry battle by solving the following equation:

$$RC_l(t) = 220 \tag{3.23}$$

Using the Wolfram Mathematica command NSolve, we get that the duration of the battle is $t = 0.345165$. We recommend using a numerical solution since there is generally no close formula for trying to solve a system of differential equations, especially when we apply the induction process. Since we will plug the results of this battle into the next battle, there will be no precise solution.

When we plug this equation $t = 0.345165$ into the function $CC_l(t)$, we find that the number of Carthaginian cavalry after the battle is equal to 5586.45, so we can compute their losses (Fig. 3.6).

3.6.4 Counterfactual of the Second Cavalry Battle

The second cavalry battle was between Hasdrubal and Hanno on the Carthaginian side, and Varro on the Romans'. In reality, however, it never happened. According to our computation, Hasdrubal had 5586 cavalry left after the first battle, and Hanno commanded 4000. We denote the Carthaginian common force by $CC_{l,r}$, We assume that altogether the combined forces had about 9586 cavalry. The Roman force was led by Varro with 4000 cavalry. The Carthaginian forces successfully attacked the Roman cavalry from two directions, forcing them to flee Fig. 3.8. The Carthaginians won without a fight (Fig. 3.7).

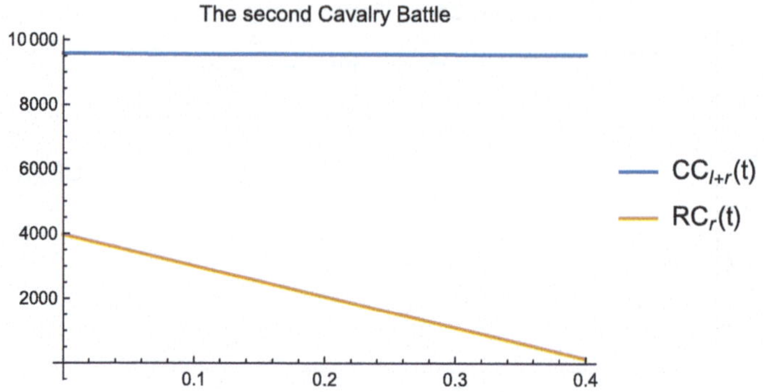

Fig. 3.7 Evolution of the battle as a function of time, where the Romans lost 90% of their forces

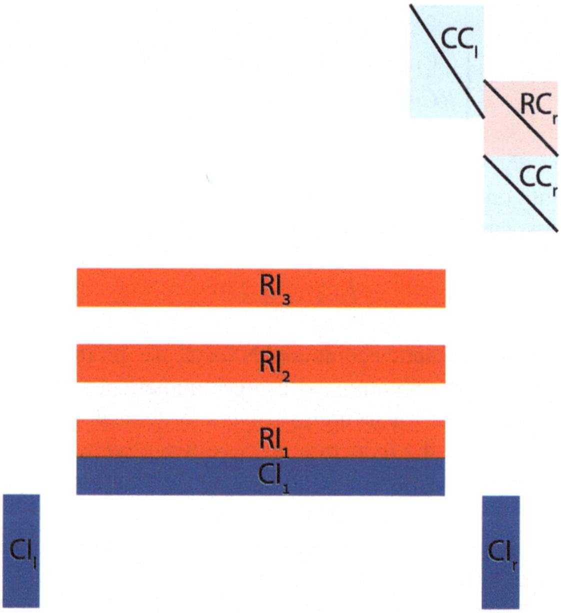

Fig. 3.8 Emulation of the second cavalry battle using Lanchester's differential equation for flanking, see 3.24. Blue is the combined forces of Carthaginian cavalry, and orange is the right squad of Roman cavalry

Our model of flanking allows us to look at what would happen if the Romans hadn't escaped and instead fought against the Carthaginian cavalry. We compute the counterfactual second cavalry battle to judge the Roman cavalry's decision to flee. In this case, the Carthaginian forces would flank the Roman cavalry, and therefore, we use Lanchester's equation for flanking (see Sect. 3.4). The equations are:

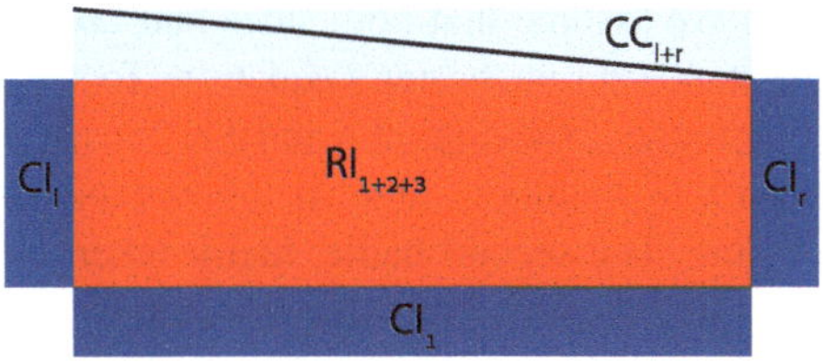

Fig. 3.9 The last stage of the battles, where the Romans were flanked and surrounded by the Carthaginians

$$\begin{cases} \frac{\partial CC_{l,r}}{\partial t} = -\sqrt{RC_l(t)} \\ \frac{\partial RC_l}{\partial t} = -CC_{l,r}(t(t)) \\ CC_{l,r}(0) = 9586 \\ RC_l(0) = 4000 \end{cases} \tag{3.24}$$

Solving these equations numerically with Wolfram Mathematica or another computing tool allows us to get the outcome and duration of the battle. In this case, the outcome shown in Fig. 3.8 would be the total annihilation of the Roman cavalry within a very short duration of the battle in 0.4-time unit—roughly 1.6 h—and minimal losses for the Carthaginian cavalry. For more calculations of time, see Sect. 3.6.7.

The battle of Cannae is very synchronized.

One may think that had the second cavalry battle been long enough, it would have given the Roman infantry time to defeat the Carthaginian infantry. We can calculate how much time the Roman infantry would have needed to win the battle—between 10 and 15 h. The flanking started 5 h after the beginning of the battle, so the Roman infantry was absent for at least 5 h. We can see that the second cavalry battle could not have lasted that long. They could have provided two or two and a half hours at most, but that was not enough.

According to our counterfactual model, even if the second battle had been fought, the Carthaginians could continue their plan without any changes, and the Romans would have lost an additional 4000 cavalry while the Carthaginian losses would have been about 20 cavalry.

After the second cavalry battle ended with the Roman cavalry's retreat, the Carthaginians moved their cavalry to attack the main Roman forces. The Romans were surrounded, and the fourth and final battle began. Although the Romans were ahead in terms of numbers, they were attacked from all four sides and could not effectively fight back Fig. 3.9.

3.6.5 The First Infantry Battle

The first infantry battle started at the beginning of the Cannae campaign and ended after the Roman cavalry was eliminated. A reasonable estimation of the duration of the battle is one unit of time, approximately four or five hours. During that period, the first line of the Roman infantry RI_1 fought the Carthaginian's main infantry forces

CI_1. We assume that both sides had 26, 666 soldiers. To describe the fight, we use the standard Lanchester's equation. To see the battlefield of the first infantry battles, refer to these two Figs. 3.5 and 3.8. We assume that the duration of the battles ranged from 0 to 1 units of time. It follows that the first battle lasted from 0 to 0.35 units of time. The second battle, had it taken place, would have taken around 0.4 units; however, it was not, and therefore we take away 0.25 for the sake of our computation. Organizing the cavalry amounts to 0.2 units of time after each battle, which is roughly 0.8 h.

When trying determine α and β in the original Lanchester's equation, we find ourselves facing two problems. One is the Romans' advantage over the Carthaginians. We assume that since both sides were professional soldiers possessing the same technology, there was no advantage. Therefore, we assume $\alpha = \beta$.

Another thing we need to take into account is how to compare infantry with cavalry. Here we compare the speed of cavalry and infantry, assuming that infantry soldiers could march at a speed of 15 km per hour and the cavalry could move at a speed of 80 km/h. Therefore,

$$\alpha = \beta = 15/80 \tag{3.25}$$

$$\begin{cases} \frac{\partial CI_1}{\partial t} = -15/80RI_1(t) \\ \frac{\partial RI_1}{\partial t} = -15/80CI_1(t(t) \\ CI_1(0) = 26666 \\ RI_l(0) = 26666 \end{cases} \tag{3.26}$$

The outcomes of the first infantry battle can be seen in Fig. 3.10 The functions are equal since the Romans and the Carthaginians had the same technology and started out with the same number of soldiers.

The outcome of the battle is approximately 4560 casualties from both sides. These were the main Carthaginian casualties for the day.

3.6.6 *The Final Battle*

The final battle began after the elimination of the Roman cavalry. We assume that the battle started after a single unit of time, and the Roman infantry was flanked and surrounded by the Carthaginian forces, as shown in Fig. 3.9. At the beginning of the battle, Carthaginian forces contained 4 squads of troops. The main force was infantry $CI_1(0) = 22106$. This infantry fought in the first battle, where it lost 4560 soldiers, as explained in the previous subsection. The soldiers were exhausted, but their technological constant remained equal to the Roman forces. Thus, we can ignore this fact in our computation. The second and third forces of the Carthaginians were the Numidian reserves, which lay hidden from the main Roman forces during the first infantry battle. The size of these forces was CI_l, $CI_r = 6666$. The fourth group

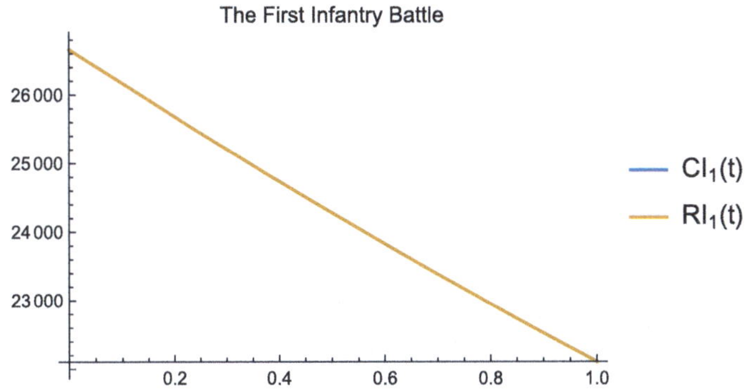

Fig. 3.10 The number of infantry during the first unit of time, which is roughly 4 h. Two functions in the graph are equal, denoted by blue and orange, and so the former is obstructed by the latter

of Carthaginian forces was the combined cavalry $CC_{l,r} = 9586$ since the second cavalry battle never really happened.

The combined Roman infantry forces are denoted by $RI_{1,2,3}$. The number of infantry at the beginning of the battle is $75438 = 22106 + 2 \times 26666$. The starting positions of this battle are shown in Fig. 3.9.

The differential equation for flanking is that the effectiveness of the surrounded army is measured by the square root of its size. Since we have four Carthaginian armies, we assume that the destruction is divided equally, i.e., the destruction for each of the Carthaginian army is $\sqrt{RI_{l,r}(t(t)}/4$. This assumption does not really matter. The first equation describes the destruction of the Roman army. We assume that the infantry works with an efficiency of 15/80 compared to the cavalry; see Eq. 3.25.

$$\begin{cases} \frac{\partial RI_{1,2,3}}{\partial t} = -15/80(CI_1(t) + CI_l(t) + CI_r(t)) - CC_{l,r}(t) \\ \frac{\partial CI_1(t)}{\partial t} = -\sqrt{RI_{l,r}(t(t)}/4 \\ \frac{\partial CI_l(t)}{\partial t} = -\sqrt{RI_{l,r}(t(t)}/4 \\ \frac{\partial CI_r(t)}{\partial t} = -\sqrt{RI_{l,r}(t(t)}/4 \\ \frac{\partial CC_{l,r}}{\partial t} = -\sqrt{RI_{l,r}(t(t)}/4 \\ RI_{1,2,3}(0) = 75438 \\ CC_{l,r}(0) = 9586 \\ CI_1(0) = 22106 \\ CI_l(0) = 6666 \\ CI_r(0) = 6666 \end{cases} \tag{3.27}$$

Looking at the solution of the differential equation (see Fig. 3.11), we can see that the purple line shows the destruction and annihilation of the Roman army at the end of the battle, while the other lines describe the Carthaginian forces whose numbers did not change during the battle. It is easy to see that the Romans suffered

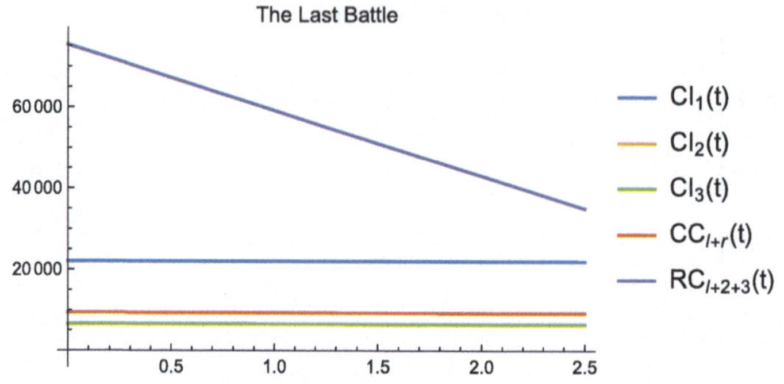

Fig. 3.11 Evolution of the final battle during the first 2.5 units of time

massive casualties during the battle, and assuming the battle ended within two units of time, the casualties would be as follows: The Romans suffered 32283 casualties in the final battle, and the remaining number of soldiers after both infantry battles was 43,155. We can compute the total infantry losses in both battles, which is 36,845. This total includes 2200 losses in cavalry, which equals 39,045. At the same time, the remaining number of Carthaginian forces after the final battle is: $CI_1 = 22083$ $CI_l = CI_r = 6544$ $CC_{l,r} = 9464$ This includes the numbers of the first Carthaginian infantry army, two reserve Numidian armies, and the remaining cavalry. The total number of remaining Carthaginian forces is 44,637.

The total casualties of the Carthaginians in all four battles was 5362. Almost all losses occurred during the first infantry battle, and only 386 casualties were counted for the final battle.

386.784

3.6.7 *Computing the Time*

Lanchester's equations measure the way a number of soldiers change through time, employing the idea of derivatives to capture a small change in the number of soldiers over a short period of time. General mathematical tools allow us to solve these Lanchester's equations either by numerical means or using exact solutions.

Lanchester's equations contain several destruction parameters which describe the number of soldiers that were killed in an instant. However, in general, it is difficult to determine these parameters exactly.

We evaluated the parameters in the Cannae battle by assuming that the parameters for the Roman and Carthaginian cavalry's destruction are one; this was an arbitrary choice.

We then set the infantry's destruction parameters. We decided that both Romans and Carthaginians possessed the same abilities; therefore, they should be equal. To set the exact value of the infantry's destruction parameters, it is necessary to compare it with the cavalry. To determine this relationship, we used the natural speed of humans versus horses.

We are left with an arbitrary choice of the cavalry's destruction parameters as one. There are several ways to overcome this problem, and ours may be the simplest. For each battle, we use Lanchester's equation to compute how many units of time are necessary to achieve a decisive outcome.

In the Battle of Cannae, we have two real battles and one counterfactual. The first cavalry battle took about 0.4 units of time; see Sect. 3.6.3. The second cavalry battle, which is a counterfactual, see Sect. 3.6.4, took another 0.4. Then we add 0.2 units of time for organizing the units. As the second battle never really happened, we define the time as 0.4 for the first battle and 0.6 for organizing cavalry. Therefore, the first infantry battles took one unit of time.

We assume that the second infantry battle took two units of time. In this case, the Cannae campaign took three units of time altogether. We note that the Battle of Cannae took place in August in Italy, when daylight was about 14 h long, with the sun rising at 6 o'clock in the morning. Assuming that (a) it took the armies two hours to organize and (b) the battle ended near sunset, it follows that the duration of the battle was 12 h. Therefore, one unit of time is equal to 4 h.

An equivalent description of the battle is to set the cavalry destruction parameters to $1/4$, and the infantry destruction parameters to $15/320 = 15/(4 * 80)$.

We set parameters by looking at each battle locally and set the global parameters to the total duration of the battle. We believe that this approach is better since the aggregation of time helps cancel noise.

In the next section, we begin to describe the general method of studying war history.

3.7 General Description

In general, when describing battles between two sides, Red and Blue, we will assume that each side has its own sequence of armies. Denote the sequence of the Red armies by $R = (R_1, \ldots, R_{nr})$, where nr is the number of different armies on the Red side. The Blue side has a sequence of Blue armies denoted by $B = (B_1, \ldots, B_{nb})$, where nb is the number of different armies on the Blue side.

Each of these armies has the following function: location, denoted by L, and number of soldiers, denoted by S.

We split the description of a campaign into several battles, $\mathcal{B}_1, \ldots, \mathcal{B}_k$, and allow the units to move in time between those battles. Note that units cannot participate in more than one battle at a time. Each unit has a position, and each battle has time and a location. In order to participate in the battle, each unit must be in the location of the battle at the time it is happening.

During the battle, each unit has a differential equation describing how the number of soldiers changes throughout the battle. We assume that this differential equation does not change during the battle. If it does, we split the battle into several battles.

For example, the Cannae campaign was split into several battles for this very reason. The differential equation before flanking the Romans was not the same as after it; therefore, we had to split the infantry battle into two.

We describe movement between battles with linear interpolation. For example, the location of the left cavalry of the Carthaginian force at the end of the first cavalry battle, which happened at 0.34 units of time, is denoted by $L(CC_l)(0.34)$. According to Fig. 3.5 the location is (−0.1, 1) i.e., we can write

$$L(CC_l)(0.34) = (-0.2, 1) \tag{3.28}$$

The location of the same unit at the beginning of the second cavalry battle, according to Fig. 3.8 is $(1, 0.8)$. The second cavalry battle was supposed to start at 0.6 units of time. Therefore, we can write

$$L(CC_l)(0.6) = (1, 0.8) \tag{3.29}$$

The location of the left Carthaginian cavalry between 0.34 to 0.6 is given by the linear interpolation for $t \in [0.34, 0.6]$

$$L(CC_l)(t) = (-0.2, 1)(t - 0.34)/(0.6 - 0.34) + (1, 0.8)(0.6 - t)/(0.6 - 0.34) \tag{3.30}$$

This function can be seen in Fig. 3.7.

Similarly, we can describe each movement during the campaign. The idea is to break down time into intervals where the description in each interval is simple and homogeneous. This means that a single equation describes the number of soldiers or the location of the units at each interval. We recommend splitting the description of the number of soldiers and their location to simplify the computation.

Note that in order to be able to write the differential equation that describes the number of soldiers during a battle (as we did in the Cannae campaign), we must set the destruction matrix ρ, i.e., the ability of each army to destroy each unit of the opposite side. In order to summarize all of the destruction coefficients in the matrix, we need to organize all of the armies that participate in the battle. For example, in the last Cannae battle, we organized the armies according to $O(CI_1, CI_l, CI_r, CC_{l,r}, RI_{1,2,3})$. In this case, the destruction matrix ρ is

$$\rho = \begin{bmatrix} 0 & 0 & 0 & 0 & \frac{3}{16} \\ 0 & 0 & 0 & 0 & \frac{3}{16} \\ 0 & 0 & 0 & 0 & \frac{3}{16} \\ 0 & 0 & 0 & 0 & 1 \\ \frac{3}{16} & \frac{3}{16} & \frac{3}{16} & \frac{3}{16} & 0 \end{bmatrix} \tag{3.31}$$

For example, according to the order O, the first line of the matrix ρ describes the destruction parameters of the Carthaginian army CI_1, which is $(0, 0, 0, 0, \frac{3}{16})$. This means that unit CI_1 cannot destroy any unit of the Carthaginian army.

3.8 Conclusion

In this chapter, we developed Lanchester's equation from the perspective of a historical battle. After explaining how to analyze the flanking, we described the Battle of Cannae using Lanchester's equation, where we were able to verify the number of casualties on both sides. Using this model, we were able to compute a more detailed description of the battle than the historical sources tell us. We were able to examine the decision of the right squad of the Roman cavalry to flee the battlefield instead of being eliminated. We then described the general framework for analyzing other historical battles. The advantage of our approach is that it allows us to question or examine the historical quantities and evolution of the battles with much more details than ancient historians could imagine. This can either add some missing data from the description or estimate if the existing data makes any sense, as ancient historians tended to exaggerate.

In the next chapter, we will describe a method that allows us to tell the historical narrative of two clocks. In general, historians tend to measure time using the Greenwich clock. However, it is well-known that we all have our own subjective clock, and historians are not exempt from this phenomenon. Using these tools allows us to compute critical time and compare historical entities.

3.9 Exercise

Lanchester's equations

1. One of the classic battles in ancient times is the first battle between the Jewish Hasmoneans and the Greeks, known as the Battle of the Ascent of Lebonah. The warring parties were the rebels, led by Judas Maccabeus, and the forces of the Seleucid Empire, led by Apollonius, governor of the Lebonah region. Apollonius' troops ventured through a curved, narrow passage, as depicted here by the red arrow. The blue mark represents the Maccabee forces, which were waiting on both sides of the passage. For more details, see [10] Describe the battle using Lanchester's equation. Model the elevation advantage the Hasmoneans had by setting the technology parameters of the Hasmoneans to be greater than Apollonius' technology parameters. Divide the Hasmoneans into four armies, each numbering 200, and divide Apollonius' forces into three armies, with the first two numbering 100 soldiers each. These armies fought against the Hasmonean forces. The third army, which contained the rest of Apollonius' forces—1800 soldiers—were used as replacements as soldiers from the first two armies were injured or killed in battle.
2. When we analyzed the Battle of Cannae in Sect. 3.6, we determined the concept of Lanchester's equation using the average speed of cavalry/horses and

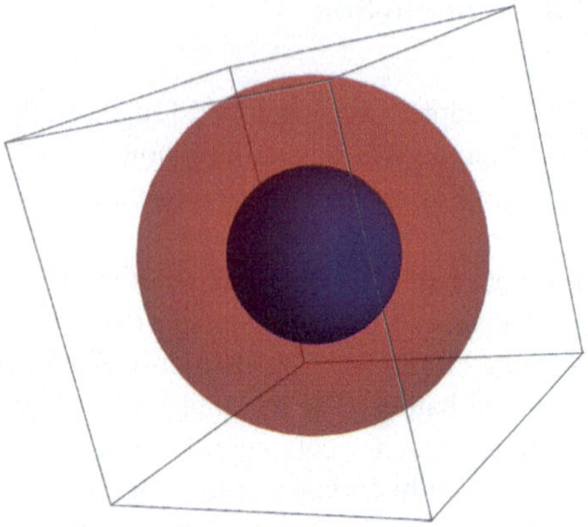

Fig. 3.12 An example of a typical 3D-flanking in air combat

infantry/human. In this exercise, we justify this methodology. We consider a non-homogeneous Lanchester model when time runs faster by a factor of 2 for the Blue army, i.e., $t_b = 2t_r$. Solve the system of equations and discuss its meaning.

$$\begin{aligned} \frac{\partial R(t_r)}{\partial t_r} &= -\alpha B(t_b) \\ \frac{\partial B(t_b)}{\partial t_b} &= -\beta R(t_r) \end{aligned} \tag{3.32}$$

Adapting Lanchester's Laws for Flanking

3. In the future, battles will be fought with flanks of drones. Imagine that we have two armies of drones, red and blue. Assume that the Red army flanks the Blue army of drones (see Fig. 3.12). Write Lanchester's equation of 3D flanking for this case. How would your answer change if, when flanked, the Blue drone army loses all communication with the headquarters, including radio connection?

Incorporate Geography for the Lanchester's Model

4. In this exercise we are going to analyze the Battle of the Kalka River. Read the evolution of the battles as depicted in the Wikipedia page [11]. Model the battle using Lanchester's equations together with geometry in a similar way to our analysis the Battle of Cannae (see Sect. 3.6).
5. In [12] you can find the list of most important battles in history. Pick your favorite and analyze it using the methodology we've developed in this chapter. Do the results of your analyses match the description of the battle? Can you find the time during which the battle was fought as we did for the Battle of Cannae? How many sub-battles comprised the entire battle?
6. In this exercise, we will derive a result equivalent to Lancaster's square law, but adapted to the case of Lanchester's Laws of Flanking (see Eq. 3.6). Specifically, we aim to prove:

$$\beta\left(R(t)^2 - R(0)^2\right) = \alpha\left(B(t)^{1.5} - B(0)^{1.5}\right) \tag{3.33}$$

Hint: Start by dividing the two equations in Eq. 3.6.

References

1. A. Rapoport, Lewis F. Richardson's mathematical theory of war. Confl. Resolut. **1**(3), 249–299 (1957)
2. A.J. Jones, *Game Theory: Mathematical Models of Conflict* (Elsevier, 2000)
3. S.J. Deitchman, A Lanchester model of guerrilla warfare. Oper. Res. **10**(6), 818–827 (1962)
4. Judy Hong Tam, Application of Lanchester combat model in the Ardennes campaign. Nat. Resour. Model. **11**(2), 95–116 (1998)
5. Daewon Chung, Byeongseon Jeong, Analyzing Russia-Ukraine war patterns based on Lanchester model using SINDy algorithm. Mathematics **12**(6), 851 (2024)
6. H.K. Weiss, Combat models and historical data: the US civil war. Oper. Res. **14**(5), 759–790 (1966)
7. Jerome Bracken, Lanchester models of the Ardennes campaign. Nav. Res. Logist. (NRL) **42**(4), 559–577 (1995)
8. C.-Y. Hung, G.K. Yang, P.S. Deng, T. Tang, S.-P. Lan, P. Chu, Fitting Lanchester's square law to the Ardennes campaign. J. Oper. Res. Soc. **56**(8), 942–946 (2005)
9. M. Zyda, D. Pratt, J. Falby, P. Barham, K. Kelleher, NPSNET and the naval postgraduate school graphics and video laboratory. Presence: Teleoperators Virtual Environ. **2**(3), 244–258 (1993)
10. Battle of the ascent of Lebonah (2023). https://en.wikipedia.org/wiki/Battle_of_the_Ascent_of_Lebonah
11. Battle of the Kalka river (2023). https://en.wikipedia.org/wiki/Battle_of_the_Kalka_River
12. Wiki. Wikisourcelists-of-battles (2023). https://en.wikipedia.org/wiki/Lists_of_battles

Chapter 4
History Through Clocks

Influential linguists Labov and Waletzky defined narrative as consisting of two temporal connected sentences [1]. Switch out "sentences" for "clocks" or "functions," and you have the essence of this chapter. Any two clocks generate a narrative; this is especially useful in digital history, where historians and machines easily understand clocks (Fig. 4.1).

4.1 Syuzhet and Fabula Times

Time is central in narrative discourse, as extensively described in [2]. Although there are many types of time in the context of a historical narrative, for now, we will focus on two: Fabula and Syuzhet, which describe exactly what happened and how, respectively. In other words, Fabula describes the thematic content of the situation and Syuzhet describes the actions in chronological order of the narrative.

Note that, in general, the way the narrative is told cannot be monotonic in time. Therefore, there is a difference between the Syuzhet and Fabula time. To capture the relationship between the two, it is helpful to represent the Fabula as a function of Syuzhet. For example, one can see the Fabula as a function of Syuzhet of the movie *Memento* [3].

For machines, it is much easier to compute the Syuzhet time than the Fabula time. In the same manner that we measure time in units of minutes, seconds, and so on, we can measure Syuzhet in units such as words, paragraphs, number of hyperlinks, etc. When moving from small units of Syuzhet time to bigger units – moving from letters to words to paragraphs—two phenomena occur simultaneously. More and more cognitive information appears, and Syuzhet time becomes increasingly parallel. This reveals that sometimes events are not points in time, but rather intervals.

We now consider a simple example of seven main historical events of WWII. We already explored them in Sect. 2.2.5. Assume that we have a narrative

Z. Lotker, *History by Algorithms*,
https://doi.org/10.1007/978-3-031-93627-2_4

Fig. 4.1 A cartoon about the problem of multiple clocks

$$\mathcal{N}ar = (p_1, p_2, \ldots, p_7), \tag{4.1}$$

that tells the history of WWII in seven paragraphs. The i-th paragraph p_i tells the story of the h_i historical event depicted in Fig. 2.2, which summarizes WWII according to [4]. Since we follow the Fig. 2.2, the Syuzhet time is

$$t_s = 1, 2, 3, 4, 5, 6, 7 \tag{4.2}$$

The Fabula time for the first paragraph, "Invasion of Poland," starts on the 1st of September, 1939 and ends on the 6th of October, 1939. The Fabula time for the second paragraph, "Battle of France," starts on the 10th of May, 1940 and ends on the 25th of June, 1940. The Fabula time for the third paragraph, "Battle of the Atlantic," starts on the 3rd of September, 1939 and ends on the 8th of May, 1945.

Therefore, the Fabula time can be seen in Fig. 4.2.

Occasionally, time is described in intervals instead of a single point, for example, a war that goes on for years. When discussing the concept of time, we can talk about several events from different periods in the same paragraph.

Syuzhet	Start	End
1	Fri 1 Sep 1939	Fri 6 Oct 1939
2	Fri 10 May 1940	Tue 25 Jun 1940
3	Sun 3 Sep 1939	Tue 8 May 1945
4	Thu 30 Nov 1939	Wed 13 Mar 1940
5	Tue 9 Apr 1940	Mon 10 Jun 1940
6	Mon 10 Jun 1940	Thu 13 May 1943
7	Sun 22 Jun 1941	Wed 9 May 1945

Fig. 4.2 Table Seven major historical events of World War II. The first column presents the syuzhet (narrative order), while the following columns represent the fabula (chronological structure), with the second column indicating the start time and the third column the end time of each event

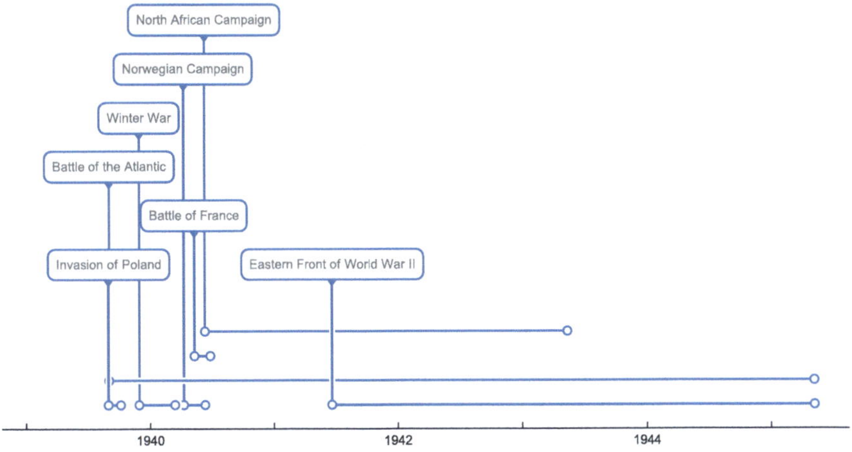

Fig. 4.3 This figure shows the timeline of the main historical events of WWII

Another problem is that the same paragraph or line can have several interpretations, especially when discussing something without context. For example, when we talk about Caesar, it is not immediately clear which namesake we are referring to: Julius Caesar or Augustus Caesar. This issue can be overcome by human advice or by putting a probability measure on each possible interpretation.

Now we explain in detail how to represent Fabula time as a function of the Syuzhet time. Our starting position is the timeline. For simplicity, we continue with a simple example of the main historical events in WWII mentioned above, see Fig. 4.3.

Each historical event receives a single paragraph while they describe different Fabula times. Therefore, Fabula time is a function of Syuzhet. We have three flashbacks (see Fig. 4.4) for the timeline of the seven main historical events of WWII. Note that in the timeline, Fabula time is an advantage, but there is no Syuzhet time. Now, if we wish to draw Fabula time as a function of Syuzhet time, both need to change. Assuming that Syuzhet time is linear, in the basic units of Syuzhet time, we need to compute the slope of the Fabula as a function of Syuzhet. This is given by the formula below:

$$\frac{\Delta Fabula(t_s)}{\Delta t_s} \tag{4.3}$$

Using the above formula (4.3), the Fabula of our WWII example can be seen in the following figure.

Now, we can transform the timeline in Fig. 4.3 using formula (4.3) to draw the Fabula function as a Syuzhet function when our narrative follows the Fig. 2.2. The outcome of this process can be seen in Fig. 4.4. Note that although each historical event in our example has a single paragraph, the slope that describes paragraphs is different because the length of the Fabula time is different.

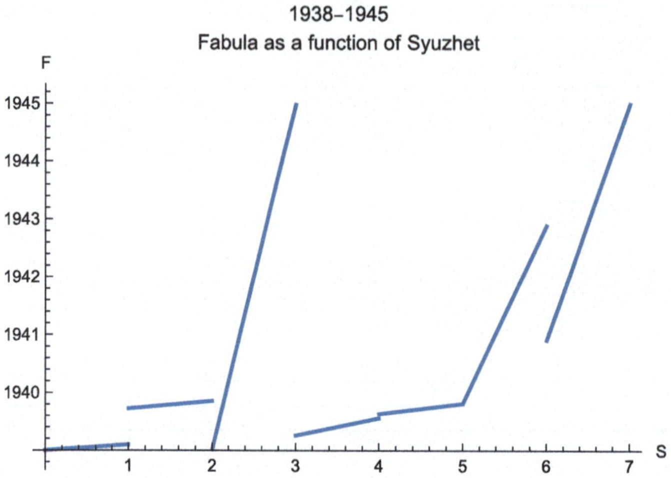

Fig. 4.4 WWII Fabula time as a function of the Syuzhet time in the simple narrative with 7 paragraphs

The link structure allows us to quarry the knowledge base for each entity, the time points, or time intervals. This will give us the mapping from Syuzhet time to Fabula time. We will discuss more about the relationships between Fabula and Syuzhet in Chap. 13.

The next section will discuss tools that are used to measure time.

4.2 Clocks

Time is a complex concept, and understanding it is essential to history. It is customary to use clocks to measure time in exact sciences, such as physics, and we follow this tradition. However, because we are interested in subjective time , we use the following definition of clocks: a mechanism for measuring time. In physics, time only moves in one direction. Clocks move only forward, and the measurement always increases. In mathematical terms, a clock

$$C : \mathfrak{T}_s \to \mathfrak{T}_m \tag{4.4}$$

is a monotonically increasing function of time, i.e., if $t_1 \leq t_2$ then $C(t_1) \leq C(t_2)$. We will assume that both $\mathfrak{T}_s$ and $\mathfrak{T}_m$ are compact, i.e., bounded and close.

Note that the Syuzhet time is actually the clock since it is monotonic, while the Fabula is not in most of the narratives, as it is not monotonic.

The clocks we use are subjective. They track whichever historical object we are interested in. When we talk about *objective* clocks in physics, we mean that they are made of two parts: an oscillator and a counter. The same may be said about *subjective clocks* or the Syuzhet $\mathfrak{T}_s$. The oscillator in a subjective clock is a generator of all historical events. It is a subjective generator, as it is a subjective decision for which events are part of the generator. The lists L off of which we work determine which events push the clock forward. As such, our starting list is the oscillator of our historic subjective clocks. The second part of the clock, the counter, keeps track of the object of interest.

Figure 4.10 demonstrates time as a key component in distributed computing, where each machine has its own clock, which is also a machine itself. The character in the figure is trying to measure time using a specific machine designed for time measurement, represented by a cuckoo clock. The figure suggests that there might be a conflict between the character's subjective time and that of the cuckoo clock.

In addition to the traditional stream of time, a counter can keep track of any physically countable progression. Different counters may be useful, depending on the narrative being processed. For example, a historian operating from a nationalist framework might sort and count nations of origin in their list, whereas a Marxist framework might call for the sorting and counting of economic classes, and a feminist framework would necessitate sorting and counting by gender categories.

Clocks are two-dimensional objects. To define a clock, we need an oscillator or event generator and a counter. In physics, this oscillator might be a pendulum, crystal, or any one of many other mechanisms. In the humanities, the events generated by the world are what drives the clock forward.

A narrative can be created when we look at two or more clocks. The question is how to compare the two clocks in order to find the critical events that make up the narrative. We use the methods developed in the 14th chapter of my first book, *Analyzing Narratives in Social Networks* [5].

The narrative that two clocks tell is formed when they deviate from each other. If they are always synchronized, they are the same and therefore produce no narrative. When they start to deviate from each other, we can ask why one clock begins to run faster than the other? The key is to compare the average speed of the clocks with their instantaneous speed. If the clocks cannot be synchronized, we deduce that there is no relationship between them. This means that there is at least one point of time t^* before which one of the clocks was running faster and the other slower with respect to their average speed. This relationship is reversed after t^*, which is the critical point where the average speed changes. We will develop the M-diagram as a mathematical tool to find these critical points.

In the context of history, we can define many subjective clocks, and any two of them tell a narrative that can be used to shed light on historical narratives. Defining subjective clocks in the context of a historical narrative is an algorithmic tool that allows historians to probe the critical points.

Clocks are central to understanding narratives. We can easily generate historical clocks from a list of people L_p—Wikipedia has many such lists on practically every subject. For example, the following link contains a list of famous physicists: [6]. Each of these people has their own Wiki page. Another example is a list of famous feminists, see [7].

We can convert each list into a clock according to the following procedures: For each person on the list L_p, we can find their date of birth as stated on their respective Wiki page. We can define clock $C_B(t)$ as the number of people in list L_p born before time t. This is a monotonically increasing function and is therefore a clock. The following diagram represents the process of computing the clock. To simplify the diagram, we assume that the archive is

$$\mathbb{A}cr_1 = \{(t, c_1(t)) : t \in \mathbb{T}ime\} \tag{4.5}$$

In Chap. 2, we developed the diagram language to describe digital history. We will now build on that by describing diagram clocks and their tales using functions. Below is a diagram of generating a clock from a list of people with a Wiki page.

$$\begin{array}{ccccc} & & \mathbb{T} & & \\ & \overset{Oscillator}{\nearrow} & \uparrow \chi & \overset{Clock}{\searrow} & \\ L_p & \xrightarrow{Oscillator} & \mathbb{A}rc_1 & \xrightarrow{counter} & \mathbb{R} \end{array} \tag{4.6}$$

In the next section, we will describe the process of generating two clocks and the narrative they tell.

4.3 Related Work

According to St. Augustine, time is an extension of the soul and, therefore, every person perceives time subjectively; the way we perceive time is a window to our soul (see [8]). Similarly, Historical entities perceive time through their collective soul, and the way they perceive time is a window to the collective soul of the entity. Therefore, it is important to construct the subjective clock of the historical entity. In this chapter, we provide several examples of how to construct those subjective clocks and discuss the methodology for analyzing them.

Time is one of the seven fundamental physical quantities in both the International System of Units (SI) and the International System of Quantities (see [9]). In physics, time is measured by clocks (see [10] for more information).

In ancient times, the best way to accurately measure time was the human heart - specifically, our pulse. The first to construct a more accurate method for measuring time was Christiaan Huygens, with his invention of the pendulum clock (see [11]). Up to Huygens, there was no difference between objective and subjective time from a practical point of view, as humans used the heart to measure both kinds of time.

Time is a central component in narratives (see Sect. 2.3.2), and can be divided into two measurements: *syuzhet* and *fabula*. Syuzhet is a sequence of events as presented in the narrative, while fabula is a sequence of events as they occurred in history. For more on this, see [3].

4.4 Comparing Two Clocks

The methodology of comparing two clocks was developed in my previous book [5]. Most of this section is taken from that book with a slight variation. For the sake of simplicity, we will assume that a clock's domain and co-domain are some close intervals. We give only the necessary information; a more extensive explanation is provided in the aforementioned book.

To compare two clocks, we must first normalize them in order to cancel the translational and rotational symmetry. This means that any comparable function Ω of two clocks $\Omega(C_1, C_2)$ must satisfy the transformation, formally: $\Omega(C_1, C_2) = \Omega(C_2, C_1)$. The normalization is done using the linear functions provided below. In general, clocks have Euclidean symmetry, meaning that we can change the unit of time and the starting point for counting it. In other words, we can measure time in seconds, minutes, hours, etc. We can start measuring time from the birth of Jesus, the Roman Revolution, or the fall of the Bastille. In other words, the symmetry of clocks is translation and rotation. This means that we can define equivalent relationships $\sim$ between clocks: two clocks C_1, C_2 are equivalent if there are two monotonically increasing linear functions f_1, f_2 such that the following equation fits for all $t \in \mathfrak{T}_s$

$$f_1(C_1(t_t) = C_2(f_2(t)) \tag{4.7}$$

The next step before we can compare two clocks we would be finding natural representatives for each equivalent relationship of $\sim$. To do so, we define the normal clock's notation as a clock where domain and co-domain are the unit intervals. Formally:

Normal-Clocks
Definition 5.1. Clock C will be considered *Normal clock* if both standard and measured time are the unit interval

$$C : [0, 1]. \to [0, 1]. \tag{4.8}$$

such that

$$C(0) = 0;\ C(1) = 1 \tag{4.9}$$

In my previous book [5] I've shown that for all clocks C, there exists a unique normal clock NC. Two clocks that are equivalent $C_1 \sim C_2$ have the same normal clock, i.e.,

$$\text{If } C_1 \sim C_2,\ \text{Then } NC_1 = NC_2 \tag{4.10}$$

Now we can compare clocks using M*-diagrams*. The diagram 5.2 captures the variance of one clock with respect to another. It is necessary to normalize both clocks before comparing them using M-diagrams. After normalizing the two clocks, the first clock should be removed from the second one. We use the technique proposed in my previous book [12]. The first clock is represented in the M-diagram by the x-coordinate, while the difference between the two clocks is shown by the y-coordinate. Formally:

M-diagram

Definition 5.2. Let C_1, C_2 be two clocks. The M*-diagram* of C_1, C_2 is the function

$$M[C_1, C_2](t) = NC_2(t) - NC_1(t) \tag{4.11}$$

This means that in the M-diagram, the function starts at 0 and ends at 0 since

$$M[C_1, C_2](0) = 0, \tag{4.12}$$

given that

$$NC_1(0) = NC_2(0) = 0 \tag{4.13}$$

and

$$M\left[C_1, C_2\right](0) = 0, \tag{4.14}$$

since

$$N(C_1)(1) = N(C_2)(1) = 1. \tag{4.15}$$

In the previous section, we described the process of generating a single clock in the function diagram 4.6. Now, we are going to describe the process of generating two clocks C_1, C_2 from two lists L_{p1}, L_{p2} of people. We change the archive to $\mathbb{A}cr_2 = \{(t, c_1(t), c_2(t))$

$$\begin{array}{ccccc} & & \mathbb{T}\times\mathbb{T} & & \\ & \overset{O_1\times O_2}{\nearrow} & \uparrow x & \overset{C_1\times C_2}{\searrow} & \\ L_{p_1}\times L_{p_2} & \xrightarrow{f_1\times f_2} & \mathbb{A}rc_2 & \xrightarrow{counter} & \mathbb{R}\times\mathbb{R} \end{array} \tag{4.16}$$

To compare different clocks, we need to normalize them. The following diagram represents a process of normalizing a single clock. Note that $\mathbb{A}rc_1$ was defined in Eq. 4.5.

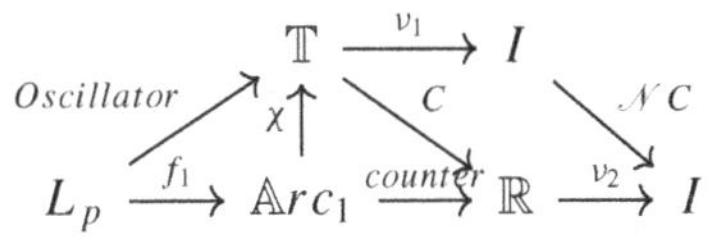

(4.17)

After normalizing the two clocks, we compute their martingale diagram $M[C_1, C_2] = M[NC_1, NC_2]$, like so:

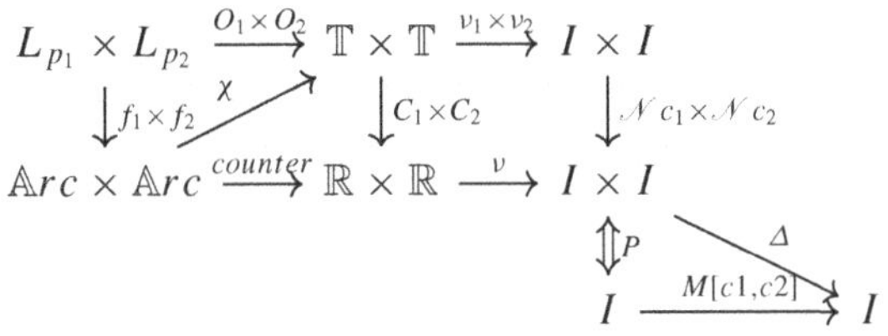

(4.18)

4.5 Programming Tips for Clocks

According to our definition, subjective clocks are a monotonous function of time. One of the advantages of subjective clocks is that there are many ways to define and compute them. This section will give some technical tips for programming and computing clocks. We will distinguish between two methods for computing subjective clocks. The first method starts with a list of historical entities and computing their subjective clocks. The second method starts with a historical text and transforms it into a sequence of positive numbers using LLM, generating a monotonic function from these numbers. In the next subsection, we discuss how to transform the list into subjective clocks.

4.5.1 Generating Subjective Clock from a List

We start with a list of objects that describe the historical entities we wish to study. For example, if we wish to study the history of physics, we can review the list of prominent physicists on Wikipedia. We denote this list by L.

First, we would like to generate a temporal order on list L. To do that, we will look for each element in L and its unique identifier for some global archives such as Wiki-data or Wolfram knowledge base archive. During this process, some of the elements in the list could be lost. We delete all elements without a unique identifier and denote the new list L_1.

Our next step is to give each element in L_1 its own time signature. In the case of people, we choose to work with the date of birth, which is P569 in Wikipedia.

One can choose other time signatures, such as date of death P570. We look for each element in L_1 to find the time signature. If the element doesn't have the needed signature, we exclude it from the list.

Next, we generate a function $key2Greenwich0$ using Association in Wolfram language or a dictionary in Python. This data structure has a key, and we have a single value for each key. Note that for each key k, the function has a unique value $key2Greenwich0[k]$. However, this function is not invertible[1] in the general case since a single time can contain many keys. Our next step is to transform the function into an invertible one. One way is to perturb the Greenwich time in a unique way. We call this function $key2Greenwich$. We will use this function to determine the Fabula time, meaning that for each two elements $x, y \in L_1$ we will say that x happens before y if, and only if

$$key2Greenwich[x] < key2Greenwich[y] \tag{4.19}$$

Let us explain how to generate general subjective clocks C. We assume that we are given a function Δ such that

$$\Delta : L_1 \to \mathbb{R}^+ \tag{4.20}$$

This means that for each element $x \in L_1$ we have a positive number $\Delta(x) > 0$. If there exists an element $x \in L_1$ for which Δ is not defined, we dispose of it. We denote the remaining element by L_2.

$\Delta(x)$ will determine how much the clock will advance once element x appears, according to the Fabula time. Again we use Association in Wolfram language or dictionary in Python to store the function Δ.

Now we have to accumulate Δ according to the Fabula time. To do that we generate for each $x \in L_2$ a two-dimensional vector

$$v(x) = (key2Greenwich[x], \Delta[x]) \tag{4.21}$$

Next, we sort all the two-dimensional vectors $\{v_x \in \mathbb{R}^2 : x \in L_2\}$ according to the Fabula time using they key $key2Greenwich$, with Greenwich time denoted by L_2. Denote those vectors sorted according to the Greenwich time by

$$V = ((v_{1,1}, v_{1,2}), (v_{2,1}, v_{2,2}), \ldots, (v_{n,1}, v_{n,2})) \tag{4.22}$$

This means that

$$v_{i,1} < v_{i+1,1} \tag{4.23}$$

Now we can define the general C clock on the set

$$\{v_{1,1}, v_{2,1}, \ldots v_{n,1}\} \subset \mathbb{R}^n \tag{4.24}$$

[1] Meaning the inverse function does not exist.

as follows:

$$C(v_{i,1}) = \sum_{j=1}^{i} v_{j,2} \tag{4.25}$$

We can use linear interpolation to transform the clock C from a discrete clock to a continuous one.

4.5.2 Generating a Subjective Clock from Historical Document

In this subsection, we discuss a general method for transforming a historical document into a sequence of non-negative numbers $\Delta(i)$ using Yes/No questions or any other quantitative questions asked to LLM on each sentence or paragraph i of the document. Once we transform the document into a sequence of numbers, we can generate a clock C_{IA} out of those numbers:

$$C_{IA}(t) = \sum_{i=1}^{t} \Delta(i) \tag{4.26}$$

In order to use the M-diagram method, we need a second clock C_2. Clearly, there are many ways to compute a second clock: either by asking AI different questions or using the standard time I_s, where I_s is the number of sentences or paragraphs. Now we can compute critical events using the M-diagram method $M\,[C_{Ai}, C_2]$.

Since this method is affiliated with using AI for history analyses, several examples are provided in the last part of the book. For more information, see Chaps. 14, 15, 16, 17, 18.

In the next section, we offer several examples of generating clocks from a list of famous physicists on Wikipedia.

In general, we generate a numerical representation of historical documents using AI analyses; we call this method AI-numerical history . Generating subjective clocks can be an example of such analysis.

4.6 The History of Two Clocks: Examples

In the following subsections, we will use the clock method to study the history of physics from several points of view. We compare the Greenwich clocks with different clocks. Each of the two clocks provides various perspectives on the history of theoretical physics relevant to the clocks in the examples. We start by explaining how to transform the list into clocks.

4.6.1 From List to Clocks

As an example of using two clocks to explore a historical narrative, we will calculate several major revolutionary events in the history of physics. We start with a list of prominent physicists L_p. This list can be found in [6].

4.6.2 Greenwich Clock

Following the methodology used in this chapter, we will define several clocks, each of which will tell us different aspects of the history of physics. The first clock, C_g, counts the number of years. We call it the Greenwich clock. It starts with the birthday of the first prominent physicist on list L_p and ends with the birthday of the last prominent physicist on the list. This clock measures the time exactly like the Greenwich clock.

4.6.3 Uniform Clock

The second clock, C_u, is the uniform clock, where each physicist is considered equal. $C_u(t)$ represents the number of physicists who were born before or on year t in list L_p, formally,

$$C_u(t) = |\{p \in L_p : B(p) \leq t\}| \tag{4.27}$$

Where $B(p)$ gives the birthday of the physicist $p \in L_p$.

The left side of Fig. 4.5 shows two functions: the blue function represents the number of prominent physicists on list L_p as a function of the Greenwich time. We call this function the uniform subjective clock of the list of prominent physicists derived from Wikipedia. It is easy to see that the blue line is not linear, since the number of prominent physicists rises rapidly following the scientific revolution. The red line on the left side of Fig. 4.5 is the expected number of prominent physicists if it were a linear function through time.

The right side of Fig. 4.5 shows the M-diagram $M\left[C_u, C_g\right]$ that compares the Uniform clock C_u with the Greenwich clock C_g. Note that the right side of the figure can be obtained from the left through simple rotation and scaling by sending the red linear line on the left side to the unit interval on the right and scaling the Y-coordinate by some factor derived from the formula. In this instance, it is 0.8, but different cases can generate different factors.

In the M-diagram of $M\left[C_u, C_g\right]$, there is one minimal extreme point around 1763, which is the critical point. From 476 CE to 1763, the Greenwich clock C_g ran much faster than the Uniform clock C_u. This can be seen in the fact that the M-diagram $M\left[C_u, C_g\right]$ is linearly decreasing in the first interval between zero and the red point

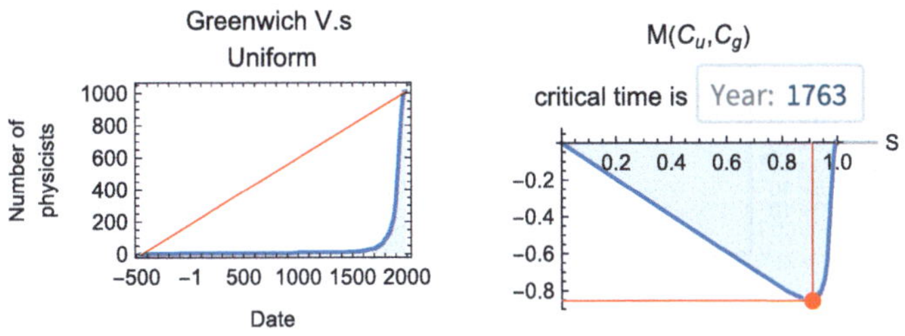

Fig. 4.5 Two graphs representing the relationship between Uniform and Greenwich clocks. In the left graph, the blue line represents the Uniform clock as a function of the Greenwich clock. Clearly, the number of physicists does not grow linearly in the time period. The red line represents the average number of physicists as a function of the Greenwich clock if the number of physicists was uniformly spread through time. The right-side graph shows the M-diagram of the Uniform clock C_u compared to the Greenwich clock C_g; the first clock accumulates the number of physicists in list L_p, and the second clock measures time according to the Greenwich clock. The red point is the minimum of the graph $M\left[C_u, C_g\right]$

in Fig. 4.5. If we just look at the first interval in the diagram, then those two clocks can be synchronized as they have a linear relationship. Around 1763, the Uniform clock started running much faster than the Greenwich clock. From that moment on, humanity began to generate more physicists than before. This makes the year 1763 a critical turning point. Looking at the history of physics, a reasonable interpretation is the beginning of research on electromagnetism. In 1767, Joseph Priestley was the first to propose an electrical inverse-square law, which propelled new discoveries in the field. Collaborative research also became more prominent.

Note that 1763 is not the exact year of the revolution. Several issues influence the critical points of the clock. The first and most important one is the list itself; the list can be argued forever, so for the sake of simplicity, we accept Wikipedia's point of view but bear in mind that it is not the optimal list. Another issue is the list's starting and ending points. Lastly, computing the marking time is another crucial aspect; we can bind the error of the critical point using the average person's life expectancy, as their contribution to science was made during their adult life.

4.6.4 *Knowledge Clock*

The third clock C_k that we will use measures the accumulated human knowledge of physics up to year t. To approximate this clock, we count the total number of words on the Wiki pages of all physicists born before year t. Formally,

$$C_k(t) = \#word(\{\ wik(p) \in \mathbb{A}rc : p \in L_p, \text{ and } B(p) \leq t\}) \tag{4.28}$$

where the function $\#word()$ counts the number of words in sets of documents or archives.

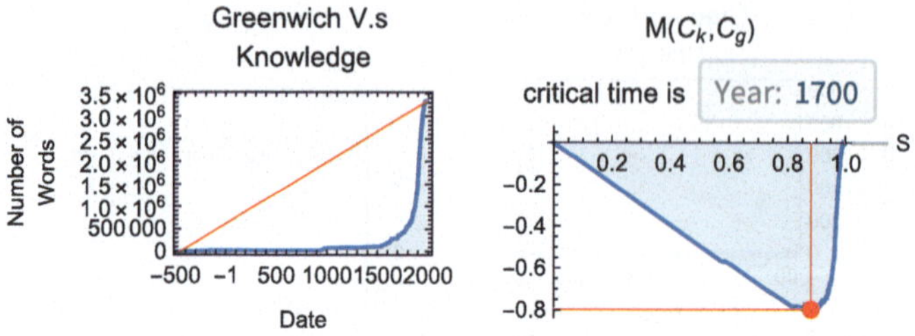

Fig. 4.6 Similar to Fig. 4.5, the left side shows the Knowledge Clock C_k as a function of Greenwich time. The red line is a linear representation of the expected number of words in Wikipedia pages about prominent physicists. The right side shows the M-diagram $M\left[C_k, C_g\right]$ of two normalized clocks; the first clock counts the number of words in Wiki pages about all scientists born before year t, which we call the Knowledge Clock $C_k(t)$ and the second one is the Greenwich clock C_g

Again, the left side of Fig. 4.6 shows two functions: the blue function represents the accumulated number of words on the Wiki pages about prominent physicists sorted according to Greenwich time. We call this function the Subjective Knowledge Clock.

The blue line is clearly not linear, since the number of prominent physicists grows much more rapidly after the scientific revolution. The red line on the left side of Fig. 4.6 is the expected number of prominent physicists if it were a linear function through time.

The right side of Fig. 4.6 shows the M-diagram $M\left[C_k, C_g\right]$. When comparing the Knowledge Clock to the Greenwich Clock, the critical point shifts to the year 1700. Human knowledge had dramatically increased due to the works of Newton and Leibniz.

4.6.5 *Wikipedia Pageviews*

Our next clock is based on Wikimedia Pageviews. Wikimedia's API provides an efficient way to access detailed data on how often specific Wikipedia articles are viewed. It allows users to analyze page views for a particular article across different time frames and categories, such as access methods (desktop, mobile web, or mobile app) and user agents (human users versus automated bots).

The third clock is a popularity clock $C_{pop}(t)$ with which we count the total number of visits to all Wiki page documents of physicists born before or at year t. Formally,

$$C_{pop}(t) = \#vist(\{\ vist(wiki(p)) \in \mathbb{A}rc : p \in L_p,\ \text{and } B(p) \leq t\}) \qquad (4.29)$$

where the function $vist$ gets a Wiki page $wiki(p)$ of a physicist p as input and output the total number of Wiki page $wiki(p)$ visitors.

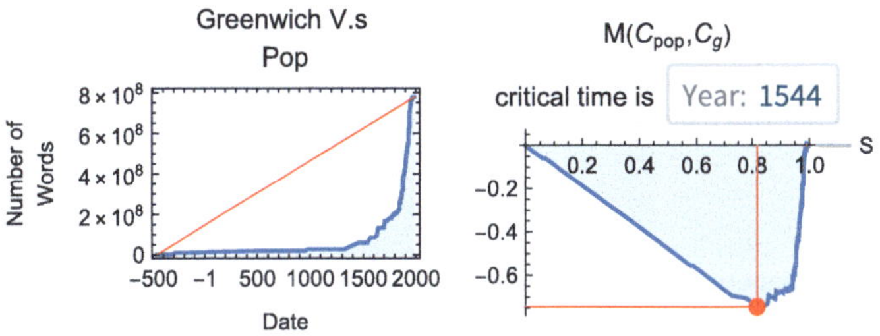

Fig. 4.7 Similarly to Figs. 4.5 and 4.6, this image has two parts. The blue line on the left shows C_{pop} clock as a function of the Greenwich Clock. The red line is the linear model of the number of page views through the Greenwich clock. The right side is the M-diagram of two normalized clocks $M\left[C_{pop}, C_g\right]$; the first counts the accumulated number of Wiki pages' visits sorted by the Greenwich clock physicists, and the second clock is the Greenwich clock

Like in the previous figure, the left side of Fig. 4.7 shows two functions: the blue function that represents the accumulated number of visits to the Wiki pages of prominent physicists sorted according to Greenwich time. We call this function the Popularity of Wiki page Subjective Clock. The red line on the left side of Fig. 4.7 is the expected number of visits to the Wiki pages of prominent physicists if it were a linear function over time.

It is easy to see that the blue line is not linear since the number of visits to Wiki pages of prominent physicists who appeared after the scientific revolution grows more rapidly. The red line on the left side of Fig. 4.7 is the expected number of visits to the Wiki pages of prominent physicists if it were a linear function over time.

The right side of Fig. 4.7 shows the M-diagram $M\left[C_{pop}, C_g\right]$. When comparing the Popularity clock with the Greenwich clock, the critical point becomes the year 1544. The reason may be that people who read about scientific history are mainly interested in scientists who operated from the start of the scientific revolution.

4.6.6 *Network Clock*

Another way to generate a clock from a list of people L_p is the Network Clock $C_{net}(t)$. To simplify the construction of a Network clock, we will assume that the list $L_p = (l_1, \ldots, l_n)$ is sorted according to the birthdays of the people on the list, i.e.,

$$\text{for all} 1 \leq i < j \leq n,\ B(i) \leq B(j) \tag{4.30}$$

In this example, we use the birthdays of people in L_p, but any other chronological order would work just as well.

The first step to constructing the Network clock is to generate a network $G(V, E)$ of the people in the list L_p, and compute the centrality χ_g for each node or person:

$$\chi_g : V \to \mathbb{R}^+ \quad (4.31)$$

We use the birthday order in L_p to define the subset V_i as follows:

$$V_i = \{l_1, \ldots, l_i\} \quad (4.32)$$

Now we can generate evolving social networks from the list L_p using the subsets V_i as follows:

$$\gamma_i = Ind(G, V_i) \quad (4.33)$$

Where $Ind_{G,V_i)}$ is the maximal induce subgraph of G in which the nodes are the subset V_i).

We can now describe how to construct the Social network G from the list L_p. The nodes in graph G are the people on the list $V = L_p$. The edges of graph G can be computed in many ways, either manually or mechanically. In this section, we are going to put a direct edge $e = (i, j)$ between nodes $i, j \in V$ if there is a link from the Wiki page $wik(i)$ to the Wiki page $wik(j)$. The giant component of the graph can be seen in Fig. 4.9.

Once we have graph G, we can use any chronological order $\leq$ on the node to generate an evolving social network γ_i from graph G, as explained in Eq. 4.33 C_{net}. We use the "born before" to generate a chronological order of the network's appearance.

Now we can define

$$C_{net}(t) = \sum_{i \in V_t} \chi_g(i) \quad (4.34)$$

Note that the domain of clock C is the set $\{1, 2, \ldots, n\}$. The domain for our Network clocks is the birthday of the nodes, and the set $\{1, 2, \ldots, n\}$. To change the domain, we use clock $B(i)^{-1}$, which transforms time $B(i)$ to i, formally:

$$B(i)^{-1} = i \quad (4.35)$$

Now we can define the Network clock :

$$C_{net}(B(i)) = C(B(i)^{-i}) \quad (4.36)$$

Figure 4.8 shows the Network clock C_{net} as a function of the Greenwich clock and their M-diagram. According to the figure, the critical point from the Wiki page social network perspective is the year 1837. This could be interpreted as Lord Kelvin's birthday (1824), which is associated with the Cambridge laboratory, the central institution of physics in Europe.

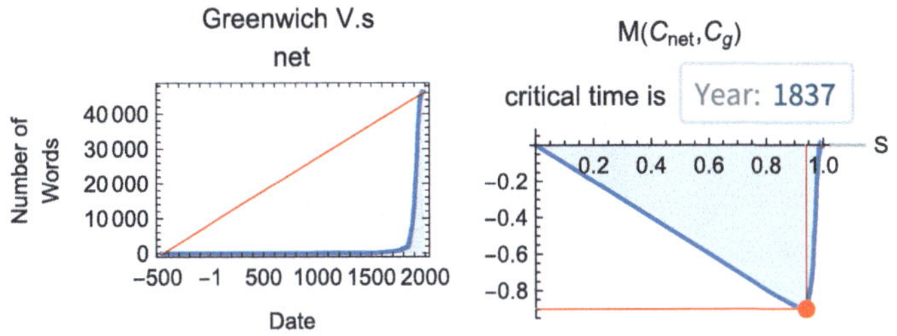

Fig. 4.8 Similarly to Figs. 4.5 and 4.6 this image has two parts. The blue line on the left side is the Network Clock C_{net} where we use the out-degree as a Kernel of the Network clock. The red line is the linear model of the expected number of out-degree links in the Wiki link graph. The right side is the M-diagram of two normalized clocks $M\left[C_{net}, C_g\right]$; the first count is the accumulated out-degree of a node sorted by Greenwich clock physicists, and the second clock is the Greenwich clock

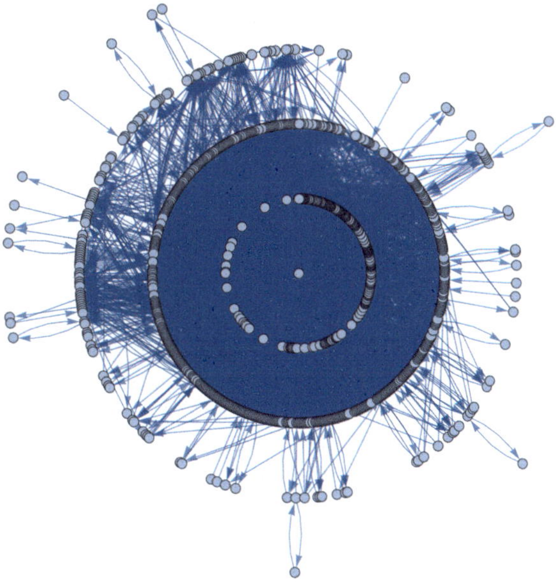

Fig. 4.9 The giant weakly-connected component where the nodes are physics that appear on list L_p. Node i connects to node j if there is a link in the Wiki page of node i to the Wiki page of node j

4.7 Time Machine

In this section, we are going to use the two-clock method to travel back in time and see what people in the past thought about the great physicists. Wikipedia provided us with two methods for doing that based on two independent questions (Fig. 4.10).

1. Which list to use?
2. Where to cut the list?

For example, the oldest list appeared on Wikipedia on 2001-10-02. This list contains 38 physicists [13].

Fig. 4.10 The complex relationship between a machine and time

It contains only two physicists born before Newton. The current list contains more than 1000 physicists, of whom 26 were born before Newton. These lists tend to grow as a function of time for three reasons: the expansion of Wikipedia and new data about old physicists, the development of physical science, and the appearance of new physicists born in the late 20th or even the 21st centuries. As time progresses, the list grows and we can use it to see how time changes the narrative of physics.

In general, all versions of the list can be found in the following link [14].

The first method is to look at the past using the current list L_p of great physicists. The second method requires going back to the old version of the list and computing the critical points of both lists to reflect the perspective from earlier times, i.e., Issac Newton's vision of the list.

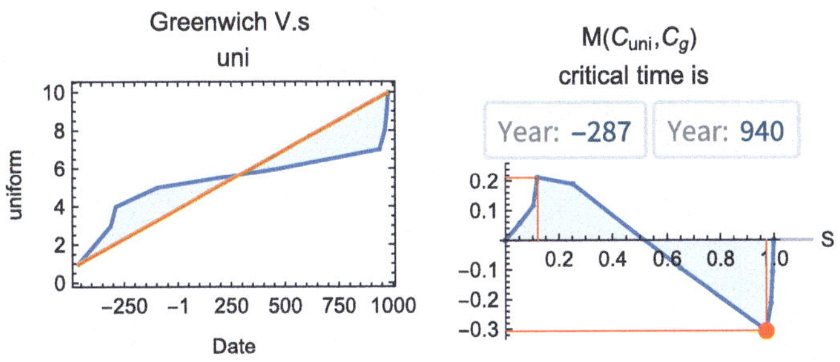

Fig. 4.11 The left side show the Uniform Clock composed from the current list L_p, which cuts off at the year 1000. On the right is the M-diagram of the two clocks that shows two critical points: The first occurred around the time Archimedes was born, and the second around the time Abu sahl Al-Quhi was born

4.7.1 The Current List

We begin with analyzing the more current list with more than 1000 great physicists. The first great physicist, Democritus, was born in 460 BC.

We are therefore going to use the current list L_p in Wikipedia and cut it at some time $t < 2023$ in the past. This allows us to generate the current point of view of what people in time t thought about the great physicists of the previous eras. For example, looking at Fig. 4.11 we can see that around the year 1000, there were two great physicists: Archimedes of Syracuse, Greece (ca. 287–212 BC) and Abu sahl Al-Quhi of İran (born 940).

To generate Fig. 4.12, we take the list L_p and cut it from the first physicist Democritus to the i-great physicists when the list $L_p$is sorted according to Greenwich time. Once we have the new sublist $L_p|_1^i$, we compute the M-diagram $M\left[C_u|_1^i, C_g|_1^i\right]$ of the Uniform clock with the Greenwich clock. The M-diagram gives us two critical points of time, $y_1(t) < y_2(t)$. We draw these two lines as a function of the birthday $b(i)$ of the (i)-th great physicist.

Figure 4.12 reveals several interesting revolutions. The first can be called the Greek era, with Archimedes and Democritus being the most important physicists; the second is the Muslim era, during which Al-Quhi translated Archimedes, while Democritus' theory of atoms was forgotten; the last one is the scientific revolution.

4.8 Many Clocks

Up to this point, we compared only two clocks and used the M-diagram to find the point at which they start to deviate from each other. We've shown that these points in time have interesting historical interpretations concerning the clocks. However, we often have many clocks to consider at once. This is especially true when we

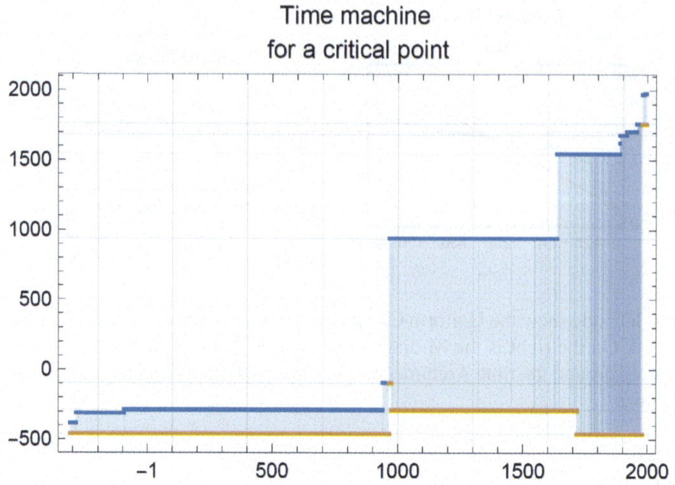

Fig. 4.12 This figure shows the time machine of the current list of great physicists. For each year we compute the M-diagram of the list that starts with Democritus and ends at that year t. The M-diagram generates two years $y_1(t) < y_2(t)$. The orange line is $y_1(t)$ as a function of t and the blue line $y_2(t)$ as a function of t

consider subjective clocks of citizens of states or any other historical entities. In this section, we will explain how to extend the M-diagram to many clocks and provide an example. The idea is to normalize all the clocks and draw the trajectory of the clocks in a unit d-dimensional box, where d is the number of clocks, i.e., we have a dimension for each clock.

Next, we transform the many clocks into a single clock SC, which measures the norm of the vector of all the clocks. Now we can use the standard time C_s, i.e.,

$$C_s(t) = t, \tag{4.37}$$

as the second clock in the M-diagram.

Formally, assume that we are given d-clocks,

$$c_i : \mathbb{T} \to \mathbb{T}_i \tag{4.38}$$

The first step is to normalize these clocks, i.e., to change their domain and co-domain into the unit interval, using a monotonic increasing linear function. Denote the normalized clock by

$$\mathcal{N} c_i : [0, 1] \to [0, 1] \tag{4.39}$$

Now we can represent all the clocks together as a function C

$$C : [0, 1] \to [0, 1]^d \tag{4.40}$$

according to the following equation:

$$\mathcal{N}C(t) = (\mathcal{N}c_1(t), \mathcal{N}c_2(t), \ldots, \mathcal{N}c_d(t)) \tag{4.41}$$

Now we compute the *Single clock SC* according to

$$SC(t) = ||\mathcal{N}C(t)||_2 \tag{4.42}$$

To find the critical event in the multiple clocks narrative, we compute the M-diagram

$$M\left[SC(t), t\right] \tag{4.43}$$

and find the extreme point of $M[SC(t), t]$, formally,

$$\tau_c = \max_{t\in[0,1]} M[SC(t), t] \tag{4.44}$$

4.9 Many Clocks Synthetic

To verify the method, we construct the following simple example: Imagine that we are given a country with n people, each with their own subjective clock

$$c_i : \mathbb{T} \rightarrow \mathbb{R} \tag{4.45}$$

To simplify our example, we assume that the clocks are generated by accumulating random streams of numbers $\Delta_i = (\delta_{i,j})_{j=1}^{t}$ which are either 0 or 1. The number 0 represents an insignificant event that will be forgotten, and therefore the subjective time is not advanced by the event. The number 1 represents a significant event and, therefore, the subjective clock will advance by a single tick. Moreover, to emphasize our point, we embed a collective significant event in the memory of all people from time t_1 to t_2 by putting 1 in the streams of all Δ_i.

Formally, for all $i = 1, \ldots, n$ and for all $j \in \mathbb{T} \setminus \{t_1, t_1 + 1, \ldots t_2\}$ $\delta_{i,j}$ there will be a Bernoulli identical independent random variable, and for all $j \in \{t_1, t_1 + 1, \ldots t_2\}$ we will have

$$\delta_{i,j} = 1 \tag{4.46}$$

Clearly, for the events to have a significant influence on the entire population, they would have enough energy, therefore:

$$t_2 - t_1 = O(\sqrt{n}) \tag{4.47}$$

since the embedded event has to overcome any random fluctuation containing $\sqrt{n}$.

Now, after defining the sequence Δ_i, we can define the clock C_i as follows:

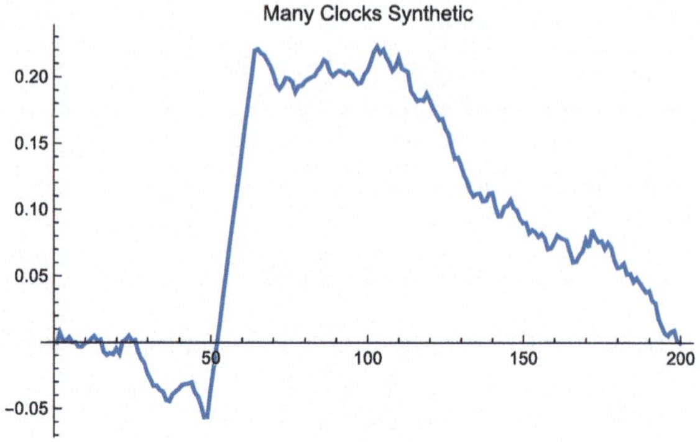

Fig. 4.13 The synthetic model of combining many clocks

Denote these numbers by

$$c_i(t) = \sum_{j=1}^{t} \delta_{i,j} \tag{4.48}$$

to normalize the clock $c_i(t)$ according to the equation

$$\mathcal{N} c_i(t) = \frac{(c_i(t|T|) - c_i(1))}{(c_i(t_e) - c_i(1))} \tag{4.49}$$

Once we have the normalized clocks $\mathcal{N} c_i(t)$, we can compute the single clock as shown in Eq. 4.42, and compute the M-diagram as shown in Eq. 4.43 (Fig. 4.13).

4.10 Combining the Physicists' Clocks

In this section, we provide a simple example of combining clocks $\mathcal{N} C_g$, $\mathcal{N} C_u$, $\mathcal{N} C_k$, $\mathcal{N} C_p$, $\mathcal{N} C_{net}$ into a single clock by taking the norm of the vector and comparing it to the Greenwich clock using the M-diagram function, see Fig. 4.14. Analyzing the extreme point of the M-diagram, we deduce that the critical time of the combined clock is the year 1818. Around that time, physicists discovered electromagnetic force, clearly a central event in the history of physics, separating the discovery of mechanics by Newton on the one hand and Einstein's quantum/relative revolution on the other.

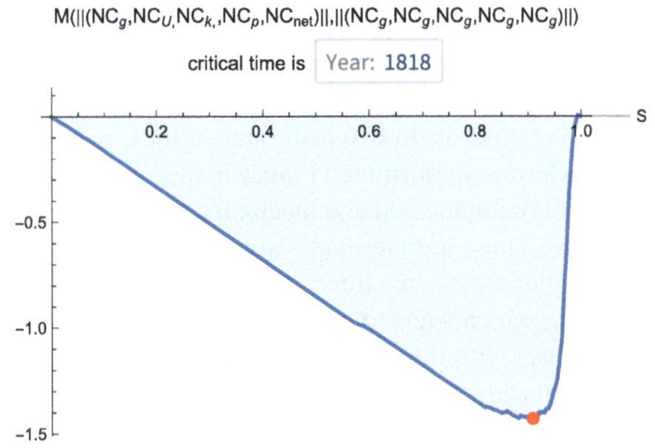

Fig. 4.14 The M-diagram of many combined clocks $M\left[||(\mathcal{N}C_g, \mathcal{N}C_u, \mathcal{N}C_k, \mathcal{N}C_p, \mathcal{N}C_{net})||, ||(\mathcal{N}C_g, \mathcal{N}C_g, \mathcal{N}C_g, \mathcal{N}C_g, \mathcal{N}C_g)||\right]$ The red point is the critical point which appears near the year 1818

4.11 Conclusion

In this chapter, we've developed tools to compare subjective clocks. The main idea is that any two subjective clocks generate a narrative that tells the story of why those clocks deviate from each other. We demonstrated how historians can use clocks to probe historical narratives. We study the list of prominent physicists in Wikipedia using several subjective clocks and identify critical events with respect to those clocks.

The idea is to start with a list of people, use their birthday as one of the subjective clocks, then use the other subjective clocks to investigate the historical phenomena we are interested in. We introduce four clocks: the Uniform Clock, the Knowledge Clock, the Popularity Clock, and the Network Clock. The Network Clock is not a single clock but a family of clocks, since it uses two variable options in its definition: the network itself, which can be changed according to the definition of an edge, and the centrality function, which can change as well.

Another way to generate a subjective clock is to start with a historical text and use a Large Language Models (LLMs) to transform the text into a sequence of positive numbers by posing a question that transforms each sentence or paragraph into a non-negative number. Then we accumulate those numbers and generate subjective clocks relevant to the question we posed.

Once we have two clocks, we compare them by using their M-diagram. After computing the M-diagram, we can compute the extreme M-point of the diagram, i.e., the critical points where the clocks start to deviate from each other. We remark that in the historical narrative, the M-diagrams tend to have between one and two critical points. This is in contrast with literature narratives, where M-diagrams tend to have between seven and ten critical points.

Surprisingly, all historians assume that time runs at the same speed when analyzing conflicts between two social entities. This is clearly wrong. Time is relative and is influenced by the technology of the social entity. In the same way, two individuals have their own subjective time, as do two historical entities.

For example, when Germany confronted France in the WWII Battle of France, the German army was fully communicated and mechanized, while the French army was technologically lagging. Thus, the Germans' subjective time moved much faster, allowing them to conquer France in three weeks. Time proves to be subjective according to technology, which leads to significant advantage at wartime.

We end up this chapter with the construction of a time machine. We have two different kinds of time machines: one that observes the past with a heavy influence on the future, since it uses the current version of the list L_p. The other uses the old version of the list L_p. The latter are reminiscent of the work of historians as they study old historical texts. For example, a historian who wishes to study the Roman era can read Julius Caesar's book on the Gallic War. In the same manner, the historians of the 25th century would be able to look at the list of prominent physicists of the 21st century as it is compiled today. Clearly, that list will not contain physicists from the 22nd century, as they have not yet been born.

4.12 Exercise

Comparing Two Clocks

1. Let

$$\Delta_1(t) = \begin{cases} 1/4, 0 \leq t \leq 1/2 \\ 7/41/2 < t \leq \end{cases} \tag{4.50}$$

   Let

$$\Delta_2(t) = \begin{cases} 7/4, \ 0 \leq t \leq 1/2 \\ 1/4, \ 1/2 < t \leq 1 \end{cases} \tag{4.51}$$

   Let $c_i(t) = \int_0^t \Delta_i(x)dx$. Prove that the two functions $c_1(t), c_2(t)$ are normal clocks. Compute the M-diagram $M_{c_1,c_2}(t)$. Find the extreme points of the function $M_{c_1,c_2}(t)$. Can you explain why the extreme point is equal to $t = 1/2$?
2. Following Sect. 2.4.2, compute the main historical event using anomaly detection for the number of deaths per year. Consider the clock that counts the number of dead people in Wikipedia up to the year t. Formally, this clock could be defined with the following equation:

$$cd(t) = \sum_{i=0}^{t} dnum_i \tag{4.52}$$

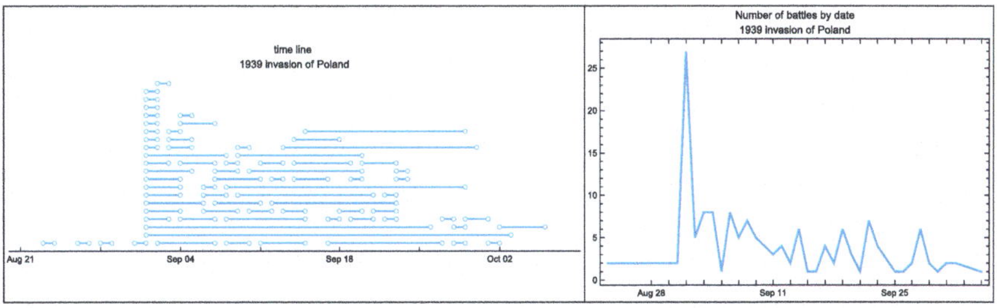

Fig. 4.15 Timeline and number of battles at a given point t-time during the invasion of Poland in 1939

where $dnum_i$ is the number of dead individuals. When we ask machine learning to find an anomaly in function cd, it will fail to recognize one since the accumulated numbers cancel the noise in the sequence. This suggests that when trying to find anomalies, it is better to remove the expectation from the data. Indeed, that is what the M-diagram does.

a Compute the anomaly on $cd(t)$
b Compute the anomaly on $(i, dnum_i)_{i=1}^{1500}$
c Compute the anomaly on $M\left[cd(i), C_g(i)\right]$ where $C_g(i) = i$ is the Greenwich clock
d Compute the anomaly on

$$(i, M\left[cd(i), C_g(i) - M\left[cd(i), C_g(i-1)\right]\right])_{i=1}^{1500} \tag{4.53}$$

3. Repeat the analyses of Sect. 4.6.1 for chemistry. Use the list in [15].

a Repeat the operations from Sect. 4.6.3 and compute the Uniform clock $C_{c,uni}$.
b Repeat the steps from Sect. 4.6.4 and compute the Knowledge clock $C_{c,k}$.
c Repeat the sequence in Sect. 4.6.5 and compute the Popularity clock $C_{c,p}$.
d Repeat the process of Sect. 4.6.6 and compute the Network clock $C_{c,net}$.
e Repeat the steps of Sect. 4.7 and compute the time machine.

4. This exercise suggests a way to measure the complexity of a battle. Figure 4.15 shows the timeline and the number of battles at a given point in time during the invasion of Poland. Use these graphs to develop a method to compare the battle complexity, then develop a method to find a point in time with the highest battle complexity. The invasion of Poland will be used to demonstrate the connection between time and place later in the book.
5. In this problem, we will compare advances in chemistry with advances in physics. We will compute two subjective clocks: one for chemistry C_c and one for physics C_p. We do that in a manner similar to what we did in Sect. 4.6.1. In

a previous exercise, we computed the clocks $C_{c,uni}$, $C_{c,k}$, $C_{c,p}$, $C_{c,net}$. To compare chemistry with physics, we need to compute the same clocks for physics $C_{p,uni}$, $C_{p,k}$, $C_{p,p}$, $C_{p,net}$. You can use the list in [6] to do that. After we've calculated these clocks:

a Compute $M\left[C_{c,uni}, C_{p,uni}\right]$ and find the extreme point. Can you explain what happened at that time?
b Compute $M\left[C_{c,k}, C_{p,k}\right]$ and find the extreme point. Can you explain what happened at that time?
c Compute $M\left[C_{c,p}, C_{p,p}\right]$ and find the extreme point. Can you explain what happened at that time?
d Compute $M\left[C_{c,net}, C_{p,net}\right]$ and find the extreme point. Can you explain what happened at that time?

References

1. W. Labov, J. Waletzky, Narrative analysis: oral versions of personal experience (1997)
2. S. Rimmon-Kenan, *Narrative Fiction: Contemporary poetics* (Routledge, 2003)
3. Wiki. wiki-fabul-syuzhet (2023). https://en.wikipedia.org/wiki/Fabula_and_syuzhet
4. Wikipedia contributors. List of timelines of world war ii (2024). https://en.wikipedia.org/wiki/List_of_timelines_of_World_War_II
5. Z. Lotker, *Analyzing Narratives in Social Networks Taking Turing to the Arts Authors* (Springer, Berlin, 2021)
6. wikisource (2023). https://en.wikipedia.org/wiki/List_of_physicists
7. wikisource (2023). https://en.wikipedia.org/wiki/List_of_feminists
8. S. Augustine, *The Confessions* (Clark, 1876)
9. Wiki. wikisource-international-system-of-units (2024). https://en.wikipedia.org/wiki/International_System_of_Units
10. Wiki. wikisource-clock (2023). https://en.wikipedia.org/wiki/Clock
11. J.G. Yoder, Christiaan huygens, book on the pendulum clock (1673), in *Landmark Writings in Western Mathematics 1640–1940* (Elsevier, 2005), pp. 33–45
12. Z. Lotker, *Analyzing Narratives in Social Networks: Taking Turing to the Arts* (Springer Nature, 2021)
13. wikisource (2023). https://en.wikipedia.org/w/index.php?title=List_of_physicists&oldid=274052
14. wikisource (2023). https://en.wikipedia.org/w/index.php?title=List_of_physicists&action=history&dir=prev
15. wikisource (2023). https://en.wikipedia.org/wiki/List_of_chemists

Chapter 5
Networks and Macrohistory

5.1 Introduction

In the previous chapter, we discussed time. In this chapter, we discuss the way we model space. In general, any finite discrete space can be modeled as a weighted graph. Therefore, this chapter's main focus is graphs and weighted graphs as used in digital history.

Economics is divided into micro- and macroeconomics. Similarly, in history, one can talk about micro- and macrohistory. Macrohistory is the act of observing historical events from a global perspective, analyzing events from the countries' or social institutions' points of view (Fig. 5.1).

Similarly to economics, we distinguish between two kinds of graphs: macro-level graphs that describe relationships 'from above', i.e., historical events, institutions, and the relationships between them, and micro-level graphs that describe social networks and relationships between people.

5.2 Macrohistory

As we discussed in Sect. 2.2, historical events $h = (t, p, m)$ are composed of three components: time, place, and meaning. Therefore, we will have three main different graphs that allow us to discuss the relationships between historical events. Interval graphs capture the "happened at the same time," geo-graphs that describe the geographical relationship between historical events, and space-time graphs that describe the relationship between historical events, taking into account the relationship between historical events through time and place. Note that, in general, time and place have local structure, meaning that close points in time and space have more influence on each other than far-away events.

We start with the definition of graphs.

Z. Lotker, *History by Algorithms*,
https://doi.org/10.1007/978-3-031-93627-2_5

Fig. 5.1 A cartoon about the division of Poland at the beginning of World War II

5.3 Graph Theory

A *graph* is a mathematical structure used to model pairwise relationships between objects. Formally, a graph $G(V, E)$ is defined as:

- V is a *set of vertices* (also called nodes), typically denoted as

$$V = \{v_1, v_2, \ldots, v_n\}. \tag{5.1}$$

- E is a *set of edges*, which are pairs of Vertices representing connections between them.

5.3.1 Types of Graphs

In this section, we discuss several types of graphs that will be useful throughout the book. For more on graph theory, see [1] or [2].

- *Undirected Graph* $G(V, E)$: The edges have no direction, meaning that if $\{u, v\} \in E$, then v is connected to u and vice versa. An example of an undirected graph with 4 nodes and 4 edges can be seen in Fig. 5.2.
- *Directed Graph (Digraph)* $G(V, E)$: The edges have a direction, meaning that $(u, v) \in E$ represents a connection from u to v but not necessarily from v to u. An example of a directed graph with 4 nodes and 4 direct edges can be seen in Fig. 5.2.

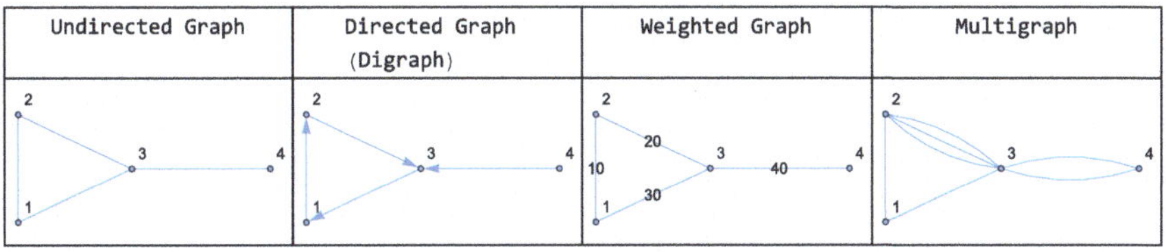

Fig. 5.2 This figure presents examples of different types of graphs

- *Weighted Graph* $G(V, E, W)$: Each edge has an associated weight or cost, where W is the weight function[1]:

$$W : E \rightarrow \mathbb{R}. \tag{5.2}$$

  An example of a Weighted Graph with 4 nodes and 4 edges can be seen in Fig. 5.2.
- *Multigraph*: A graph $G(V, E, W)$ that allows multiple edges between the same pair of Vertices. One may think of a multigraph as a Weighted Graph where the function W is defined on the natural numbers, formally:

$$W : E \rightarrow \mathbb{N}. \tag{5.3}$$

  An example of a multigraph with 4 nodes and 7 multi edges can be seen in Figure

Special Families of Graphs

Next, we describe some important families of graphs:

In graph theory, a *path* is a sequence of Vertices connected by edges, where each edge connects consecutive Vertices in the sequence.

A *path* of length k in a graph $(G(V, E)$ is a sequence of Vertices:

$$P = (v_0, v_1, v_2, \ldots, v_k) \tag{5.4}$$

such that:

$$\{v_i, v_{i+1}\} \in E \quad \text{for all } 0 \leq i < k. \tag{5.5}$$

An example of a path graph with 4 nodes and 3 edges can be seen in Fig. 5.3.

In the case of a *directed graph*, the edges must respect the direction:

$$(v_i, v_{i+1}) \in E \quad \text{for all } 0 \leq i < k. \tag{5.6}$$

Next, we define a *cycle graph*:

[1] Sometimes we talk about weighted matrix $W = [w_{i,j}]$.

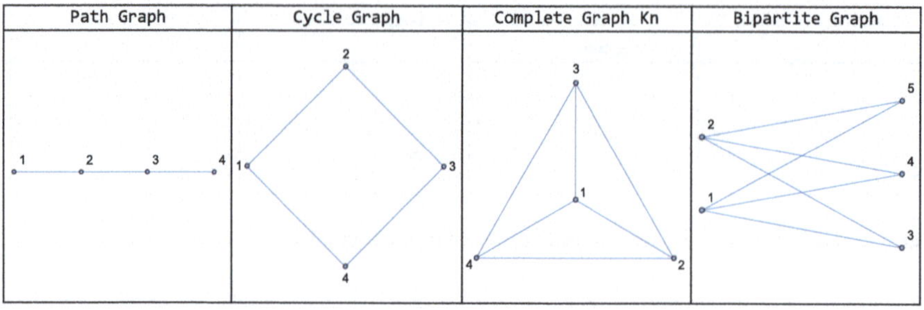

Fig. 5.3 This figure presents examples of types of families of graphs

- *Cycle graph*: A path that starts and ends at the same vertex without repeating any edges.

An example of a cycle graph with 4 nodes and 4 edges can be seen in Fig. 5.3.

While paths and cycles contain a linear number of edges, the graph that maximizes the number of edges is the *complete graph* and has a quadratic number of edges.

- **Complete Graph** $K_n = G(V, E)$: A graph in which every pair of distinct vertices is connected by an edge. Formally,

$$E = \{\{i, j\} \mid i \in V, j \in V, i \neq j\}. \tag{5.7}$$

- **Bipartite Graph** $G(V, E)$: A graph where the vertex set can be divided into two disjoint subsets $V = A \cup B$ such that no two Vertices within the same subset are adjacent. That is,

$$\forall \{i, j\} \in E, \quad \text{either } (i \in A \text{ and } j \in B) \text{ or } (i \in B \text{ and } j \in A). \tag{5.8}$$

An example of a complete graph K_4 with 4 nodes and 6 edges, as well as a complete bipartite graph $K_{2,3}$ with 5 nodes and 6 edges can be seen in Figure 5.3.

5.3.2 *Functions of a Graph*

When analyzing graphs, we often define functions on the graph structure. A function f maps Vertices to real values:

$$f : V \to \mathbb{R}. \tag{5.9}$$

Such functions help capture important properties of the graph $G(V, E)$. Examples include:

- *Degree of a vertex* denote by d_v: The number of edges connected to a vertex v. In direct graph, the degree is split into indegree and outdegree. The indegree counts the number of edges going into the node, and the outdegree counts the number of edges coming out of the node.
- *PageRank*: Originally developed for ranking web pages, it assigns importance to nodes based on the number and quality of links pointing to them, capturing influence beyond immediate neighbors.
- *Centrality metrics* of the graph g, χ_g: These include betweenness, closeness, and eigenvector centrality, which quantify various aspects of a node's significance in terms of shortest paths, reachability, or influence propagation.

In the next subsection, we describe some data structures relevant to graphs that we use throughout the book.

5.3.3 Data Structures for Graph Representation

When dealing with digital information, the natural question is how the data is stored on the computer. In this subsection, we explain two main methods for storing graphs on the computer: adjacency matrix and adjacency list. To simplify, we assume that the set of nodes is $V = \{1, \ldots, n\}$ where n is the number of nodes in the graph $G(V, E)$.

- *Adjacency Matrix*: $A \in \mathbb{M}_{n,n}[\{0, 1\}]$ matrix where the entry at (i, j) indicates whether an edge exists between i and j. i.e.,

$$a_{i,j} = \begin{cases} 0, & \text{if } (i, j) \notin E \\ 1, & \text{if } (i, j) \in E \end{cases} \tag{5.10}$$

- *Adjacency List*: A list where each vertex i has a list of adjacent vertices l_i.

The next A is the Adjacency Matrix of the Complete Bipartite Graph $K_{2,3}$, see Fig. 5.3.

$$A = \begin{bmatrix} 0\,0\,1\,1\,1 \\ 0\,0\,1\,1\,1 \\ 1\,1\,0\,0\,0 \\ 1\,1\,0\,0\,0 \\ 1\,1\,0\,0\,0 \end{bmatrix} \tag{5.11}$$

and the Adjacency List of the $K_{2,3}$ is

$$\begin{array}{c|c} \text{Node} & \text{adjacent to} \\ 1 & (1,2,3) \\ 2 & (1,2,3) \\ 3 & (1,2) \\ 4 & (1,2) \\ 5 & (1,2) \end{array} \tag{5.12}$$

Graph theory is one of the main building blocks of contemporary technology and is discussed extensively throughout this book. We assume that the reader is familiar with it, as teaching the graphs is beyond the scope of this book. The Appendix offers a list of prompts that allow the reader to familiarize themselves with the topic Appendix A.5.

One of the most important families of graphs in digital history is interval graphs, which are the subject of the following section.

5.4 Interval Graphs

An *interval graph* is a type of graph that represents the intersection relationships between intervals on the real number line. Interval graphs are useful for modeling temporal relationships between historical events. Formally, an *interval graph* is an undirected graph $G = (V, E)$, where:

- Each vertex $v \in V$ corresponds to an interval I_v on the real line, and
- There is an edge $(u, v) \in E$ if and only if the intervals I_u and I_v intersect.

For more on interval graphs, see [3]. We can construct the *timeline interval graph* using only time interval of the historical events. To understand this concept, it is useful to compare Figs. 4.3 and 5.4, both of which represent seven major historical events of WWII. The former illustrates the intervals, while the latter depicts their relationships by treating each interval as a node and connecting two nodes if and only if their corresponding intervals intersect. For further details, refer to Sect. 2.2.5 and Figs. 4.3 and 4.4.

Looking at the figure, we see that the interval graphs of seven major WWII historical events and the time interval graphs are connected. However, if the Battle of the Atlantic were removed, the graphs would become disconnected. This reveals that indeed the Battle of the Atlantic was the longest and most difficult due to the fact that no one really controlled the sea. The Allies had a material advantage in ships and aircraft, while the Axis had a material advantage in submarine fleet.

Next, we explain how to compute the Fabula time from weighted interval graphs.

In Sect. 4.1 we discussed how to compute Fabula time as a function of Syuzhet. In this section, we explain how to compute Fabula time as a function of Syuzhet from the weighted interval graphs.

In general, to construct the function of Fabula versus Syuzhet from the weighted historical events $\mathcal{H}$, we need the following input:

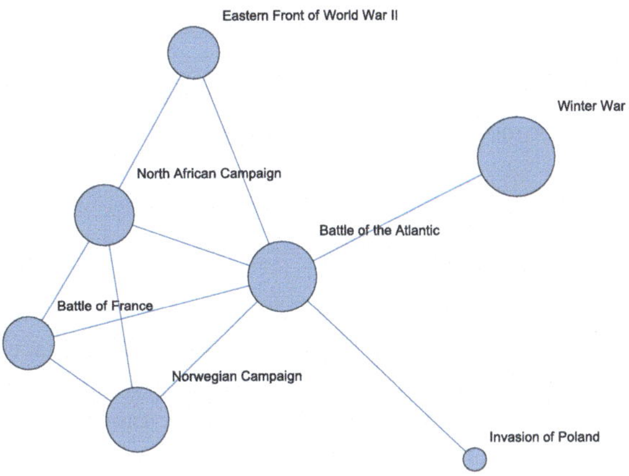

Fig. 5.4 This figure shows the intersection graph of the time interval of the seven main events of WWII. The size of the node is proportional to the time length of the historical events. An edge connects two historical events if the time intervals of these events intersect

1. A set of interval timelines $I = \{\mathrm{i}_i\}_{i=1}^n$ which we can compute from the historical events.
2. A weighted function $w : I \to \mathbb{N}$ which can be computed from the meaning of the historical events.

Now we can construct a Weighted interval graph $G(V, E, W)$ from the input, where from each interval $I_i \in I$ we have a corresponding vertex $i \in V$. Two nodes $i, j \in V$ are connected by an edge $\{i, j\} \in E$ if intervals I_i, I_j have non-empty intersection. The weighted function $W : V \to \mathbb{N}$ of the vertex $i \in V$ is equal to

$$W(v_i) = w(I_i) \tag{5.13}$$

In general, any weighted interval timeline $\{\mathrm{i}_i\}_{i=1}^n$ can be told as a narrative in many ways and, therefore, can have many different Fabula as a function of Syuzhet time. The weights of the intervals represent the number of words used to describe each interval. Therefore, the slope in which we draw each interval is proportional to the ratio between the Fabula time of the interval divided by the number of words needed to describe the interval.

5.5 Spacetime Intersection Graph

In this section, we will discuss how to analyze geographic data. In the next section, we will explain how to combine time and space. In order to demonstrate the tools we developed in this subsection, we analyze the German invasion of Poland [4], one of the seven main historical events in WWII. The reason we move from analyzing the main historical events of WWII to the invasion of Poland is that the geography of the battles in this historical event is easier to analyze. In general, there are 80 historical events that describe the invasion of Poland. It is not easy to determine the geographical position of all of them. Of the 80, we were able to find the geographical position of 74.

We import data structure from computational geometry to analyze geographic data. For more about computational geometry, see [5], or [6].

5.5.1 Convex Hull

The *convex hull* of a set of points in a Euclidean space is the smallest convex set that contains all the given points. Geometrically, it can be visualized as the shape formed by stretching a rubber band around the outermost points and letting it snap into place. The convex hull is widely used in computational geometry, where it serves as a fundamental structure for solving many problems. For more on convex hull, see [7]. In computational history, convex hull is used to define the position of a set of historical events and drawing area with a common historical entity. This is especially useful in studying battles and wars, since violating the convexity makes one of the armies vulnerable after being surrounded.

Formally, given a finite set of points S in $\mathbb{R}^n$, the convex hull is the intersection of all convex sets that contain S. In applications, convex hulls are crucial in computational geometry, machine learning (for classification boundaries), and computer graphics.

We can use them to compute the border of Poland during the war; see Fig. 5.5 or [8].

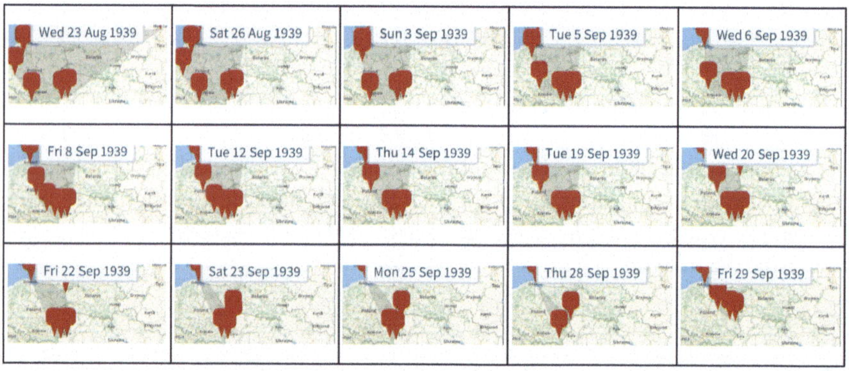

Fig. 5.5 This figure shows the position of the convex hull battles during the German invasion of Poland at 15 key points in time. The grey area represents the area of free Poland. The red points are the major battles during the invasion

5.5.2 Delaunay Graph and Minimum Spanning Tree

Another important data structure for computational geometry is the Delaunay triangulation. In digital history, the Delaunay triangulation can be used to combine temporary data with geographic data, since the Delaunay graph captures the structure of the shortest local path. Assuming that movements are efficient, the Delaunay graph allows us to reduce the number of edges from quadratic to linear, tremendously simplifying the graph structure. We will use the Delaunay triangulation to reduce the number of edges in the casual graphs, see Sect. 5.8.

The *Delaunay triangulation* of a set of points in the plane is a triangulation such that no point is inside the circumcircle of any triangle. It maximizes the minimum angle among all possible triangulations, avoiding skinny triangles and ensuring a well-shaped mesh. Therefore, it is useful to model the army movements over the Delaunay triangulation, where nodes are the battles and edges are the main roads that the army travels. Clearly, this is an approximation, but useful nonetheless (Fig. 5.6).

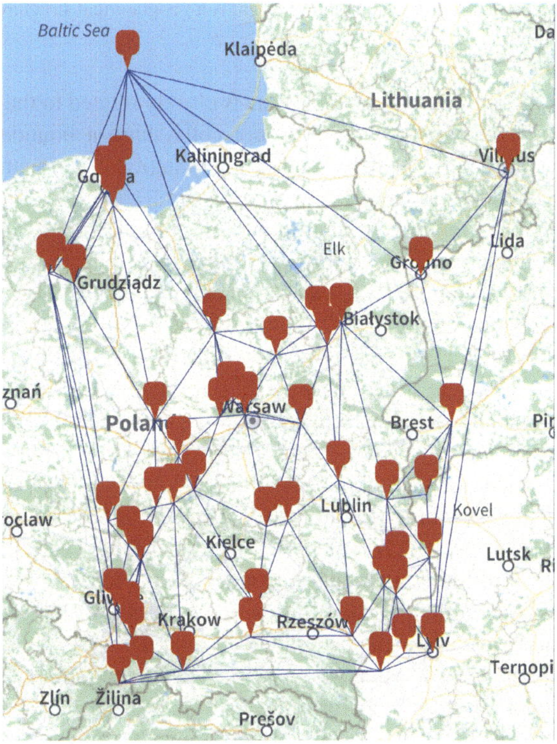

Fig. 5.6 This figure shows the position of the main battles during the German invasion of Poland. The red points are the locations of the battles and the blue lines are the Delaunay edges

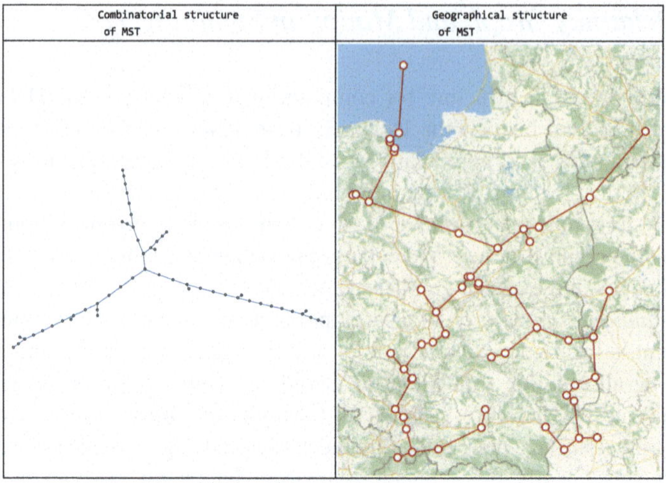

Fig. 5.7 This figure shows the MST of the main battles during the German invasion of Poland. The red points are the locations of the battles and the blue lines are the MST edges. The left is a combinatorial structure, and the right is a geographical structure

The Delaunay triangulation is closely related to the Voronoi diagram, as its edges connect points whose Voronoi cells share a boundary. It can be computed efficiently and is widely used in computational geometry. For more on the Delaunay triangulation, see [9].

The Delaunay triangulation and the minimum spanning tree (MST) of a set of points in the plane are closely related in graph theory. The MST is always a subgraph of the Delaunay triangulation, meaning that every edge in the MST appears in some Delaunay triangle. This relationship is useful for computing the MST efficiently. For more, see [10]. Figure 5.7 shows the MST of the invasion of Poland in 1939.

Both data structures are useful for many geographical analyses. For example, the minimum spanning tree is useful for analyzing the geographical structure of battles since the objective of battle is to reach critical points as quickly as possible and conquer them. As can be seen in the left of Fig. 5.7, the minimum spanning tree has a single point where all three legs of the MST tree converge. The center of the tree is Warsaw, which can be seen from the right. The fall of Warsaw appears only after three different locations surrounded it.

According to the definition of a historical event, see Sect. 2.2, the position of any historical event can be embedded in $mathbbG$.

In general, when we wish to analyze geographical data of a historical event, we recommend to transform the historical event into nodes and use the geographical data of historical events to determine if there is an edge between two historical events. Formally, the node $V = 1, \ldots, \tau$ of the graph represents the historical events $\{h_1, \ldots, h_\tau\}$ The edges of the graph represent a relationship R between two sets of the plane

$$R \subseteq \mathbb{G} \times \mathbb{G}, \tag{5.14}$$

We can now transform the relations R into a graph $G(V, E)$. For example, we can say that nodes $i, j \in V$ are connected by an edge $\{i, j\} \in E$ if the two historical events h_i, h_j are at a distance less than some constant c.

5.6 Knowledge Graphs

Knowledge graphs are structured representations of information that capture relationships between entities in graph format. They consist of nodes, which represent entities such as people, places, concepts, or historical events, and edges, which define the relationships between these entities. Unlike traditional relational databases, knowledge graphs provide a more flexible and interconnected way of storing and retrieving data, making them particularly useful for representing complex domains, such as historical processes.

By leveraging semantic relationships, knowledge graphs allow for more meaningful querying and reasoning, enabling us to capture local structures mechanically.

One of the primary advantages of knowledge graphs is their ability to integrate information from multiple sources while preserving the semantic meaning of relationships, such as combining time and space as needed in historical process understanding. This makes them valuable for search engines, recommendation systems, and artificial intelligence applications. For example, Google's Knowledge Graph enhances search results by linking related concepts, allowing users to find relevant information more efficiently.

Building and maintaining a knowledge graph involves multiple challenges, including data integration, scalability, and consistency. Since knowledge graphs aggregate data from heterogeneous sources, ensuring accuracy and resolving conflicts between contradictory information is a key concern. This is crucial when using knowledge graphs in history, since the quality of the analyses is based on the truthfulness of the data.

Moreover, as knowledge evolves, the graph must be dynamically updated to reflect new discoveries and changes in relationships. Advances in machine learning and natural language processing have contributed to automated knowledge graph construction and refinement, but human oversight is often necessary to maintain quality and reliability. Despite these challenges, knowledge graphs continue to be a crucial technology for organizing and understanding complex information in an increasingly data-driven world.

In this section, we use the Wikipedia structure to construct knowledge graphs of historical events. We will construct our knowledge graph using a hyperlink. Typically, knowledge graphs are directed graphs where each node represents a page, and a directed edge from page A to page B exists if A contains a hyperlink to B.

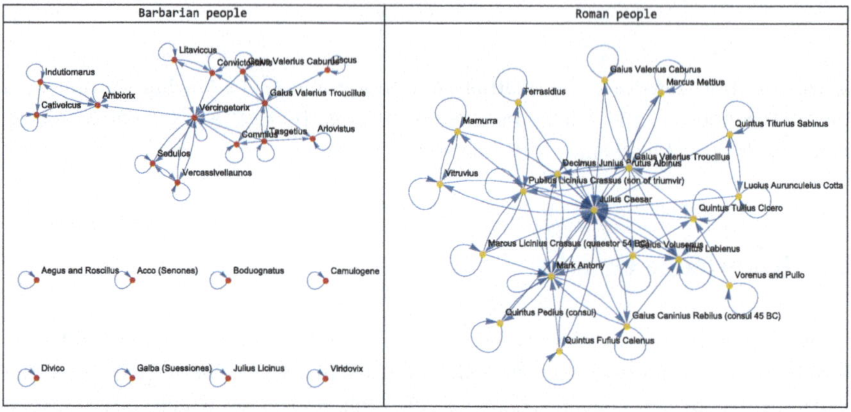

Fig. 5.8 This figure shows the social network of the parties involved in the Gallic War. The left shows the Barbarians; the right are the Romans. It is clear that the Roman social network is strongly connected and controlled by Caesar, while the Barbarian network is separated and disconnected, a fact believed to be one of the reasons for their defeat

In the *Wiki page knowledge graphs*, the node V_h in the hyperlink graph $G_h(V_h, E_h)$ is a webpage and a direct edge $(v_i, v_j) \in E_h$ between two nodes $v_i, v_j \in V_h$ if there is a hyperlink in the webpage v_i that points to the webpage v_j.

We look at the Gallic War to demonstrate how the Wiki page knowledge graphs can be used in history study. We compare two knowledge graphs: one for the Romans and one for the Barbarians. To construct the two graphs, we use the Wiki categories [11, 12]. Now, we can construct two Wiki page knowledge graphs: one for the Barbarians and one for the Romans, see Fig. 5.8.

Analyzing the Wiki page knowledge graphs between the people may help us compute the main characters in historical conflicts and the structure of the social network during the conflict. From Fig. 5.8 we can infer that the central figure in the Roman social network is Julius Caesar, while the central figure in the Barbarian camp is Vercingetorix. We can also see that the Roman network is well connected, while the Barbarian network is disconnected, which was probably one of the key reasons for the Gallic defeat. Our conclusions are well-known historical facts. However, we were able to compute them mechanically, which enables us to verify our method.

5.7 Dynamic Network

Dynamic graphs are graphs that change over time. Formally, dynamics networks are either a sequence of graphs or a function of time to the space of all graphs/matrices. In fact, a better model for studying historical events would be evolving network or dynamic networks. An dynamic networks is a finite sequence of graphs:

$$\gamma = (G_1(V_1, E_1), \ldots, G_1(V_\tau, E_\tau)) \tag{5.15}$$

The duration of the sequence is τ, and it measures the historical period in question, where time units depend on the application. Time can be measured in days, years, or centuries. For more about dynamic networks and dynamic graphs, see [13] and 3.

5.8 Causality

When we combine time and place, we can infer some kind of causality using a causality graph. In this section, we explain how this can be computed. Consider the German invasion of Poland as a case study. During the invasion, the main ground forces of the German army were tanks, which could travel about $speed = 40$ kms/day kilometers per day. We consider the battles to be historical events, see ..., so, if we have two historical events (battles) $h_i = (t_{i,s}, t_{i,e}, goe_i), h_j = (t_{j,s}, t_{j,e}, goe_j)$ such that the battle h_i ended at time $t_{i,s}$.Let the battle h_j ended at time $t_{j,e}$. Assume that

$$t_{i,e} < t_{j,e}. \tag{5.16}$$

We can ask if the tanks from battle h_i can participate in battle h_j. Clearly, if the tanks did not have enough time to travel before battle h_j ended, then they could not influence the outcome of the battle. Therefore, we can place the causality link between battles h_i and h_j if the tanks had enough time to participate in battle h_j. Formally, this is equivalent to the equation

$$speed(t_{j,e} - t_{j,e}) > d(goe_i, goe_j) \tag{5.17}$$

Note that Eq. 5.17 is in the spirit of the Minkowski metric equation. In relativity, we assume that information can't travel faster than the speed of light c. In the case of the German invasion of Poland, we assume that tanks could not travel faster than 40 km per day.[2] For more about causality structure, see [14]. We remark that this is a weak version of causality, since the fact that the tanks could arrive didn't mean that they did arrive during the actual battles.

In the Poland invasion, speed was a deciding factor. One way to demonstrate the importance of speed is by our causality graph, see Fig. 5.10. The train was first used in war during the Crimean War in 1855, when the British built the Grand Crimean Central Railway to transport supplies. The American Civil War (1861–1865) marked the first large-scale use of railroads for moving troops and equipment. However, it was the Franco-Prussian War of 1870–1871 that truly demonstrated the strategic power of rail transport in warfare. The Prussian military, under Helmuth von Moltke, used a meticulously planned railway system to mobilize and deploy troops rapidly,

[2] In WWII, it was possible to transport tanks faster, for example, by train; however, we assume that it was not done.

Fig. 5.9 This figure demonstrates how trains can change the speed of causality graph

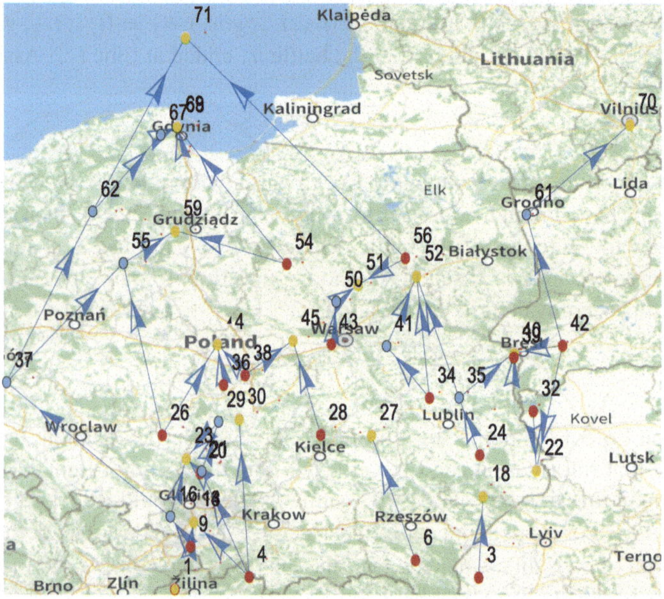

Fig. 5.10 This figure shows the causality graph of historical events restricted to the Delaunay edges. Yellow nodes have an outdegree equal to 0, and red nodes have indegree equal to 0

giving them a decisive advantage over France. This marked the beginning of modern military logistics, where railroads became central to planning and execution see Fig. 5.9.

Now that we have an equation to test causality, we would like to compute a causality graph between the different battles in the invasion of Poland. Here, we face at least two possibilities. We can compute the causality between all pairs of

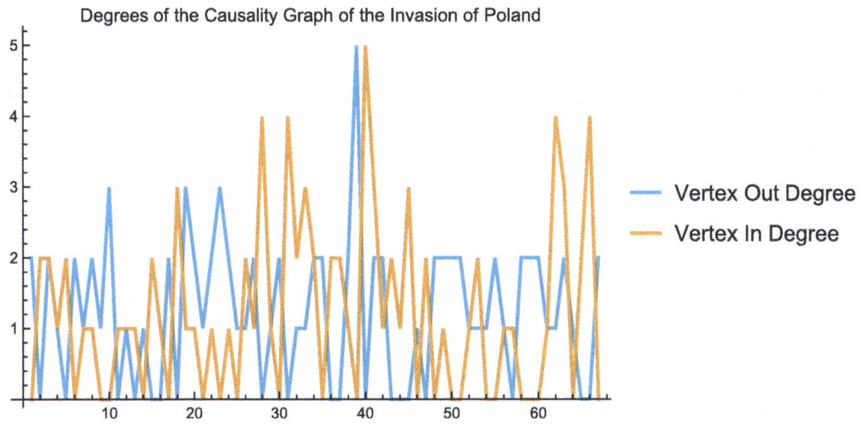

Fig. 5.11 This figure shows the in- and outdegrees of each of historical events in our causality graphs

battles. In this case, we will have too many edges in the graph. It ignores geography and uses only distances. To reduce the number of edges, we recommend checking causality only on the Delaunay edges. The causality graph that was based only on the Delaunay edges can be seen in Fig. 5.10. Once we compute the causality graph, we can look at the indegree and outdegree sequences of the causality graphs, see Fig. 5.11. Now, we can examine the maximums of those sequences. The max indegree is the historical event which depends on the outcome of all the battles. In our case, this is the German–Soviet military parade in Brest-Litovsk. The max outdegree turns out to be Battle of Brześć Litewski. At first glance it looks like an insignificant battle in the Poland campaign, which ended on September 17, 1939; but from a political standpoint, this is a major battle since the German army conquered Litewski, part of the Soviet Union according to the Molotov-Ribbentrop Pact. One may argue that there is a tactical reason as to why Stalin invaded Poland. September 17, 1939 is the maximum time until which Stalin could postpone the invasion to Poland without losing the territories.

5.9 Conclusion

In this chapter, we introduced the concept of graphs that allow us to represent the relationship between historical events. We introduced the concept of interval graphs, which allow us to describe the relationship between concurrent historical events. We extended the notion of interval graphs into intersection graphs. We will use intersection graphs extensively in Chap. 13 where we discuss historical machines.

Since historical events contain information about geography, we introduced several graphs from computational geometry, such as minimum spanning tree, Delaunay triangulation, and convex hull. We demonstrate how the convex hall can compute the size of Poland during the German invasion of 1939.

We ended this chapter by showing how to combine the information about space and time to causality graphs and use it to study the German invasion of Poland. Our analyses provided some explanation as to why the Soviet Union invaded Poland precisely on September 17, 1939—not before or after.

We also explained how to transform a list of webpages into directed graphs using the hyperlinks in each of the webpages. Those hyperlink graphs will be used extensively throughout the rest of the book.

This chapter deals with macrohistory where nodes are usually historical events or historical entities and the edges are the nodes that represent relationships between two historical events or two historical entities.

In the next chapter, we will discuss microhistory where graphs tend to be social networks and the nodes are the people of the past. Since we generally do not have the social network of ancient societies, we will develop some rules of thumb that will help us estimate the properties of these networks.

5.10 Exercise

1. k_n is the complete graph with n nodes and all possible edges.
2. The **complete graph** K_n is an undirected graph with n Vertices where every pair of distinct Vertices is connected by a unique edge. Formally, it is a simple graph (no loops or multiple edges) with the maximum number of edges possible for n Vertices.

   a How many edges are there in K_n?
   b What is the degree of any node K_n? How does this relate to the number of edges?
   c Consider the adjacency matrix A of K_n. What is the rank of A? Is A invertible for $n > 1$? Why or why not?
   d Define the *diameter* of a graph as the longest/shortest path between any two vertices. What is the diameter of k_n?

3. In Sect. 5.4 we explained how to compute the interval graph of the seven main events of WWII. Pick your favorite timeline from [15] and compute its timeline interval graphs.
4. In Sect. 5.5 we explain how to compute the Delaunay graph and MST of the invasion of Poland. Pick your favorite war timeline from [15] and compute the Delaunay graph and the MST. What can these graphs teach you about the war?
5. In Sect. 5.8 we compute the causality graph of the Delaunay graph. Pick your favorite war, and repeat what was done in Sect. 5.8. What does the causality graph teach you about that war.

References

1. Wikipedia contributors. Graph (discrete mathematics) (2024). https://en.wikipedia.org/wiki/Graph_(discrete_mathematics). Accessed 2024
2. R. Diestel, *Graph Theory* (Springer, Berlin, 2006)
3. Wiki. wiki-interval-graph (2024). https://en.wikipedia.org/wiki/Interval_graph
4. Wiki. wikisource-timeline-of-the-1939-invasion-of-poland (2024). https://en.wikipedia.org/wiki/Timeline_of_the_1939_invasion_of_Poland
5. M. De Berg, M. Van Kreveld, M. Overmars, O. Schwarzkopf, *Computational Geometry* (Springer, Berlin, 1997), pp. 1–17
6. Wikipedia contributors. Computational-geometry (2024). https://en.wikipedia.org/wiki/Computational_geometry. Accessed: 2024
7. Wikipedia contributors. wikipediaconvex-hull (2024). https://en.wikipedia.org/wiki/Convex_hull. Accessed: 2024
8. Z. Lotker, 1939: Poland during the war. YouTube video (2025)
9. Wiki. wikisource-delaunay-triangulation (2024). https://en.wikipedia.org/wiki/Delaunay_triangulation
10. Wikipedia contributors. wikipedia-euclidea-mst (2024). https://en.wikipedia.org/wiki/Euclidean_minimum_spanning_tree. Accessed: 2024
11. Wiki. wikisource-category-barbarian-peopleofthegallicwars (2023). https://en.wikipedia.org/wiki/Category:Barbarian_people_of_the_Gallic_Wars
12. Wiki. Category:roman-people-of-the-gallic-wars (2023). https://en.wikipedia.org/wiki/Category:Roman_people_of_the_Gallic_Wars
13. Z. Lotker, *Analyzing Narratives in Social Networks Taking Turing to the Arts Authors* (Springer, Berlin, 2021)
14. Wiki. wikisource-causal-structure (2024). https://en.wikipedia.org/wiki/Causal_structure
15. Wikipedia contributors. List of timelines of world war ii (2024). https://en.wikipedia.org/wiki/List_of_timelines_of_World_War_II

Chapter 6
Networks and Microhistory

6.1 Introduction

The previous chapter discussed macrohistory. In this chapter, we discuss microhistory. Microhistory analyzes historical events from the perspective of individuals. Traditionally, this is done from the narratives of the leaders. However, when discussing microhistory from a mathematical perspective, we model society using a social network $G(V, E)$ where the set of nodes V in the graph are people and edges in the graph are connections between people. For more on social network, see [1] and Sect. A.6. In this chapter, we will discuss the structure of social networks through a simple rule of thumb. This will allow us to compute the structure of the social network, knowing the size of the social network. In many important historical events, we know the size of the society during these events. Therefore, our rule of thumb will allow us to do the estimation of the historical social network (Fig. 6.1).

6.2 Social Networks

Social networks are one of the building blocks of computational history. Generally speaking, an introductory discussion of social networks involves mathematical machinery, such as the Erdős–Rényi random graph, preferential attachment random graphs, and social network harvesting via real data or an online source like Wikipedia.

Once we construct a relevant social network, we can then apply several standard algorithms such as community detection [2] or conflict function [3] to analyze the social networks. This analysis can pinpoint interesting interpretations of the network narrative.

In this chapter, we use a different approach to introduce a social network. We develop rules of thumb on how a typical social network behaves. The reason for

Z. Lotker, *History by Algorithms*,
https://doi.org/10.1007/978-3-031-93627-2_6

Fig. 6.1 A cartoon about the historical social networks

choosing such an approach is that the rules of thumb themselves are simple to understand, making it easier to avoid unnecessary mathematical complexity. They are easy to apply to historical data and are, in general, flexible enough to capture the essence of a social network.

Although we will be using advanced mathematics as an argument to convince the reader, the rules of thumb themselves can be applied without knowing the mathematical tools.

Once we introduce the rules of thumb, we can apply them to ancient society and tackle the question of how to compute the information about extinct and demolished social networks. This can be very useful when studying big historical social networks, such as the social network of ancient Rome or Carthage.

We approach this problem with five rules of thumb for social networks, then apply them to two events from the past and modern times: the first Punic War and the Syrian refugee crisis.

The rest of the chapter is organized as follows: In Sect. 6.2 we briefly introduce the use of denotation in social networks. Then we discuss the nature of the rule of thumb in Sect. 6.2. In Sect. 6.2.1 we discuss how to calculate the number of edges knowing the number of vertices in common social network. This is a very useful rule of thumb when studying extinct social networks where we do not know the number of edges and only an estimation of the number of nodes. In Sect. 6.3, we discuss the structures of social networks and develop the Core-Periphery rule of thumb. In Sects. 6.4, we use the Core-Periphery model to study the statistical structure of the social network fabric and the Small World phenomena with just the number of nodes. In Sect. 6.5, we develop the rule of thumb for the maximum network flow problem, which is important for understanding the ability of a single player to manipulate the network. In Sect. 6.6, we give two examples that use our rule of thumb to study extinct and forgotten social networks from the past and the refugee crisis in the modern era.

Figure 6.4 shows an artistic interpretation of the statistic of term counts in the chapter, where the size of the words reflects the frequency of their appearance.

A rule of thumb is a principle or guideline that provides practical advice or recommendations. A rule of thumb in social networks refers to a general principle that helps evaluate the social network's property. These rules of thumb allow us to resurrect the extinct social network, not unlike Ezekiel's vision of the dry bones. The rule of thumb changes the network from an untouched, unimaginable, and empty structure to a tangible one we can imagine, study, and argue about. This gives us a new perspective on studying historical facts and tendencies. Usually it is possible to find better analytical instruments that fully utilize all the information and get more accurate results, but in cases where accurate data is lacking, such as when analyzing the extinct social networks of the past, using the rule of thumb turns out to be the best choice.

The most famous rule of thumb is the Pareto principle, widely used in business, economics, and project management. Also known as the 80/20 rule, it states that just 20% of the causes generate roughly 80% of the effects. This rule was originally proposed by Italian economist Vilfredo Pareto, who observed that 20% of landlords in Italy owned 80% of all land. The Pareto rule has since been applied to many different areas of life.

Pareto's rule

20% of customers generate 80% of a company's profit.

In business, the Pareto rule can help companies prioritize their efforts and allocate their resources more effectively. For example, a company might realize that 20% of its customers generate 80% of its profit. The company can potentially increase sales and profitability by focusing its marketing efforts on those 20%. The Pareto rule can also identify areas of inefficiency, such as products that consume excessive resources but could be more profitable. A company can optimize its operations and improve overall performance by addressing these areas.

Another example of a rule of thumb is Price's square root law, also known as Price's law [4]. It is devoted to the estimation of the number of authors publishing academic literature. If there are m publications written by n authors, then half of them, i.e., $m/2$, were written by square root, i.e., $\sqrt{n}$-authors. For example, if 25 authors write 100 papers, then 50 papers are written by 5 authors.

Empirical data suggests that Price's law needs to be a better fit and was wrongly formulated. The right formulation for the square root rule would be that a liner fraction of the paper is written by a subliner fraction of the authors [5].

Price's law

There exists $1 > \alpha > 0$, such that $\Theta(m)$ papers are written by $\Theta(n^{\alpha})$ authors.

Note that Price's law is related to Lotka's law [6].

There are many other rules of thumb in science, and all share the property of providing easy calculation for analyzing complex social phenomena. Rules of thumb are equivalent to the separation of wheat from the chaff. But despite their convenience, they usually sacrifice accuracy for ease of use.

6.2.1 Number of Edges in a Social Network

The relationship between the number of edges and the number of nodes in a social network can vary depending on the structure of the network and the meaning of the edges. For example, let us look at the social network where nodes are all real people who connect to others with Facebook accounts, nearly 3 billion people. Clearly, such a graph would have huge cliques, and the number of edges in this particular graph is

$$|E| = \binom{|V|}{2} \tag{6.1}$$

Following this, we can see that it is easy to come up with many examples where the number of edges is square to the number of vertices; however, the graph in this example is trivial and not interesting. Generally speaking, it seems that when we have too many edges, the social graph becomes trivial and not interesting.

In a social network with n nodes, the number of edges is directly related to the number of connections between the nodes. The more connections there are, the more edges there will be in the network. In a fully connected network, where every node is connected to every other node, the number of edges is equal to $n(n-1)/2$. However, in most real-world social networks, not every node is connected to every other node, and the actual number of edges is lower than this maximum value.

The number of edges in a social network with n nodes is often found to be proportional to $nlog(n)$. It can be explained as the result of the network growing when the number of connections between nodes increases, but not equally for all nodes. Some nodes may have a higher degree of connectivity than others. Moreover, the underlying structure of social networks often exhibits characteristics of small-world networks, which are characterized by a high clustering coefficient and short average path lengths between nodes. These properties can be modeled using graph theory, and the resulting analysis shows that the number of edges in a social network with n nodes is typical of the order $nlog(n)$.

In this section, we are going to present several fragments of evidence that support the assumption that the number of edges in the majority of **interesting** social networks is bounded from below by the number of nodes n and from above by $O(nlog(n))$. The rule of thumb in Eq. 6.10 gives a good indication if the definition of the social network properly captures the essence of the problem. If the equation is not satisfied, then we distinguish between two disjoint cases. The first is if we have too few edges, i.e., the network is not connected, in which case the Eq. 6.10 tells us that we need to concentrate on a single connected component, meaning that we need to reduce the number of nodes. The second case is where we have too many edges. The Eq. 6.10 then tells us that we should increase the number of nodes. For example, consider a core-periphery social network: If we concentrate only on the core, then the network will be too dense. When we add the periphery to the network, we come back to the original network. If we concentrate only on the periphery, then the network breaks down into many connected components, each of them linear in the number of nodes for the majority of the time. To conclude, by saying a social network is **interesting** we mean that there is a balance between the number of nodes and the number of edges, following Eq. 6.10.

We present three different arguments in support of this statement. The nature of the first one is the connectivity argument, the second one comes from the heavy tail of degree distribution, and the third one is experimental.

6.2.2 Connectivity Argument

For the network to be connected, the number of edges must satisfy

$$|E| \geq |V| - 1 \tag{6.2}$$

Therefore, in most cases, the number of edges in the social network is at least linear in the number of nodes.

Leskovec et al. [7] have shown that the number of edges is super-linear in the network, i.e., as the number of nodes increases, the ratio between the number of edges and the number of nodes also increases.

6.2.3 Heavy-Tail of Degree Distribution

One of the main models that describes the evolution of the social network is preferential attachment. In the basic preferential attachment model, nodes arrive one after the other; to connect to the network, they choose the random edge according to the uniform distribution and connect to one of its nodes in the same manner. The preferential attachment model attempts to explain the heavy tail of the degree distribution in the social network.

The preferential attachment models are known for their ability to generate networks with a power law degree distribution [8], i.e., a distribution where the fraction of nodes with degree k is proportional to $k^{-\beta}$ for some fixed β which is called the power law distribution parameter. The power law degree distribution has been observed in many real-life networks and complex systems.

In a simple preferential attachment model, Chung et al. have shown [8] that β the power law degree distribution parameter satisfies

$$2 < \beta < 3 \tag{6.3}$$

Clearly, in this case, the number of edges is linear to the number of nodes.

More general models were introduced by Avin et al. in [9], dubbed the most general preferential attachment model. In the general model in each event, an edge arrives and therefore the sum of the degrees increases by two in each time step t.

However, there are three different events, where the number of arriving nodes could be zero, one, or two. When zero new nodes arrive, we say that an edge event $p_e(t)$ occurs. This edge event occurs with some probability. If one node arrives, we say that a single-node event occurs with a probability of $p_v(t)$. If new connected components arrive, i.e., two nodes connected by a single edge, we can say that a component event occurs with a probability of $p_c(t)$.

In this most general preferential attachment model, it was shown that the degree distribution parameter β could achieve the full spectrum of the degree distribution power law, i.e.,

$$1 \leq \beta < \infty \tag{6.4}$$

When considering the number of edges as a function of β, we distinguish between four categories of values:

1. $1 \leq \beta < 2$
2. $\beta = 2$
3. $2 < \beta \leq 3$
4. $3 < \beta$

To achieve the first two cases, we need to have more and more edge events occurring as time progresses, which means that the connective component event probabilities tend to zero:

$$\lim_{t \to \infty} p_c(t) = 0 \tag{6.5}$$

To achieve the last case, we need to have more and more component events happening. The graph in this case will not be connected.

In the first case, $1 \leq \beta < 2$ the number of edges is

$$|E| = \Theta\left(n^{2-\beta}\right) \tag{6.6}$$

Most edges are connected to a constant number of nodes, making it wholly not interesting.

In the second case, the node events happen with a probability of

$$p_v(t) = \Theta\left(\frac{1}{log(t)}\right) \tag{6.7}$$

The number of edges is therefore

$$|E| = \Theta(nlog(n)) \tag{6.8}$$

Here, the preferential attachment model and the core-periphery model coincide at the symmetry point. For more information, see Sect. 6.4.2. In the next two regimes, the number of edges is linear. To summarize, when considering the degree distribution and the preferential attachment, it follows that the most interesting case is when the number of edges is between linear and $n\log(n)$

$$\Omega(n) = |E| = O\left(n\log(n)\right) \tag{6.9}$$

6.2.4 Experiments

It's worth noting that in real-world networks, the relationship between the number of edges and the number of nodes is usually not a simple linear one, see Table 6.1. Social networks tend to have a complex structure, and the number of edges can be affected by many factors such as network density, number of communities, and the distribution of connections between the nodes. In the following table, we present seven well-known networks, the data of which can be found in Wolfram Mathmatica's data repository, see [10].

Table 6.1 contains four columns—the first one is the name of the network, the second the number of nodes in the net, the third the number of edges in the network and the last column is the ratio between the number of edges and our rule of thumb to the number of edges, i.e., $|V|\log(|V|)$. As can be gleaned from the last column, the ratio is between 0.5 and 1.41. This gives us a good indication that the rule of thumb is correct.

To conclude the number of nodes and the number of edges in the **interesting** social network, we suggest the following rule of thumb:

Law of meaningful social network

Let $G(V, E)$ be a meaningful social network, so that

$$\Omega(|V|) \leq |E| \leq O(|V|log(|V|)) \tag{6.10}$$

In the next section, we will discuss the simple structure of the social network and use it as a rule of thumb.

Table 6.1 Table summarizing the relationship between number of nodes and number of edges

Name of the network	$\|V\|$	$\|E\|$	$\|E\|/(\|V\|log(\|V\|))$
Davis Southern Women	32	89	0.8
Dolphin Social Network	62	159	0.62
Family Gathering	18	31	0.59
Friendship	9	13	0.65
Marvel Universe Social Graph	19428	96662	0.5
Online Social Network	1899	20296	1.41
Zachary Karate Club	34	78	0.65

6.3 Structures of a Social Network

Human society consists of many layers and different social classes. These layers branch in and out, connecting to each other directly and indirectly.

When developing a rule of thumb for the structure of society, we have two contradictory goals that generate a conflict and force us to balance them delicately. On the one hand, we want to make the mode as simple as possible; on the other hand, we try to make the model as close to reality as possible. The simple model will allow us to compute easily, while the complex model makes computing difficult. The success of the model, therefore, depends on the balance between these conflicting goals.

In this section, we will discuss two major simple models that describe the social network. The first is Erdős–Rényi model and the second is the core-periphery model.

6.3.1 Erdős–Rényi Model

Perhaps, the simplest structure assumes that the social network is homogeneous. This model is called Erdős–Rényi model, see [11, 12]. The Erdős–Rényi model was the first random graph model, and it revolutionized science when it was proposed a century ago.

Erdős–Rényi Model Definition

There are several great books on the Erdős–Rényi random graph model, see for example [13–15].

The Erdős–Rényi model is based on a simple mathematical model of probability theory that describes the behavior of random graphs. The model is also called a random graph model, and it defines a random graph as a graph in which edges between vertices are formed randomly and independently.

In the Erdős–Rényi model, a graph with n vertices is constructed by selecting each of the $n(n-1)/2$ possible edges with a probability of p and adding them to the graph. This means that a graph generated by the model will have an average of $pn(n-1)/2$ edges. By varying the value of p we can obtain different types of random graphs, ranging from sparsely connected graphs with few edges to highly connected graphs with many edges.

The Erdős–Rényi model had different applications, for example, the study of phase transitions in random graphs, which occurs when the graph structure changes abruptly as a function of the parameter p. It has been shown that when n is large enough, a random graph generated by the Erdős–Rényi model will almost surely have a large connected component if p is greater than a certain critical value p_c, see [13, 15].

The Erdős–Rényi model was used to model relationships in real-world networks, and new, relevant features were subsequently added—for instance, the presence of power-law degree distributions or community structure, see [16]. With these extensions, the model provides a deeper understanding of the structure and function of complex networks in fields ranging from biology to sociology.

The Diameter of Erdős–Rényi Model

One of the most famous universal properties of a social network is known as Small-world. Milgram's experiment was the first attempt to empirically prove small-world concepts. It was conceived as a means to finding the answer to the question of what is the structure of the social personal connections network.

Stanley Milgram conducted his experiment in order to determine the number of steps needed to deliver a message from a random person in the United States to a predetermined human target in another part of the country. In the experiment, Milgram provided participants with the message and name of the target and asked them to deliver the message to someone whom they knew personally and believed might be connected with the target. The process was repeated in several iterations until the message was delivered to the end target or until it became clear that the chain of acquaintances is at a dead end and cannot be continued.

This experiment had unexpected results, showing that on average the messages that arrived to the target person were delivered in just six steps. The expected delivery time before the experiment was much longer as none of the participants was personally connected to the target person, and due to the distance between them, the number of steps was assumed to be much bigger. The experiment showed a great level of interconnections between people in such a large country as the United States. This led to the introduction of the "small-world phenomenon" concept, earning deeper inspection in further research. The Milgram experiment had a huge impact on the analysis of social network structure.

In fact, Stanely Milgram's Small-world experiment is closely related to studying the average diameter of the social network. Those of you who wish to learn more can try [17] for a comprehensive review. Many authors have tried to estimate and explain why the social network has a small diameter, most notably [18–20]. However, all works on the Small-world phenomenon try to prove the upper bound on the diameter with one exception—Fraigniaud's [21] which proves a lower bound in the Kleinberg model. In this section, we use the Erdős–Rényi model to show that the diameter must be bigger than two. The argument is simple: For any Erdős–Rényi random graph with constant p when the number of nodes tends to infinity, the diameter is 2 with high probability. To show that it is enough to find two nodes that are not connected to each other, pick any node v and the degree d_v of node v shall be a random variable with a binomial distribution formula:

$$\Pr[d_v = k] = \binom{n-1}{k} p^k (1-p)^{n-1} \tag{6.11}$$

Therefore with high probability, the node will not be connected to at least one other node, if not more.

As pointed out by Kleinberg [19], the participants in Milgram's experiment successfully found a short route in the network without knowing its structure. This ability was defined as the navigability of the social network, and Kleinberg proposed a model according to which some social networks are global, i.e., everyone knows them, while other edges are known only to the local points of the links. He then defined the grid-routing strategy and showed that the average distance when using this strategy is $O(\log(n)^2)$, which is an attempt to explain how the message found its way to the source. In this section, we are going to provide a different explanation. Our assumption is that the social network is described by the Erdős–Rényi random graph model, where the probability of having an edge p is a constant. Later we will study the same question under the core-periphery model.

Our next step is to define a strategy—i.e., an algorithm—to route the message. Clearly, the worst strategy we can use is to simply let the message travel at random without any global knowledge of the graph. This routing algorithm is called a simple random walk.

A simple random walk on a graph $G(V, E)$ has a starting position $v_0 \in V$, and is defined by induction: at time $t \in \mathbb{N}$ the walk is positioned in location $v_t \in V$. The next location can be one of the neighbors of V_t i.e., N_{V_t}. The walk chooses one of these nodes according to uniform distribution. In other words, the next location is selected with the probability of $1/N_{v_t}$.

The starting point of a simple random walk routing strategy is the origin or source of the message.

If the node knows the target of the message, it sends the message directly. Otherwise, it selects a random node from its connections and sends the message there. Now it is easy to calculate the expected distance $Ndis$ that the message travels.

$$E[Ndis] = \sum_{i=1}^{\infty} ipq^{i-1} = 1/p \tag{6.12}$$

Note that in this simple model, the number of edges is $\Theta(pn^2)$. An interesting variation is when the number of edges is $\Theta(n\log(n))$, which happens when $p = \log(n)/n$. In this case, the average distance will be $1/p = n/Log(n)$. Clearly, this is wrong. To fix that, we will move the Erdős–Rényi random graph model to the core-periphery model.

6.3.2 Core-Periphery

Should we want to have a more realistic model with two social classes, then the next structure to use would be the core-periphery structure. The core-periphery concept is a theoretical framework that is used to understand the hierarchical relationships between elements or groups within a larger system. The basic assumption is that society is not homogeneous, but can be decomposed into two homogeneous sets: one is the core, and the other is its periphery, see Fig. 6.2.

The core is a group that possesses a higher level of evolution and resources, while the periphery consists of less developed elements that are structurally dependent on the core. The core-periphery framework has been widely used in the social sciences, particularly in economics, political science, and geography, to analyze the complex relationships between countries and regions, as well as to understand the dynamics of inequality and power within societies. In economics, the core is a central region characterized by good communications and high population density, which contribute to its prosperity. Such a region lies in contrast with the periphery, which consists of less developed regions with poor communication and sparse populations. The core-periphery concept was used in a significant number of papers in the field of economy, for example, [22]. In political science, the core can refer to countries or regions that hold the most political power and influence in the international system. They often have the most advanced economies and military capabilities, and their policies and actions can significantly impact the rest of the world. See the book [23], where the authors analyze core-periphery relations in the scope of EU politics. In geography, the core can refer to the central regions of a country or continent that are more urbanized and developed, while peripheral areas are typically rural and less developed. This can lead to disparities in access to resources and opportunities between core and periphery. The classic Krugman's core-periphery model contains two factors that are used to input resources—the first is spatially immobile and the second is spatially mobile [24, 25].

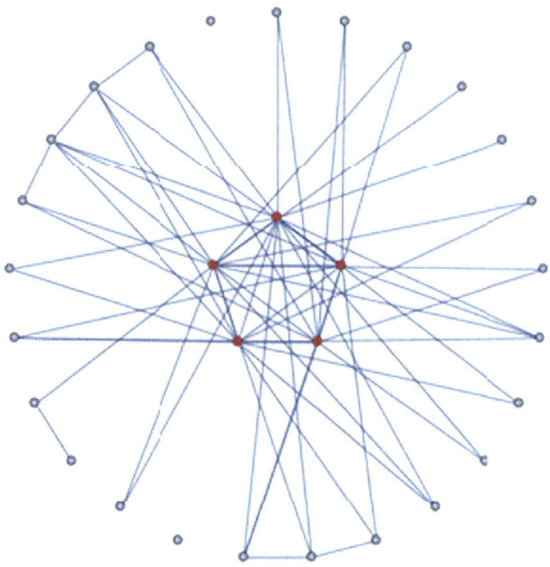

Fig. 6.2 A core-periphery random graph with 5 red nodes in the core and 25 nodes in the periphery

Law of meaningful social network

Let $G(V, E)$ be a meaningful social network, so that

$$\Omega(|V|) \leq |E| \leq O(|V|log(|V|)) \tag{6.13}$$

In the core-periphery model, core nodes have a high number of connections to both periphery and core nodes. Periphery nodes, on the other hand, have a small number of connections to periphery and core nodes. We assume that each node in the periphery is connected to a constant number of core nodes, and at the very least one.

The core-periphery model can be described by the block model, which is widely used in the statistical analysis of complex network structures. The block model partitions the nodes of a network into multiple blocks, and the edges between nodes are generated randomly within and between blocks according to certain probabilities. The random block model studies the formation of groups or communities according to characteristics such as demographics, interests, or behaviors.

A social network can be divided into blocks according to age, gender, or political affiliation. The edges between nodes can be generated based on the likelihood of connections between individuals with similar characteristics. The random graph block model has a significant advantage as it provides a flexible framework for generating synthetic networks with a structure similar to a real-world social network. The model allows testing hypotheses about the formation of social networks, comparing the results of these simulations with real-world data, and estimating evolution probabilities. The random graph block model has been used in a wide range of applications, including the study of social networks, recommendation systems, and epidemic spread. For more information on the random graph block model, see the paper *Stochastic Block Models and Community Structure in Networks* by Karrer and Newman [26]. We are going to use the block model as a sandbox for developing a rule of thumb. To achieve this purpose, we will use the core-periphery block model, where the nodes will be divided into two blocks: one will be referred to as the core and the other as the periphery.

Formally, the core-periphery model has two numbers that describe the population sizes n_c, n_p of the core and the periphery, respectively. We partition the nodes of the random graph into two sets $v_1, v_2..., v_{n_c}$: the nodes belonging to the core and the nodes $v_{n_c+1}, v_{n_c+2}..., v_{n_c+n_p}$ belonging to the periphery. The last element in our core-periphery model is the stochastic matrix $\rho \in \mathbb{M}_{2,2}[\mathbb{R}]$ called communities; a symmetric ρ of edge probabilities. See, for example, Fig. 6.2, where the red nodes represent the core and the blue nodes represent the periphery. To summarize: the model needs five numbers to determine. Although there are six numbers in the example below, since we assume that the matrix ρ is symmetric, we need only five to fully describe the model.

1. $n_p \in \mathbb{N}$
2. $n_c \in \mathbb{N}$

3. $\rho = \begin{bmatrix} \rho_{1,1} & \rho_{1,2} \\ \rho_{2,1} & \rho_{2,2} \end{bmatrix}$

Typically, we suggest implementing the rule of thumb for the core-periphery random model with one parameter m, which has many meanings in the model instead of just five. This will significantly simplify the calculation. $O(m)$ can be thought of as the size of the periphery, the number of edges in the core, the number of edges in the periphery, and the number of edges in the graph. The rule of thumb for interesting social networks can be used when we know the number of edges m. We denote a sequence of graphs that satisfy Eq. 6.14 by $G_{CP}(m)$.

The rule of thumb for core-periphery in interesting social networks

$$1.\quad n_p = \Theta(m)$$

$$2.\quad n_c = \Theta(\sqrt{m})$$

$$3.\quad \rho = \begin{bmatrix} \Theta(1/n_{n_p}) & \Theta(1/n_c) \\ \Theta(1/n_c) & \Theta(1) \end{bmatrix} \tag{6.14}$$

Next, we would like to examine whether the sequence of graphs $G_{CP}(m)$ has Small-world properties. In other words, we would like to compute the expected distance between two random nodes in $G_{CP}(m)$, see Fig. 6.3. The dashed red lines are the edges that connect the periphery to the core, and the solid lines connect the edges inside the core in the Small-world navigation path. We will analyze the distances

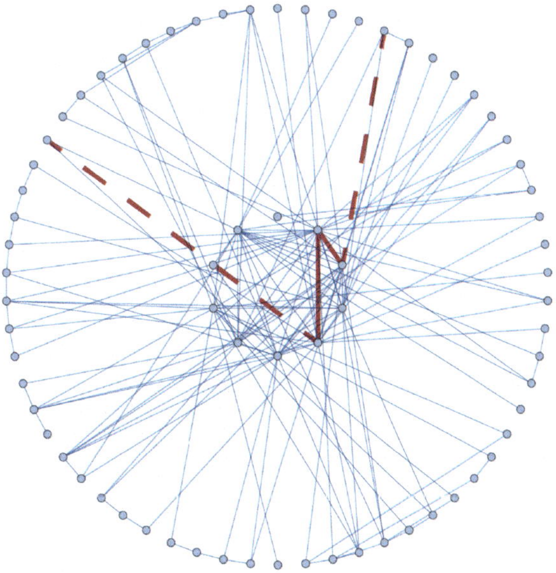

Fig. 6.3 A core-periphery random graph with 8 red nodes in the core and 64 nodes in the periphery. The red path demonstrates the navigation distance between two node in the periphery. The first two dashes are on the path from the periphery to the core, and the solid three are on the path from the core to the second periphery dash

which appear with high probability, i.e., we eliminate low probability before computing so that we can ignore the case of one of the nodes belonging to the core. This happens with a vanishing probability $\lim_{m\to\infty} O(1/\sqrt{m}) \to 0$.

Therefore, when we pick two random nodes u, v, they are usually in the periphery. This is true with high probability. The shortest distance between these two nodes passes through the core with high probability. We assume that each u, v is connected to a single node in the core. Denote the connection of node u to the core by c_u. In the same manner, denote the connection of node v to the core by c_u. Since the core is an Erdős–Rényi random graph with constant probability, i.e., $0 \ll p < 1$, the distance between c_u and c_v is 2 with high probability.

Therefore, the expected distance between any two nodes in the periphery is at least 4 $G_{CP}(m)$. In the above calculation we assume that the nodes know how to find the shortest distance in the graph $G_{CP}(m)$. It is clear that this is not the case with social networks, meaning that 4 is the lower bound. To calculate the upper bound of the distance that the message travels, we apply the following assumptions:

1. Upon receiving the message, each node in the periphery

$$v \in \{v_{n_c+1}, \ldots, v_{n_c+n_p}\} \tag{6.15}$$

sends it directly to its representative in the core.
2. Each node in the core first checks if it can send the message directly to its target. If so, it does; otherwise, it uses a simple random walk routing strategy inside the core.

Following the description above, the message travels 1 edge from the periphery to the core. According to Eq. 6.12 it travels $1/p$ edges inside the core, and 1 edge outside the core. Therefore, the entire route can be formally represented by the equation

$$Ndis = 1 + 1/p + 1 \tag{6.16}$$

Since Milgram's experiment taught us that the average distance is 6, we can use this information to reveal the upper bound of the density of the core $p > 1/4$. Equation 6.17 summarizes the rule of thumb for the Small-world phenomenon in the social network.

Small-world in social network

Let $G(V, E)$ be a social network, so that

$$3 \leq Ndis(u, v) \leq 2 + 1/p \tag{6.17}$$

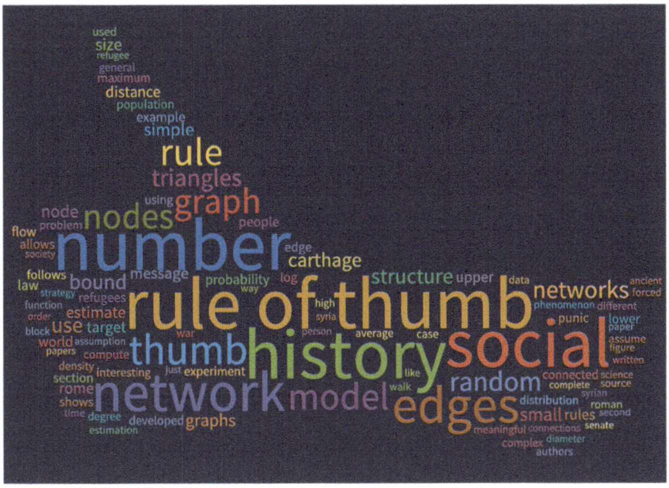

Fig. 6.4 Counting words: the chapter's word cloud

6.4 Counting Motifs in Social Network

In my previous book [27], the destruction of the social fabric is demonstrated as one of the ways to represent evil in literature. The social fabric is created by making members of the social network interact with each other, thus weaving the threads together. Therefore, the minimal structure that generates social fabric in social networks is a triangle. A complete graph with t nodes is generally denoted by K_t. We denote the number of copies of triangles in the graph $G(V, E)$ by κ_3. In the same way, the number of copies of a complete graph k_t is denoted by κ_t. While an edge is a one-dimensional object, a triangle is a two-dimensional object, and—generally speaking—k_t is a $t - 1$ dimensional object (Fig. 6.4).

The next rule of thumb we would like to enforce is the relationship between the number of edges in the social network and the number of triangles in the graph $G(V, E)$. To put it differently, a general rule of thumb that connects the number of edges m to the number of copies of the complete graphs K_t, i.e., κ_t. Of course, the answer depends on the definition of the edge in the social network. However, some of the results will be general for any graph, and some of them will be assumed for the core-periphery.

In this section, we will present evidence that support the rule of thumb by showing that the number of edges in most **interesting** social networks is bounded from below by the number of nodes:

$$\kappa_3 = \Theta(m^{1.5}) \tag{6.18}$$

In other words, the number of copies of the complete graphs k_t in the core-periphery in the social network is

$$\kappa_t = \Theta(m^{t/2}) \tag{6.19}$$

We present three different arguments in support of this statement. The nature of the first one is some combinatorial argument and is an upper bound which is true for all graphs. The second argument comes from the core-periphery model and is a low bound, and the third is experimental.

6.4.1 Combinatorial Argument

We start with the upper bounds on the number of triangles. After developing the upper bound for triangles, we develop upper bounds for complete graphs k_t in general.

Let's assume that the number of edges in the social network is m. We can bound from above the number of triangles using the number of edges, following the thesis of Debsoumya Chakraborti published in 2020, see [28].

To understand the structure of the graph that has m edges and maximize the number of triangles, we look at the colex (colexiographic) order and graphs. This allows us to maximize the number of triangles in a graph with m edges.

Colex order on the finite subsets of the natural number set $\mathbb{N}$ is defined as follows: for $A, B \subseteq \mathbb{N}$, we have that $A < B$ if and only if

$$\max((A \setminus B) \cup (B \setminus A)) \in B \tag{6.20}$$

The colex graph L_m on m edges is defined as the graph with vertex set $\mathbb{N}$, and edge set consisting of the first m 2-element subsets of $\mathbb{N}$ in colexicographic order, denoted by $<$.

It turns out that the colex L_m graph has m edges and the maximal number of triangles, see Fig. 6.5 for an example of a graph with 10, 11, ..., 13 edges that maximize the number of triangles. It follows that given a graph with

$$m = \binom{\lfloor \frac{1}{2}\left(\sqrt{8m+1}+1\right) \rfloor}{2} + r \tag{6.21}$$

edges, the maximum number of κ_3 triangles we can have is

$$\kappa_3 = \binom{\lceil \frac{1}{2}\left(\sqrt{8m+1}+1\right) \rceil}{3} + \binom{r}{2} \tag{6.22}$$

For example, the number of triangles in $l_{10} = 10, l_{11} = 10, l_{12} = 11, l_{13} = 13$

Therefore, we have the following theorem which was proven in [28]:

Theorem 6.1 *Let $G(V, E)$ be a graph where $|E| = m$, so that*

$$\kappa_3 = O\left(m^{1.5}\right) \tag{6.23}$$

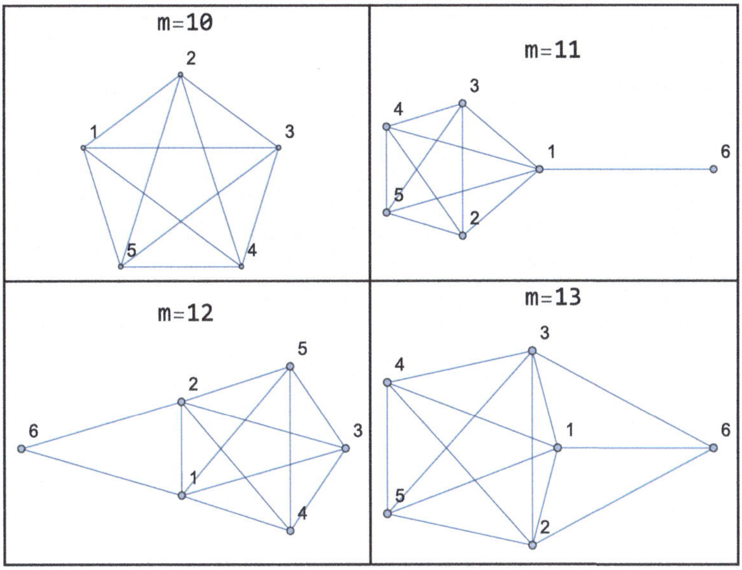

Fig. 6.5 The L_m graphs for $m = 10, 11, 12, 13$

Now we can apply Theorem 6.1 to estimate the number of triangles in the social network. It follows that the number of triangles is less than $O(m^{1.5})$. According to the previous Sect. 6.2.1, the number of edges in a social network with n nodes is $\theta(n \log(n))$. Therefore, the number of triangles in the social network is $O((n \log(n))^{1.5})$.

The same can be proven for general k_t for any fix t. The graph that has m edges and maximizes the number κ_t is L_m.

Theorem 6.2 *Let $G(V, E)$ be a graph where $|E| = m$, so that*

$$\kappa_t = O\left(m^{t/2}\right) \tag{6.24}$$

In the next subsection, we will use the core-periphery structure and the symmetry point to derive a lower bound of the number of triangles. We rely on the paper [5].

6.4.2 Core-Periphery and the Number of Triangles

At first, we need to explain the symmetry point. According to [5], the core-periphery social network is divided into two parts: Core and periphery. At the symmetry point, those parts have the same number of edges, like so:

1. Quarter $m/4$ of the edges are periphery edges, i.e., edges, where both end-points belong to the periphery.

2. Quarter $m/4$ of the edges are core edges, i.e., edges, where both end-points belong to the core.
3. Half $m/2$ of the edges are core-periphery edges, i.e., edges, where one of the end-points belongs to the core, and the other to the periphery.

According to Theorem 3.6 in [5], the size of the elite at the symmetry point is $\Theta(\sqrt{m})$. Moreover, the number of edges within the elite is $\Theta(m)$. Now, to achieve a lower bound on a number of triangles, we use another assumption that claims that the structure of the elite is a Erdős–Rényi random graph with some constant p. Therefore, the expected number of triangles within the elite is

$$\kappa_3 = p^3 m^{1.5} = \Theta(m^{1.5}) \tag{6.25}$$

In the same way, to achieve a lower bound on a number of triangles, we use another assumption which claims that the structure of the elite is an Erdős–Rényi random graph with some constant p. Therefore, the expected number of complete graphs with t nodes in the elite is

$$\kappa_t = p^{\binom{t}{2}} m^{t/2} = \Theta(m^{t/2}) \tag{6.26}$$

The next sections support the theoretical claims by studying some examples, using the databases of famous social networks. We will concentrate only on counting the number of triangles.

6.4.3 *Counting the Number of Triangles: Experimental*

In this section, we examine four real social networks; for each of them we present the name of the network, the number of E edges, the number of κ_3 triangles, and the ratio between the number of triangles and the power $\kappa_3/E^{1.5}$. All of these numbers are summarized in Table 6.2. The social networks that we examine are all well-known, and the data can be found in [29]: Dolphin social network [30], Family Gathering [31], Friendship [32], Zachary's Karate Club [33].

Table 6.2 The first column is the name of the network, the second is the number of edges, the third is the number of triangles, and the last shows the ratio which is the number of edges divided by the rule of thumb

Name of the network	E	K_3	$K_3/E^{1.5}$
Dolphin social network	159	95	0.28
Family Gathering	31	20	0.67
Friendship	13	4	0.51
Zachary's Karate Club	78	45	0.39

Therefore, to estimate the number of triangles, it is enough to know the number of m edges. According to our previous rule of thumb, the number of edges is roughly linear n in the number of nodes up to some $log(n)$ factor. So, if we combine those two sections it follows that the number of triangles in **interesting** social networks is

$$\Omega(n^{1.5}) = \kappa_3 = O((nlog(n))^{1.5}) \tag{6.27}$$

The rule of thumb for number of κ_t

For an interesting social network

$$\kappa_t = \Theta\left(m^{t/2}\right) \tag{6.28}$$

In the next section, we will discuss the rule of thumb for flow problems in social networks. This rule of thumb is extremely useful since social networks tend to be huge, and it is becoming impossible for humans to compute them themselves. The rule of thumb allows them to get answers much faster and much more simpler.

6.5 Network Flow

One of the central problems in graph theory is network flow, see for example [34, 35]. In network flow problems, we explore directed graphs $G(V, E, C)$ with V nodes and directed E edges. We also explore a capacity function C for each directed edge $e \in E$ that describes an upper bound on the flow capacity $C(e)$ that can pass through edge e, which is a positive number.

$$C : E \to \mathbb{R}^+ \tag{6.29}$$

When the graph is not directed, we add two opposite directed edges to the undirected edges. In addition, we also have a source s and a target t.

Until now, we've described the infrastructure of the flow, i.e., the pipelines that allowed us to pass fluid from one node to another. Now we will discuss the flow itself.

A flow f is a function

$$f : E \to \mathbb{R} \tag{6.30}$$

that describes how much fluid flows on each edge. Clearly, we are not allowed to overflow the capacity of the edge. Formally this can be done using the system of inequality, as described in (6.31) for each edge $e \in E$:

$$f_e \leq C(e) \tag{6.31}$$

We call the above inequality capacity restriction.

In addition to the capacity restriction (6.31), the flow function f needs to satisfy Kirchhoff's law of conservation of flow at each node, i.e., for all $v \in V \setminus \{s, t\}$

$$\sum_{i \in N_v} f_{(v,i)} = 0 \tag{6.32}$$

Note that the flow $f_{v,u}$ can be negative numbers. In this case, it means that the flow goes from u to v.

Clearly, the flow from the source s is negative and the flow to the target t is positive, therefore

$$\sum_{i \in N_s} f_{(s,i)} + \sum_{j \in N_t} f_{(j,t)} = 0 \tag{6.33}$$

The value of the flow is denoted by $|f|$, and is equal to

$$|f| = \sum_{j \in N_t} f_{(j,t)} = -\sum_{i \in N_s} f_{(s,i)} \tag{6.34}$$

The maximum flow problem is to find a maximum flow value:

$$max|f| \tag{6.35}$$

This means that we need to compute the function f that maximizes the flow from the source s to the target t.

There are many algorithms to compute the maximum flow, and the one involves time complexity $O(E^{1+o(1)} \log U)$, see [36]. We will provide a very simple rule of thumb that will allow us to compute the value of the flow using a simple formula that can be applied in most cases.

In the social network, the capacity function is 1 for all the edges, and 0 otherwise, i.e.,

$$C(e) = \begin{cases} 1, e \in E \\ 0, e \notin E \end{cases} \tag{6.36}$$

A key ingredient in computing flow is the max-flow min-cut Theorem. An s-t cut $Cut = (S, T)$ is a partition of node V such that $s \in S$ and $t \in T$. Cut-set X_{Cut} is the set of edges that have a source in S and a target in T, formally

$$X_{Cut} = \{(v, u) \in E : v \in S \text{ and } u \in T \tag{6.37}$$

The max-flow min-cut Theorem ensures that once we know the min-cut, we can compute the max-flow. The structure of social networks allows us to find the min-cut very easily by looking at the neighborhood of the source N_s or the neighborhood of the target N_t. This follows from the fact that r-local neighborhood in social networks, i.e., the set of all the nodes at a distance r from the given nodes, grows rapidly, thereby forcing the cut to be either near the source s or near the target t.

To formulate our rule of thumb, we need the following variation of the degree d_v of a node v for a direct graph. Let d_v^{out} be the number of edges that go out from node v:

$$d_v^{out} = |\{(v, u) \in E\}| \tag{6.38}$$

In the same way, we can define d_v^{in} as the number of edges that enter node v.

The rule of thumb for max flow in social network

$$|f| = \min(d_s^{out}, d_t^{in}) \tag{6.39}$$

In the next section, we will describe several applications for these rules of thumb.

6.6 Application to History

Up until now, we developed several rules of thumb that allowed us to estimate and compute many properties of the social network just from the number of edges or nodes. Rules of thumb have many applications when applying common wisdom; they allow us to apply critical thinking. When applied to history, they offer a huge advantage since they allow us to resurrect the social network from the ashes of history. We can suddenly talk about the history of society without having complete data on the network. In this section, we will demonstrate the methodology of social network resurrection through several examples.

6.6.1 The First Punic War

The First Punic War was a conflict between the Roman Republic and the city-state of Carthage that lasted from 264 BC to 241 BC. Carthage was located in present-day Tunisia and fought the Romans over control of Sicily, which was a valuable source of grain and other resources.

The war was long and costly for both sides, ending in a Roman victory. As part of the peace settlement, Carthage was forced to pay a large indemnity to Rome, cede its territories in Sicily, and limit its navy. The First Punic War marked the beginning of Rome's expansion beyond Italy and paved the road to the centuries-long Roman domination of the Mediterranean region [37, 38].

Carthage

The exact population of Carthage during the First Punic War is difficult to determine, as there are no reliable census records from that time period. The numbers heavily depend on the counting methodology, as some ancient sources counted only male citizens, while others counted the whole population. However, ancient sources suggest that Carthage was a large and prosperous city with an estimated population of 200,000 to 700,000 people [39].

Estimating the Number of Edges

Now we can estimate the number of social edges in Carthage using our rule of thumb for the meaningful social network, see (6.13).

$$200,000 \leq |E| \leq\approx 5,000,000 = 700,000 * \log_{10}(700,000) \tag{6.40}$$

Note that the uncertainty in estimating the number of social ties increases as our rule of thumb is superlinear. Moreover, for the lower bound, we use the lower bound of the number of nodes, and for the upper bound, we use the upper bound of the number of nodes.

Estimating the Core of the Carthaginian Social Network

Using these rough numbers we can estimate the structure of the Carthaginian social network. In this case, we have two rules of thumb: one that uses the number of edges and the other that uses the number of nodes. It is better to use the number of nodes, since these are already known. A simple calculation shows that the size of the Carthaginian city's elite $Elic_C$ was

$$447.21 = \sqrt{200,000} < |Elic_C| < \sqrt{700,000} = 836.66 \tag{6.41}$$

Note that the estimation of the core/elite of the society is much more accurate than the estimation of the total population since the relationship between the total size of the population and the elite/core is sublinear. Therefore, the deviation tends to shrink. Using the rule of thumb (6.14) it follows that the core of Carthage was approximately 700 people.

Small-world Estimation of the Carthaginian Social Network

Concerning the Small-world phenomenon, there is a problem in estimating the density of the core p, since the density highly affects the diameter. To estimate the density at the core, we use the size of the Carthagian Senate which during the Punic wars numbered between 200 and 300 members. Therefore, the density of the core could be estimated as follows: the number of people in the Senate divided by the size of the Core, formally

$$200/700 < p < 300/700 \tag{6.42}$$

Now we can use the Small-world phenomenon rule of thumb (see (6.17)) to estimate the average distance in Carthage's social network $Ndis_C$

$$4.35 < Ndis_C < 5.5 \tag{6.43}$$

Estimating the Number of Triangles in the Carthaginian Social Network

Applying the rule of thumb for the number of triangles (see (6.28)), we conclude that the social network of Carthage was around

$$\kappa_3 = 125,000,000 \tag{6.44}$$

Estimating Flow in the Carthaginian Social Network

The last rule of thumb that we developed was the maximum flow rule of thumb (see (6.39)). For Carthage, our estimation is the maximum degree of order for the magnitude of the Core.

$$|f| \simeq 500 \tag{6.45}$$

Rome

The exact population of Rome during the First Punic War is not known, and estimates vary among ancient sources. According to T. Cornell, the population of the city of Rome on the eve of the Punic Wars was approximately 200, 000 people [40]. For the sake of our calculations, we will assume this number to be the lower bound.

According to T. Frank, who cited ancient authors such as Livy, the population of Rome in 234 BC was 270,000 people [41]. For the sake of our calculations, we will assume this number to be the upper bound.

Estimating the Number of Edges

Now we can estimate the number of social edges in Rome using the rule of thumb of the meaningful social network, see (6.13).

$$200,000 \leq |E| \leq\approx 1,500,000 = 270,000 \log_{10}(270,000) \tag{6.46}$$

We can see that the difference between the lower and the upper bounds is not as great as in the case of Carthage. When comparing the estimations for Rome and Carthage, it is clear that we know much more about Rome, as the calculations prove.

Estimating the Core of the Roman Social Network

The population estimation for Rome is much tighter. The calculation shows that the size of the Roman elite Eli_R was

$$447.27 = \sqrt{200,000} \leq Eli_R \leq \sqrt{270,000} = 519.61 \tag{6.47}$$

When comparing Eqs. 6.41 to 6.47 we see that the lower bound in Rome is the same as the lower bound in Carthage, but the upper bound in Rome is much smaller than in Carthage. If we believe the narrative that Carthage was the bigger city at the beginning of the conflict, then the size of core/elite for Carthage must also have been larger.

Small-world Estimation of the Roman Social Network

Concerning the Small-world phenomenon, we can use the same technique as we did for the Carthage case. To estimate the density in the core, we use the size of the Senate in Rome which numbered approximately 100 and 200 at the time of the Punic Wars. Therefore, the density of the core could be estimated as follows: the number of people in the Senate divided by the size of the core, formally:

$$100/500 < p < 200/500 \tag{6.48}$$

Now we can use the Small-world phenomenon rule of thumb (see (6.17)) to estimate the average distance in the Carthagian social network $Ndis_C$

$$4.5 < Ndis_C < 7 \tag{6.49}$$

It appears that the real number of Senate members should be closer to 200, as 100 is definitely too small.

Estimating the Number of Triangles in the Roman Social Network

The estimation for the Roman case is the same as Carthage's, as the size of the core is the same. We use the rule of thumb for the number of triangles (6.28) to estimate the social network of Rome was around

$$\kappa_3 = 125,000,000 \tag{6.50}$$

Estimating Flow in the Roman Social Network

The last rule of thumb that we developed was for maximum flow (see (6.39)). For Rome, our estimation is the maximum degree on order for the magnitude of the core:

$$|f| \simeq 500 \tag{6.51}$$

6.6.2 *Refugees*

Refugees can serve as another example of how to calculate when we have no data on the social network. We can use our rules of thumb to resurrect the social network and estimate its parameters as we did in the Punic War section.

Over the past few centuries, tens of millions of people have been forced to leave their homes, leading to the dispersion of entire societies. This happens all over the world, at different times and for various reasons.

The Refugee Problem

The current ways to measure the impact of refugee crises are either through economics or narrative expression (such as journalism, literature, and cinema). If we equate it with our previously developed rules of thumb, we can analyze the degree of destruction to the social fabric.

The global refugee crisis is a complex and long-standing issue with deep historical roots. According to the United Nations High Commissioner for Refugees (UNHCR), a refugee is defined as "someone who has been forced to flee his or her country because of persecution, war, or violence."

Throughout history, there have been numerous instances of mass displacement and forced migration due to war, persecution, and other forms of violence. One of the earliest recorded instances of mass displacement occurred during the 18th century, when colonization and the transatlantic slave trade drove millions of people out of Africa and other parts of the world.

In the 20th century, the scale of displacement reached unprecedented levels, with WWI and WWII resulting in the displacement of millions of people. In the aftermath of WWII, the international community established the 1951 United Nations Convention Relating to the Status of Refugees, which defined the rights of refugees and the legal obligations of states to protect them.

However, despite this legal framework, the number of refugees and internally displaced persons (IDPs) has continued to grow in recent decades, as a result of conflicts, persecution, and human rights abuse. According to the UNHCR, as of 2019, there were more than 26 million refugees, over 41 million IDPs, and more than 3 million asylum-seekers worldwide.

The problem of refugees has also been exacerbated by factors such as climate change and economic inequality, which have contributed to natural disasters and economic crises. As a result, many people have been forced to flee their homes in search of safety and a better life.

Related Work on Refugees

The problem of refugees and forced migration has been widely reviewed in recent scientific papers. This is a complex matter, and to understand it deeply requires further reading. The theoretical approach to the problem is presented in the book *Refugees: a Very Short Introduction* [42]. The economic insight of the refugee crisis is shown in [43], where the author talks about the cost of refugee migration focusing on the current situation. The refugee problem was the main research topic in the Handbook written by Fiddian, [44].

The Rule of Thumb for the Refugee Crisis: The Case of Syria

According to Wikipedia, the number of refugees from Syria is about 5 million. Before the crisis, the population of Syria was around 22.7 million in 2011, and 18.8 million in 2018.

Estimating the Number of Edges

Now we can estimate the number of social edges in Syria that were destroyed by the crisis using the rule of thumb for the meaningful social network (see (6.13)):

$$|E| \simeq 25,000,000 \tag{6.52}$$

Estimating the Core of Syrian Social Network

The size of the elite before the Syrian civil war was

$$Eli_S \leq \sqrt{22,000,000} = 4690.42 \tag{6.53}$$

When trying to analyze the size of the Syrian refugee core Eli_{SR}, we assume that the structure of refugee society is homogeneous, which allows us to use the core-periphery rule of thumb (see (6.14)). The calculation shows that the size of the Syrian refugee core Eli_{SR} was

$$Eli_{SR} \leq \sqrt{3,900,000} = 1974.84 \tag{6.54}$$

Assuming that it was easier for the rich to run away from the war, we can conclude that the Syrian elite made up a significant portion of the refugees.

Small-world Estimation of the Syrian Social Network

Before the crisis, the Small-world of the Syrian social network could be estimated using the same approach and was around 4, but after the civil war many of the old routes and connections were destroyed, making path calculations difficult. The simple random walk routing strategy will probably fail under these circumstances.

Estimating the Number of Triangles in the Syrian social network

We use the same technique to estimate the number of triangles in the Syrian social network. Using the rule of thumb for the number of triangles (6.28), we find that the number of triangles is approximately

$$\kappa_3 = 7.70188 * 10^9 \tag{6.55}$$

Estimating Flow in the Syrian Social Network

The last rule of thumb that we developed was the maximum flow rule of thumb (see (6.39)). In Syria, the flow changed from 4690.42, to 4242.64.

6.7 Conclusion

Most historians ignore the question of past social networks. At first glance, it would appear that we simply do not have data and therefore cannot discuss the ancient social networks.

To overcome this problem, we developed simple rules of thumb that allow us to compute many social network parameters simply from the number of nodes and edges. Clearly, these rules of thumb are not true science, and should therefore be taken with a grain of salt.

Scrutiny is necessary because these are just rules of thumb and are not accurate. They seem like some kind of magic, and magic warrants suspicion and scrutiny. However, these rules of thumb are based on a simple model, like core-periphery, which allows us to understand why they are true and in what sense. They can be applied to past, present, and future social networks. In general, they should be understood in the language of the big-O notation and not as an exact measurement.

These rules of thumb are to be respected, as they allow us to resurrect ancient and extinct and forgotten social networks. That makes history more vivid and lively and provides the narrative of those forgotten networks.

We developed five rules of thumb:

1. The law of meaningful social network, which connects the number of nodes with the number of edges.
2. The core-periphery rule of thumb that determines the relationship between the core and the periphery.
3. The Small-world phenomenon of social networks allows us to estimate how quickly and in what way news spread in the ancient social network.

4. The number of triangles rule of thumb allows us to evaluate the fabric of the ancient social network or its destruction.
5. The flow rule of thumb allows us to estimate the ability of exceptional individuals to manipulate information.

These five rules of thumb are just examples of how we can recover information about extinct or forgotten social networks. We hope that others will follow and develop other methods.

This is the last chapter of the first part of the book. In the next part, we will discuss theoretical history, which focuses on simple models that describe historical phenomena.

6.8 Exercise

1. In this exercise, we use literature as a source for constructing the social relationship between ancient famous people. Use Shakespeare's *Julius Caesar* [45] to generate the social network of the play using a "who speaks after whom" graph. For more on these graphs, see [3]. Compute the node that is farthest away from Caesar.
2. In this exercise, we use Wikipedia as a source for constructing relationships between famous people. For example, the Wikipedia category People of the Gallic wars (see [46]) contains two subcategories: one for the Romans, and the other for the Barbarians.
    a. Use the Barbarians category of the Gallic wars (see [47]) to construct a social network $G(V, E)$ of the Barbarians during the war. Use the Wiki pages of each of the people mentioned in the subcategory as a node of the graph. Place a direct edge $(i, j) \in E$ between node i and node j if person j is mentioned/linked in the Wiki page of person i.
    b. Use the Romans category of the Gallic wars (see [48]) to construct a social network $G(V, E)$ of the Roman people at the time of the war. Use the Wiki pages of each of the people mentioned in the subcategory as a node of the graph. Place a direct edge $(i, j) \in E$ between node i and node j if person j is mentioned/linked in the Wiki page of person i.
    c. Compare the diameter and spectrum of these networks. According to Wikipedia, the diameter d of a graph is the maximum eccentricity of any vertex in the graph. That is, d is the greatest distance between any pair of vertices. For more on the spectrum of the graph, see [49].
3. The size of the city of Baghdad near the end of the Abbasid Caliphate was nearly one million people, according to Wikipedia. Using our rule of thumb, estimate the social network of Baghdad at this time.
4. Repeat the previous calculation for London and Paris and compare the cities' population size to that of the countries'.

5. In this question, we analyze the history of Russia through the population of two main cities: Moscow and Saint Petersburg. You can use Wikipedia to find the population size of Moscow https://en.wikipedia.org/wiki/History_of_Moscow and Saint Petersburg https://en.wikipedia.org/wiki/Demographics_of_Saint_Petersburg.
   a. Calculate the social cost of Napoleon's war in both cities. Estimate the amount of social destruction using the number of triangles in the social network of each city. Calculate the size of the core in each city before and after the war.
   b. Calculate the social cost of the Civil War in both cities. Which city suffered more during the civil war?
   c. Calculate the social cost of WWII in both cities.
   d. Compare these three colossal wars and their impact on the cities. Pay attention to which city was Russia's capital during those events.
6. Estimate the number of social triangles that were destroyed in WWII.

References

1. Wiki. wikisource-social-network. https://en.wikipedia.org/wiki/Social_network (2024)
2. Wiki. wikisource-community-structure. https://en.wikipedia.org/wiki/Community_structure (2023)
3. Z. Lotker, *Analyzing Narratives in Social Networks - Taking Turing to the Theater* (Springer, 2021)
4. J.B. Peterson, *12 Rules for Life: An Antidote to Chaos* (Random House Canada, 2018)
5. C. Avin, Z. Lotker, D. Peleg, Y.-A. Pignolet, I. Turkel, Elites in social networks: an axiomatic approach to power balance and price's square root law. PloS One **13**(10), e0205820 (2018)
6. P.D. Allison, D. de Solla Price, B.C. Griffith, M.J. Moravcsik, J.A. Stewart, Lotka's law: a problem in its interpretation and application. Soc. Stud. Sci. **6**(2), 269–276 (1976)
7. J. Leskovec, J. Kleinberg, C. Faloutsos, Graphs over time: densification laws, shrinking diameters and possible explanations, in *Proceedings of the Eleventh ACM SIGKDD International Conference on Knowledge Discovery in Data Mining* (2005), pp. 177–187
8. F. Chung, F.R.K. Chung, F.C. Graham, L. Lu, et al., *Complex Graphs and Networks*. Number 107 (American Mathematical Society, 2006)
9. C. Avin, Z. Lotker, Y. Nahum, D. Peleg, Improved degree bounds and full spectrum power laws in preferential attachment networks, in *Proceedings of the 23rd ACM SIGKDD International Conference on Knowledge Discovery and Data Mining* (2017), pp. 45–53
10. datarepositorywol framcloud. https://datarepository.wolframcloud.com/all (2023)
11. P. Erdős, A. Rényi et al., On the evolution of random graphs. Publ. Math. Inst. Hung. Acad. Sci. **5**(1), 17–60 (1960)
12. E.N. Gilbert, Random graphs. Ann. Math. Stat. **30**(4), 1141–1144 (1959)
13. B. Bollobás, B. Bollobás, *Random Graphs* (Springer, 1998)
14. V.F. Kolchin, V.F. Kolchin, *Random Graphs*. Number 53 (Cambridge University Press, 1999)
15. A. Frieze, M. Karoński, *Introduction to Random Graphs* (Cambridge University Press, 2016)
16. M.E.J. Newman, The structure and function of complex networks. SIAM Rev. **45**(2), 167–256 (2003)
17. C. Arney, Linked: how everything is connected to everything else and what it means for business, science, and everyday life. Math. Comput. Educ. **43**(3), 271 (2009)
18. S. Milgram, The small world problem. Psychol. Today **2**(1), 60–67 (1967)

19. J.M. Kleinberg, Navigation in a small world. Nature **406**(6798), 845–845 (2000)
20. D.J. Watts, S.H. Strogatz, Collective dynamics of 'small-world' networks. Nature **393**(6684), 440–442 (1998)
21. P. Fraigniaud, E. Lebhar, Z. Lotker, A lower bound for network navigability. SIAM J. Discret. Math. **24**(1), 72–81 (2010)
22. R. Forslid, G.I.P. Ottaviano, An analytically solvable core-periphery model. J. Econ. Geogr. **3**(3), 229–240 (2003)
23. J.M. Magone, B. Laffan, C. Schweiger, *Core-Periphery Relations in the European Union: Power and Conflict in a Dualist Political Economy* (Routledge, 2016)
24. P.R. Krugman, *Development, Geography, and Economic Theory*, vol. 6 (MIT Press, 1997)
25. S.P. Borgatti, M.G. Everett, Models of core/periphery structures. Social Networks **21**(4), 375–395 (2000)
26. B. Karrer, M.E.J. Newman, Stochastic blockmodels and community structure in networks. Phys. Rev. E **83**(1), 016107 (2011)
27. Z. Lotker, *Analyzing Narratives in Social Networks: Taking Turing to the Arts* (Springer Nature, 2021)
28. D. Chakraborti, *Topics in Extremal and Random Discrete Structures*. PhD thesis, Carnegie Mellon University (2020)
29. MultiMedia LLC. Datarepository (2023)
30. D. Lusseau, The emergent properties of a dolphin social network. Proc. R. Soc. London. Ser. B: Biol. Sci. **270**(suppl_2), S186–S188 (2003)
31. M. Newman, *Networks* (Oxford University Press, 2018)
32. N. Eagle, A. Pentland, D. Lazer, Inferring friendship network structure by using mobile phone data. Proc. Natl. Acad. Sci. **106**(36), 15274–15278 (2009)
33. W.W. Zachary, An information flow model for conflict and fission in small groups. J. Anthropol. Res. **33**(4), 452–473 (1977)
34. T.H. Cormen, C.E. Leiserson, R.L. Rivest, C. Stein, *Introduction to Algorithms* (MIT Press, 2022)
35. D.P. Williamson, *Network Flow Algorithms* (Cambridge University Press, 2019)
36. L. Chen, R. Kyng, Y.P. Liu, R. Peng, M.P. Gutenberg, S. Sachdeva, Maximum flow and minimum-cost flow in almost-linear time, in *2022 IEEE 63rd Annual Symposium on Foundations of Computer Science (FOCS)* (IEEE, 2022), pp. 612–623
37. A.K. Goldsworthy, *The Punic Wars* (Cassell, 2000)
38. A. Goldsworthy, *The Fall of Carthage: The Punic Wars 265-146BC* (Weidenfeld & Nicolson, 2012)
39. S. Lancel, J.W. Betlyon, Carthage: a history. Hist.: Rev. New Books **24**(1), 43–44 (1995)
40. T.J. Cornell, *The Beginnings of Rome: Italy and Rome from the Bronze Age to the Punic Wars (c. 1000–264 BC)* (Routledge, 2012)
41. T. Frank, Roman census statistics from 225 to 28 bc. Class. Philol. **19**(4), 329–341 (1924)
42. G. Loescher, *Refugees: A Very Short Introduction*, vol. 673 (Oxford University Press, 2021)
43. C. Dustmann, F. Fasani, T. Frattini, L. Minale, U. Schönberg, On the economics and politics of refugee migration. Econ. Policy **32**(91), 497–550 (2017)
44. E. Fiddian-Qasmiyeh, G. Loescher, K. Long, N. Sigona, *The Oxford Handbook of Refugee and Forced Migration Studies* (OUP Oxford, 2014)
45. Shakespeare. Shakespeare-julius-caesa. https://docs.google.com/spreadsheets/d/1fVUCvqIil4WqBmO0AfJeIqoBdt1qqadb9urJssJmSmc/edit?usp=sharing (1599)
46. wikisource. https://en.wikipedia.org/wiki/Category:People_of_the_Gallic_Wars (2023)
47. Wiki. wikisource-category-barbarian-peopleofthegallicwars. https://en.wikipedia.org/wiki/Category:Barbarian_people_of_the_Gallic_Wars (2023)
48. Wiki. Category:roman-people-of-the-gallic-wars. https://en.wikipedia.org/wiki/Category:Roman_people_of_the_Gallic_Wars (2023)
49. Wiki. Spectral-graph-theory. https://en.wikipedia.org/wiki/Spectral_graph_theory (2023)

Part II
Metahistory: Computational Models for History

Chapter 7
What Is a Computational Model for History

Extensive literature exists on the philosophy of history; for a thorough and insightful overview, see Aviezer Tucker's Companion to the Philosophy of History and Historiography [1]. It is said that history is a narrative of the past [2, 3]. Therefore, in order to understand what history is, we must first understand what narrative is. One of the definitions of a narrative is two or more sentences that share a temporal relationship and can be interpreted in several, often conflicting, ways (Fig. 7.1).

Consider, for example, Hemingway's shortest story, *For Sale: Baby shoes, never worn*. One question that comes up upon reading this story is why are these shoes for sale? Perhaps the parents had two of the same pair, or maybe the baby died and the parents have no need for the shoes. The two conflicting interpretations transform the sentences into a narrative that ignites our imagination. Similarly, a historical narrative should contain at least two conflicting interpretations that spark our imagination.

Up until now, only humans could write narratives. Today, it is a task that can be performed by machines as well. I cover this at length in my book, [4], specifically Chap. 13 "Machine Narrative." You can also ask any advanced Large Language Models (LLMs) to write a short story about your favorite subject. This forces us to rethink what history is. We claim that a mathematical model could also be a narrative, and therefore, a mathematical model of history should be regarded as a historical text.

Such models can shed light on historical phenomena, processes, and events. In this part of the book, we provide several mathematical models that can be regarded as historical texts since they can be considered narratives about the past.

According to Wikipedia [5], "theoretical physics is a branch of physics that employs mathematical models and abstractions of physical objects and systems to rationalize, explain, and predict natural phenomena. This is in contrast to experimental physics, which uses experimental tools to probe these phenomena." Similarly, theoretical digital history can be viewed as a branch of history that implies

Z. Lotker, *History by Algorithms*,
https://doi.org/10.1007/978-3-031-93627-2_7

Fig. 7.1 A cartoon about how models help us understand history

mathematical models and abstractions of historical objects and systems to rationalize, explain, and predict historical phenomena. Note that prediction is not a part of history since it does not have an interpretational aspect.

Our main tool in developing theoretical digital history is mathematical modeling. According to Wikipedia [6], "A mathematical model is an abstract description of a concrete system using mathematical concepts and language. The process of developing a mathematical model is termed mathematical modeling. Mathematical models are used in applied mathematics and in the natural sciences (such as physics, biology, earth science, chemistry) and engineering disciplines (such as computer science, electrical engineering), as well as in non-physical systems such as the social sciences (such as economics, psychology, sociology, political science)." In this book, we add history to the list of applications of mathematical modeling.

We give examples of historical social structure, such as core and periphery and their relationship from a historical perspective Chap. 8, historical social network Chap. 6, consensus in history Chap. 9, collective memory formation game Chap. 10, propaganda Chap. 11, and stochastic terrorism Chap. 12. Our approach differs from the previous method of explaining historical phenomena by suggesting a mathematical model instead of a philosophical argument. Although our models are written in mathematical terms, we consider them to be a narrative about the past and can therefore be viewed as historical texts. Since they have different flavors compared to the standard historical narrative, they shed a different light on it.

Clearly, these are just examples, and mathematical modeling for history is a vast area for research.

We are not the first to do that. For example, battles were analyzed using Lanchester's equations [7, 8], and conflicts such as the Cuban Missiles crisis were analyzed using Game theory [9–11]. While these analyses were of specific historical events, we suggest a methodology for studying historical processes and rules that can be applied to the philosophy of history. For example, in Chap. 8 we explain why it is possible to assume that the history of the core is a good approximation of the history of the periphery. In Chap. 6 we provide a rule of thumb that allows us to estimate the social network of any historical society and, therefore, to compare different societies from different times. In Chap. 12 we suggest a model that explains stochastic terrorism phenomena that are used in today's politics. In Chaps. 9, 10, and 11, we deal with the question of how historical memory is formed, why canonical texts are needed, and why propaganda is crucial in history. All these examples can be applied to many instances in history. Therefore, they belong to the philosophy of history rather than history itself.

Usually, conflicts are described as good versus bad, two forces trying to eliminate each other. In a similar way, when we use positive and negative numbers, they try to eliminate each other and present the opposite behavior. Therefore, it turns out that the mathematical model that is used to describe conflicts tends to use positive and negative probabilities. We interpret negative probabilities that could happen to the bad guys in historical events. The idea of negative probabilities appears in several chapters in this part. In the last chapter, we define historical entities and show how their definition can eliminate negative probabilities.

References

1. Tucker, A.: *A Companion to the Philosophy of History and Historiography* (Wiley, 2011)
2. S. Lipscomb, H. Carr. *What Is History, Now?* Orion (2021)
3. J.R. Rayfield, What is a story? Am. Anthropol. **74**(5), 1085–1106 (1972)
4. L. Zvi, *Analyzing Narratives in Social Networks Taking Turing to the Arts Authors* (Springer, 2021)
5. wikisource (2023). https://en.wikipedia.org/wiki/Theoretical_physics
6. wiki mathematical model (2023). https://en.wikipedia.org/wiki/Mathematical_model
7. T.W. Lucas, J.A. Dinges, The effect of battle circumstances on fitting lanchester equations to the battle of kursk. Military Oper. Res. **9**(2), 17–30 (2004)
8. Judy Hong Tam, Application of lanchester combat model in the ardennes campaign. Nat. Res. Model. **11**(2), 95–116 (1998)
9. F.C. Zagare, *Game Theory, Diplomatic History and Security Studies* (Oxford University Press, 2018)
10. F.C. Zagare, A game-theoretic history of the cuban missile crisis. Economies **2**(1), 20–44 (2014)
11. J. Elster, Marxism, functionalism, and game theory. *Debates in Contemporary Political Philosophy*, p. 22 (2003)

Chapter 8
History Through the Core

8.1 Introduction

In the past, our history was mainly concerned with the core of society. Humankind did not preserve information about the periphery, and so historians didn't have access to the history of the periphery. In the 21st century, this is no longer true. Historians of the 21st century will have access to the materials written by the periphery. In fact, we now have an excess of information to the point that historians cannot read all the texts written by the periphery, and have to turn to machines for help. Table 8.1 presents the schematic difference between life in the core and life in the periphery.

This chapter studies the relationship between core and periphery through a mathematical model. We tackle one of history's biggest questions: Is it possible to describe a revolution with a magnifying glass of the core? Suppose society goes through a merger change. Can we study only the core and conclude that the periphery has also changed during that period of time? For most of our history, we know much more about the core than the periphery. The reason for this is a lack of written sources. For example, when the Roman religion changed from Paganism to Christianity, there was no way of knowing whether or not it happened simultaneously in the core and the periphery. The general rule is that we get the narrative only from the sampling of the society, see [1, 2]. Our history is that of those who knew how to write or were important enough to be written about. The stories we know tend to be those of powerful and influential historical figures. However, most people who lived in the past have not been recorded and are not accessible for research today. Most records of the peripheries of older societies are also written from the perspective of the elite [3].

How much of the periphery's narrative history can we figure out using only information offered by the elite? Is it statistically viable to sample the elite to produce a narrative for the whole society? We'll focus on periods of revolution in history.

To answer these questions, we use a mathematical model and try to understand when we can say that the timeline of the core is a reliable source for the timeline of the periphery. We try to read "against the grain" using mathematical modeling

Z. Lotker, *History by Algorithms*,
https://doi.org/10.1007/978-3-031-93627-2_8

Table 8.1 The difference between the core and the periphery as a way of living. Life at the periphery is usually more quite and less crowded, while life at the core is more crowded and with more traffic. Figures by Oded Lotker

Image of the periphery	Image of the core

instead of text. We start with a homogeneous model to prove the concept and then elaborate on its uses.

Our homogeneous model uses the fluid model, and a more complicated model using stochasticity could be developed and studied. There are no major differences between the more realistic stochastic model and other, less realistic fluid models, and therefore we choose the simpler version of these models to overcome unnecessary technical obstacles.

This section starts with a homogeneous model of a historic event in the core-periphery network, where the network simulates some historical shift in the network's opinion. The model's purpose is to help us better understand the structure of past revolutions and evaluate the historical method of examining a society through its core.

We assume society is a construct of two hierarchical entities: core and periphery (Fig. 8.1). While societies in the real world are incredibly multi-layered in matters such as status and influence, we will be using only two levels of society—core

Fig. 8.1 A cartoon depicting the history of the periphery as seen through the core

(elite) and periphery (non-elite)—to simplify the model. We study two models, the homogeneous core-periphery revolutionary fluid model, and the non-homogeneous core-periphery revolutionary fluid model.

We will explain the homogeneous core-periphery revolutionary fluid model in the next section.

8.2 The Homogeneous Core-Periphery Revolutionary Fluid Model

Unlike the Chap. 6, we assume that the nodes in the core and periphery are homogeneous, i.e., the neighborhood of two nodes in the core is isomorphic, as is the neighborhood of two nodes in the periphery, see for example Fig. 8.2. The degree of any node that belongs to the core node $i \in V_C$ is denoted as

$$d_i = d_c. \tag{8.1}$$

In the same way, the degree of any node belonging to the periphery, whereas the degree of the relativity of a periphery node $j \in V_P$, is denoted as

$$d_j = d_p \tag{8.2}$$

Moreover, we split each neighborhood into nodes belonging to the core and nodes belonging to the periphery according to the following matrix:

$$[dig_{c,p}] = \begin{bmatrix} d_{c,c} & d_{c,p} \\ d_{p,c} & d_{p,p} \end{bmatrix} \tag{8.3}$$

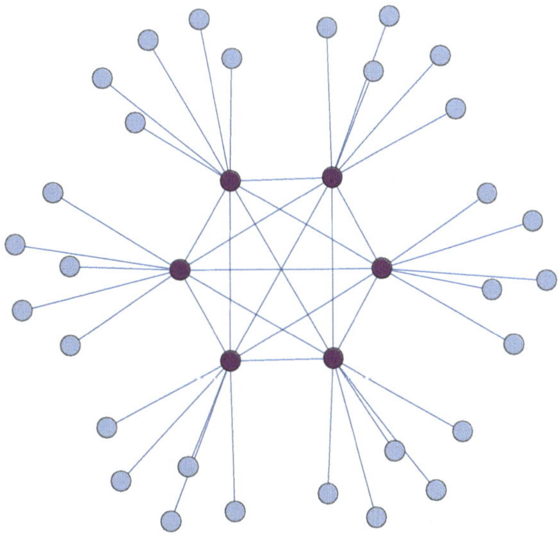

Fig. 8.2 A simplistic model for a community unit within a core-periphery society. The blue nodes of the graph represent individual periphery members connected to their community's representative: a central, purple, member of the core. Note that each peripheral node connects to one node in the core, while the core node connects to all five

Many graphs can satisfy this assumption. Later, we will explain how it is possible to change the assumption and switch to a more complex model.

Each node $i \in V$ has its opinion $o_i(t)$. At the beginning of time $t = 0$ the value of $o_i(t) = 0$. At the end of said time, the value of its opinion is $o_i(\infty) = 1$. When the opinion of all nodes is 1, we will say that the revolution has happened and all nodes had joined the revolution. Similarly, when all nodes of the core(periphery) have joined the revolution, we can say that the core(periphery) had joined the revolution, respectively.

Let's describe how nodes change their opinion. Each node $i \in V$ receives a revolutionary fluid at a constant rate f_i. The rate is different for the core $fc \geq 0$ and the periphery $fp \geq 0$. In general, we use f_i to denote the rate node i receives the revolutionary fluid:

$$f_i = \begin{cases} fc & i \in V_C \\ fp & i \in V_P \end{cases} \tag{8.4}$$

For all nodes, there is a social pressure constant $0 < \alpha \leq 1$. Nodes change their opinion in two ways: if the value of the revolutionary fluid they receive equals 1, or if the average revolutionary fluid in their neighborhood is above the social pressure α. Denote the first time node i supports the revolution by t_i^*. Therefore, the value of the opinion of node i is:

$$o_i(t) = \begin{cases} tf_i & t < t_i^* \\ 1 & t \geq t_i^* \end{cases} \tag{8.5}$$

Since all nodes at the core have the same neighborhood structure, they will all change their opinion simultaneously. Denote this time by t_C^*. Following the same argument, all nodes at the periphery will change their opinion simultaneously. Denote this time by t_P^*. However, the relationship between those two times depends on the relationship between the core and the periphery and the rate fc, fp. Note that social pressure can not exceed one, since the average opinion of the nodes in its neighborhood is maximized when all nodes have joined the revolution. In this case, the ratio should be 1.

Let us denote the social pressure node $i \in V$ receives at time t by $\alpha_i(t)$. Using the opinion function, we can write a social pressure as a function of time:

$$\alpha_i(t) = \frac{\sum_{j \in N_i} o_j}{d_i} \tag{8.6}$$

8.2.1 *The Revolution Starts at the Periphery Before the Core*

Let's start with the case when the periphery joins the revolution first. This happens if $t_P^* < t_C^*$. A simple calculation shows that the time at which the periphery changes its opinion is

$$t_P^* = \min\left\{\alpha\left(\frac{d_p}{(f_p d_{p,p} + f_c d_{p,c})}\right), \frac{1}{f_p}\right\} \tag{8.7}$$

Since the periphery changes its opinion before the core, we know that the amount of revolutionary fluid at time t_P^* is $t^* f_P$. The periphery can change its opinion if one of the following conditions are met: First, the average revolutionary fluid in the neighborhood N_i of any node i belonging to the periphery is equal to α. This happened at time t, thereby satisfying the following equation:

$$\frac{t(f_p d_{p,p} + f_c d_{p,c})}{d_p} = \alpha \tag{8.8}$$

The other condition is that the fluid injected into a node at the periphery reaches a value of 1. This happens at time $1/f_p$. The min in Eq. (8.7) is needed since the node in the periphery changes its opinion as early as possible.

We repeat a similar calculation, this time assuming that the core joins the revolution first.

8.2.2 The Revolution Starts at the Core Before the Periphery

Let's move to the case when the core joins the revolution first. This happens if $t_C^* < t_P^*$. We can repeat the same calculation as in the previous subsection

$$t_C^* = \min\left\{\alpha\left(\frac{d_c}{(f_p d_{c,p} + f_c d_{c,c})}\right), \frac{1}{f_c}\right\} \tag{8.9}$$

Note that the graph which describes society in the homogeneous core-periphery model could go on to infinity and the degree of the nodes at the core can grow to infinity as well. Therefore, the time of joining the revolution can be arbitrarily big and can be dependent on the size of the graph.

The next theorem shows that if the parameters fc, fp are constant, then the revolution happens in constant time both in the periphery and the core. Therefore, we can conclude that according to the homogeneous model, once the parameters are positive and constant, there is no major difference in describing the time the revolution happens at either the core or the periphery.

Theorem 8.1 *In the homogeneous core-periphery model, if fc, fp are constant then there are some constant $c_1, c_2 > 0$ s.t.*

$$c_1 t_c^* \leq t_p^* \leq c_2 t_c^* \tag{8.10}$$

or in the asymptotic notation

$$t_c^* = \Theta(t_p^*) \tag{8.11}$$

Proof Note that if both f_c, f_p are constant, then according to two Eqs. (8.7) and (8.9), t_c^*, t^*p are also constant.

When the parameters fc, fp are not constant we need to distinguish between four different cases. Both parameters fc, fp can go to either zero or infinity. When the parameter goes to zero, it means that the revolution is progressing very slowly. When the parameter goes to infinity, it means that revolution happens immediately.

The previous chapter introduced the core-periphery model. The relationship between the size of the core and the size of the periphery is not linear, but rather polynomial; therefore, we do not recommend and support using percentage and linear models when studying economics and historical events in the social network context. A more suitable ruler of time will be big O or Θ as was explained in Chap. 1 and was proven by Theorem 8.1 above.

8.2.3 Categories of Revolution Using a Homogeneous Model

Our homogeneous model suggests two kinds of revolution: one that starts at the core and spreads to the periphery, and the other that starts at the periphery and moves into the core.

To fully characterize the different kinds of revolution, we need the following denotations: Denote the first time the social pressure of node i went above the social pressure threshold α by t_i^α, formally:

$$t_i^\alpha = \min\{t \in \mathbb{R} : \alpha_i(t) \geq \alpha\} \tag{8.12}$$

Clearly, in the homogeneous core-periphery revolutionary fluid model, t_i^α is dependent only on the social status of the node, i.e., for all $i, j \in V_C$, $t_i^\alpha = t_j^\alpha$ and for all $i, j \in V_P$, $t_i^\alpha = t_j^\alpha$ as well. Therefore we can denote $t_c^\alpha/(t_p^\alpha)$ to be the first time the social pressure of the node at the core/(periphery) exceeded the social pressure threshold α correspondingly.

Now we can analyze both points of view: core and periphery. According to Eq. 8.5, the point of view of the node i depends only on t_i^*. To compute t_i^* we need to distinguish between two cases, depending on which part of society joins the revolution first.

In our basic model, there are 6 parameters:

$$f_c, f_p, \alpha, d_{c,c}, d_{c,p}, d_{p,c}, d_{p,p} \tag{8.13}$$

Two parameters describe the speed of the revolution f_c, f_p in the core and in the periphery respectively, α which is social pressure in society, and four parameters describe the topology of the core-periphery graph. The first three parameters create six kinds of different revolutions according to the relationships between the parameters, f_c, f_p, α:

1. $1/f_c < 1/f_p < \min\{t_c^\alpha, t_c^\alpha\}$
2. $1/f_p < 1/f_c < \min\{t_c^\alpha, t_c^\alpha\}$
3. $1/f_c < \min\{t_c^\alpha, t_c^\alpha\} < 1/f_p$
4. $1/f_p < \min\{t_c^\alpha, t_c^\alpha\} < 1/f_c$
5. $t_c^\alpha < t_p^\alpha < \min\{1/f_c, 1/f_p\}$
6. $t_p^\alpha < t_c^\alpha < \min\{1/f_c, 1/f_p\}$

We only categorize the cases of strong inequalities and ignore the cases where we have equality between the parameters.

We will briefly discuss each of these cases. We have six different kinds of revolutions divided into three groups with respect to social pressure. Each of these three groups is divided into two subgroups: either the revolution starts at the periphery or at the core.

Social pressure has three options. It can be not important at all, which happens under conditions 1 and 2. The second option is that the revolution stems without social pressure, but its second stage develops through social pressure. The last option is that the revolution is generated purely by social pressure.

In the next section, we are going to explain how to move from a homogeneous core-periphery model to a general core-periphery model. The advantage of the general model is that it allows us to study the revolution at the level of individuals instead of social classes.

8.3 General Core-Periphery Model

In this section, we explain how to work with general fluid models where each node sees its own environment, and we no longer have the homogeneous structure of core-periphery. This section aims to prove the famous rule of thumb that revolutions happen much faster than people realize. Our model will assist in explaining this phenomenon.

We now assume that we have the core and the periphery, similar to the model discussed in the previous chapter Eq. 6.14. We no longer assume that the nodes at the core or periphery are similar. According to the rule of thumb Chap. 6, all nodes at the core have a high degree and are highly connected to each other, and all nodes at the periphery have a low degree, and the graph is connected.

Similar to the previous section, we assume that the speed of the revolutionary fluids is either f_c or f_p according to the position of the node, which can be at the core or the periphery.

In the previous section, we had two equations: one for the core, and one for the periphery, since we assumed that our model was homogeneous. Now we do not assume that the model is homogeneous, so we have an equation for each node. To describe the equation we need the time t_i^* the node i joins the revolution. Clearly, by the definition of t_i^* and f_i, the opinion of each node i follows the Eq. 8.5.

Therefore, the only remaining thing is to compute t_i^*. We compute t_i^* by induction sorted on the earliest nodes that completely support the revolution. To be able to write the system of equations, we need a sequence increasing sets of nodes

$\varnothing = S_0 \subset S_1 \subset, ..., \subset S_k = V$ which the induction process will define. We can not say the exact number of induction steps since two or more nodes can join the revolution at the same time. What we do know is that the number of steps in the induction process is less than n, and therefore the process will end after a finite number of steps.

8.3.1 The Base of the Induction

Now we must explain how to compute the first node that supports the revolution t_i^*. To do that we write for each node $i \in V$ the following linear equation of the variable t_i:

$$t_i \frac{\sum_{j \in N_i} f_j}{N_i} = \alpha \tag{8.14}$$

Note that we have n linear equations with n variables; however, the i equation contains a single t_i variable, which appears only in its own equations. Therefore we can solve each equation separately and compute all t_i.

Clearly, the first time when the node joins the revolution is given by the following equation:

$$\tau_1 = \min\{1/f_c, 1/f_p, t_1, t_2, ..., t_n\} \tag{8.15}$$

Once we compute the first time the node joins the revolution, we need to find which node was the first to do so. Here we distinguish between three cases, depending on the source of the minimum (8.15).

Assume that $\tau_1 = 1/f_c$. In this case, all nodes belonging to the core change their opinion to support the revolution at time τ_1 and $S_1^c = V_C$, otherwise $S_1^c = \varnothing$.

If $\tau_1 = 1/f_p$, all nodes belonging to the periphery change their opinion to support the revolution at time τ_1 and $S_1^p = V_p$, otherwise $S_1^p = \varnothing$.

The last case is when $\tau_1 = t_i$ then all nodes satisfy

$$S_1' = \{i \in V : t_i = \tau_1\}. \tag{8.16}$$

Now we can define the set S_1

$$S_1 = S_1^c \cup S_1^p \cup S_1' \tag{8.17}$$

Clearly, S_1 is a non-empty set, and therefore we have

$$S_0 \subset S_1 \tag{8.18}$$

Now we compute S_1, which is the base of the induction.

Let's move to the general case.

8.3.2 General Case of Induction

In the general case, we assume that the set $S_1, S_2, \ldots S_j$ was already computed. Note that if

$$S_j = V \tag{8.19}$$

the claim has already been proven since we computed the revolutionary time for all nodes. Therefore we may assume that

$$S_j \subset V \tag{8.20}$$

i.e., there exists a node that does not join the revolution. Therefore we wish to compute S_{j+1} and prove

$$S_j \subset S_{j+1} \tag{8.21}$$

Again we write for each node that did not join the revolution the linear equation that computes the time it joined the revolution, i.e, for all $i \in V \setminus S_j$

$$t_i \frac{|N_i \cap S_j| + \sum_{l \in N_i \setminus S_j} f_l}{N_i} = \alpha \tag{8.22}$$

Note that in the system of Eq. 8.22, for each equation there is only one variable. Therefore we can compute t_i for all $i \in V \setminus S_j$.

Again we need to compute the sets S^c_{j+1}, S^p_{j+1}, S'_{j+1}, similar to what we did above, but now is defined as τ_{j+1}

$$\tau_{j+1} = \min\{1/f_c, 1/f_p, \min_{i \in V \setminus S_j} \{t_i\}\} \tag{8.23}$$

If the core has already joined the revolution, i.e.,

$$V_C \subseteq S_j \tag{8.24}$$

then we ignore the nodes at the core, and therefore $S^c_{j+1} = \varnothing$. Otherwise, there exists a node at the core that did not join the revolution, and therefore, the set S^c_{j+1} is

$$S^c_{j+1} = \begin{cases} V_C & \tau_{j+1} \leq \frac{1}{f_c} \\ \varnothing & \tau_{j+1} > \frac{1}{f_c} \end{cases} \tag{8.25}$$

Again we need to distinguish between two cases. The first is when the entire periphery has already joined the revolution, i.e.,

$$V_P \subseteq S_j \tag{8.26}$$

wherein we ignore the nodes at the periphery, and therefore $S^p_{j+1} = \varnothing$. Otherwise, there exists a node at the periphery that did not join the revolution, and therefore, the set S^p_{j+1} is

$$S^p_{j+1} = \begin{cases} V_P & \tau_{j+1} \le \frac{1}{f_p} \\ \varnothing & \tau_{j+1} > \frac{1}{f_p} \end{cases} \tag{8.27}$$

The last set we need to define before computing S_{j+1} is the set S'_{j+1}:

$$S'_{j+1} = \{i \in V : t_i = \tau_{j+1}\}. \tag{8.28}$$

Now we can define the set S_{j+1}

$$S_{j+1} = S^c_{j+1} \cup S^p_{j+1} \cup S'_{j+1} \cup S_j \tag{8.29}$$

By the definition of α_{j+1} it follows that

$$S^c_{j+1} \cup S^p_{j+1} \cup S'_{j+1} \neq \varnothing \tag{8.30}$$

and therefore $S_j \subset S_{j+1}$.
This completes the proof of the induction.

8.3.3 Categories of Revolution Using the Non-homogeneous Model

In this section, we use a homogeneous model to explain the phenomenon of revolutions that sometimes occur faster than most people expect.

The main difference between homogeneous and *non-homogeneous revolutionary models* is the resolution of questions we may ask. In the homogeneous model, we can only separate the core and the periphery, as individuals inside the core and at the periphery have the same neighborhood and therefore behave similarly. In the non-homogeneous model, we can zoom in on each individual and investigate his own opinion through time.

Therefore, the outcome of the non-homogeneous revolution is one of the following options. The first scenario is that the model behaves exactly as the homogeneous one since the parameters are not delicate enough to separate individuals.

In the second scenario, at some point in time, the first individual groups—perhaps single people—change their opinion and then join the revolution. These groups can be either at the core or the periphery.

At the beginning of the induction process at time 0, each node i can estimate the time it will join the revolution. This estimate is equal to $1/f_i$. Now, by definition, the time that node i joins the revolution is t_i^*. If $t_i^* < 1/f_i$, then node i joins the revolution earlier than expected. We remark that if social pressure is strong enough, this will happen to most nodes.

8.3.4 *Example*

In the previous section, we approved that evolution happened sooner than expected. In this section, we provide two examples of such cases. The first is global warming, and the second is AI.

Global warming tends to accelerate for several critical reasons. The first is that the human population grows exponentially and with it grows the use of fossil fuels. However, researchers often ignore the population growth factor. Another reason is that there are economic forces that stand to lose if they acknowledge global warming and, therefore, try to sabotage public opinion and information about global warming. As a result, environmental scientists tend to have optimistic assumptions, resulting models that always underestimate the truth of global warming.

Another example is the ability of AI to perform human tasks. In this case, underestimation follows the general assumption that humans are "the chosen ones," a notion deeply rooted in our culture. Human culture, which has been slowly developing over the course of ten thousand years or more, repeatedly suggests that humans are special—and indeed, they used to be. Therefore, it is hard for us to accept that we no longer are.

8.4 Accessing the Core Through Wikipedia

In the previous section, we studied the core-periphery revolutionary fluid model and showed that the core can tell the periphery's narrative of the revolution. In this section, we study the core of Western civilization using Wikipedia. We focus on four groups: Romans, Jews, Christians, and Muslims, during the period ranging from 700 BC to 1000 AD. For the Romans, we look at the patricians as the elite of Roman society. For the rest, we simply look at their beliefs. Our definition of core is people who have their own entries on English Wikipedia.

Wikipedia houses lists of known prominent individuals organized by year of death. The names from each year can be cross-referenced with other categories to find how many prominent members of a historical social group died annually.

What emerges is a narrative of shifting power dynamics. The changing influences between the religions are seen through the number of notable figures present over the centuries.

Starting in 700 BC, Jewish influence decreases and comes to a minimum around 586 BC. Around 500 BC we have Shivat Zion, the return to Zion—a period of Jewish population increase. This is followed by a decline until 168 BC, with the arrival of Antiochus III the Great. Then the Hasmonean dynasty rises, culminating in the Bar

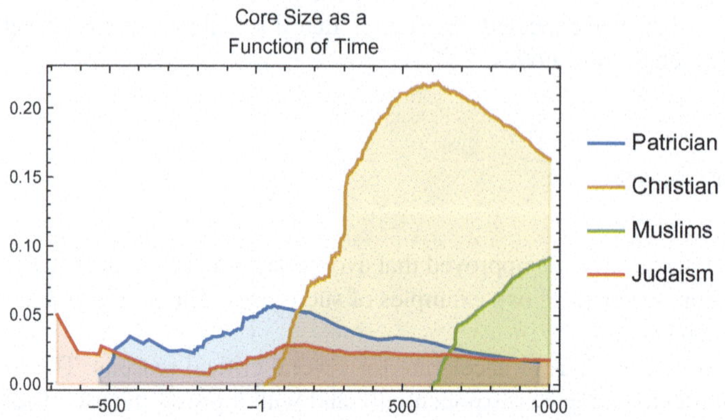

Fig. 8.3 The percentile relationship between elites of prominent religious and social groups, sampled through recorded death dates of prominent individuals on Wikipedia and charted over time

Kokhba revolt of 132-136 AD, after which comes another slow but steady decline. Around 900-1000 AD the Jews finally earn an influential role in society, even more than the Patricians. There is also a small bump with the birth of Islam (Fig. 8.3).

The Patricians only show up around 500 BC, after Rome turned from monarchy to republic. The local minimum at the time of their initial rise, 390 BC, is their defeat at the hand of the Gauls. The following minimum at 200 BC comes with the Second Punic War against Hannibal (the 260 BC plateau is attributed to the First Punic War). The republic becomes an empire in 27 BC and experiences a steep rise until 71 AD, when the Second Temple is destroyed. Then there is a slow decline as the empire declines and divides in 395 AD, while Rome is sacked in 410. The small rise in 500 AD is possibly attributed to prominent figures claiming to be Patrician-born, with the rise of the new Frank, British, and Italian empires.

It is important to note the relationship between Jews and Romans and the intersection of Jews and Christians at the time of the Bar Kokhba revolt. It is at this point that Christianity overpowered Judaism following the wavering Patrician influence. Also, note the inverse rise and decline of Christianity and Islam.

The first steep rise in Christian influence occurs around the destruction of the Temple. Christians first surpass Jews at the time of the Bar Kokhba revolt. Then they outnumber the Patricians. They experience a steep rise during the era of Constantine the Great and Constantinople, and peak around 600 AD before their influence starts to decline. The year 638 AD marks the maximum, with the capturing of Jerusalem and the surplus deaths of countless Christians.

The Muslims experience a significant rise in 632 AD with the death of Muhammad. Another steep incline occurs in 692 AD with the Siege and Fall of Mecca and the Muslim unification under the Umayyad Caliphate, continued by a constant rise in significance.

To conclude: Jews were the primary influence until Christianity took over the Middle East. Islam started to grow at the same rate.

For the time being, we use numbers as our elements, but we are moving towards geographical elements, or history hypergraphs. This is meant to serve as a pedagogical tool to enable students to search for the narrative within history on their own.

8.5 COVID-19

Up to this point, we have proven that the core narrative is similar to that of the periphery through a simple mathematical model. However, we would also like to find some data in support of this claim. As mentioned earlier, it is very difficult to find the numbers describing the narrative of the periphery and the core simultaneously. COVID-19 provides us with a rare opportunity to examine this claim from a real data perspective. To do that, we need to find the real estimation of the number of deaths as a function of time in society as a whole, as well as estimate the number of deaths at the core.

The narrative of COVID-19 is that there was a pandemic. We can see this just by studying the dramatically increasing number of deaths in 2019–2022.

In the next two subsections, we explain how to evaluate the narrative numerically from perspectives of the Periphery and the Core.

8.5.1 Estimating the Number of Deaths at the Core

To estimate the number of deaths at the core, we use Wikipedia lists of dead people. For each month, there is a list of all the people who have a Wiki page and died on that month. See, for example, the list of people who died in January 2015 [4]. The list is broken down into dates. For each person, we are given some data: name, occupation, nationality, and a link to their personal Wiki page. From this data we can generate a list of deaths according to nationality to record, for instance, the number of famous Americans who died each month.

8.5.2 Estimating the Number of Deaths Periphery

A good estimation of the number of deaths at the periphery is usually provided by the government. In the case of the United States, the data can be found on [5]. The number of deaths at the core is negligible compared to the number of deaths at the periphery.

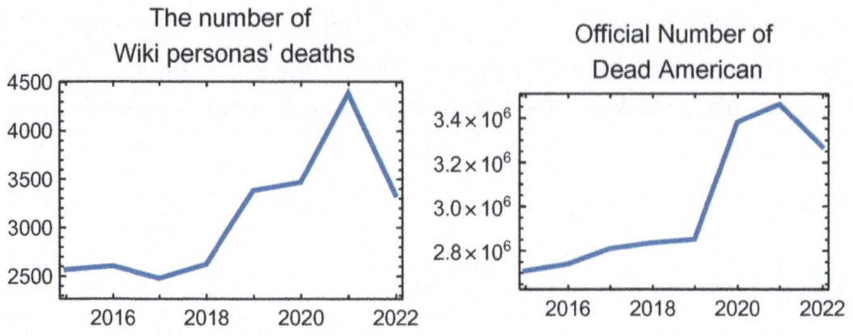

Fig. 8.4 The relationship between the narrative of the COVID-19 pandemic told by the core versus the scenarios of the periphery/entire society

8.5.3 Numerical Results

Figure 8.4 shows that both the core and the periphery tell a similar narrative. The number of dead has grown drastically since 2019. The peak was 2021, and the pandemic ended in 2023. However, there are several nuances. For example, the peak at the core is much more distinct than at the periphery.

8.6 Conclusion

The main question in this chapter is whether the core can tell the story of the periphery during revolutionary periods, as historians tend to do. Due to the lack of information about the periphery, it is difficult to answer this question using standard history tools.

To overcome this difficulty, we tackled this question with mathematical tools. We presented the mathematical "Core-Periphery Revolutionary Fluid" model. This is a very simple and naive model where society is static and composed of two social hierarchies: Core and periphery. This is clearly not representative of real-life society, which is dynamic, composed of many more levels, and is individually intricate. The model does, however, enable the connection between the core and the periphery. This model uses the concept of revolutionary fluid to describe a society that goes through a revolution.

We used a simple homogeneous model to categorize different kinds of revolutions in terms of social pressure and the core-periphery relationship.

Later, we described the non-homogeneous core-periphery revolutionary fluid model and used it to explain why revolutions tend to happen faster than expected. We did not discuss scenarios in which revolutions fail.

We ended this chapter by examining the relationships between different cores in ancient Rome.

8.7 Exercise

The Homogeneous Core-Periphery Revolutionary Fluid Model

1. In this problem, we are going to generate the Homogeneous Core-Periphery Revolutionary Fluid Mode.

   a. Construct a core-periphery graph with 256 nodes at the periphery and 16 nodes at the core, using the following parameters:

$$[dig_{c,p}] = \begin{bmatrix} 10 & 1 \\ 1 & 1 \end{bmatrix} \tag{8.31}$$

   b. Use the fluid model with the following parameters:

$$f_p = 0.1,\ f_c = 0.2, \alpha = 0.4. \tag{8.32}$$

   Determine the time the core joined the revolution t_c^*, and the time the periphery joined the revolution t_p^*

2. Write an algorithm that receives $|V_C|$, $|V_P|$ and the relative degree of each of the nodes according to the degree matrix below

$$[dig_{c,p}] = \begin{bmatrix} d_{c,c} & d_{c,p} \\ d_{p,c} & d_{p,p} \end{bmatrix} \tag{8.33}$$

   and determine if there is a graph that satisfies those numbers or not.

   General Core-Periphery Mode

3. Compute for each node in the graph the time it joined the revolution with the parameters of the Problem 1. Note that this is not a homogeneous graph; the parameters have expected values.
4. In Sect. 8.5 we use the number of dead people in Wikipedia to show the connection between the number of deaths in the core and the number of deaths in the periphery. Use this connection to determine the influence of the Spanish Flew on the ending of the First World War.
5. Consider the social network in Fig. 8.2 where the six purple nodes are the core and the 30 blue nodes are the periphery. Assume that the core nodes are denoted by 1 to 6, and the blue nodes are devoted by 7 to 37, where the highest node in the periphery is node 37 with a degree of 3. The rest of the periphery has degree 1.

Assume the revolution fluid is given by

$$f_p = 0.1,\ f_c = 0.2. \tag{8.34}$$

and the social pressure is

$$\alpha = 0.4. \tag{8.35}$$

Estimate the time each node is joining the revolution at time $t = 0$, compute the time each node joins the revolution and verify that almost all nodes joined the revolution before their estimated time.

References

1. J. Fetterley, *The Resisting Reader: A Feminist Approach to American Fiction*, vol. 247 (Indiana University Press, 1978)
2. T. Bewes, Reading with the grain: a new world in literary criticism. Differences **21**(3), 1–33 (2010)
3. C.A. McKenzie, S. Jarvie, The limits of resistant reading in critical literacy practices. Engl. Teach.: Pract. Crit. **17**(4), 298–309 (2018)
4. Wiki. wiki deaths of January 2015. https://en.wikipedia.org/wiki/Deaths_in_January_2015 (2023)
5. gov. cdc gov. https://www.cdc.gov/nchs/nvss/deaths.htm (2023)

Chapter 9
History After the Death of the Witnesses

In the previous chapter, we showed how we can construct a rough approximation of the social network from the number of people in a society. In this chapter, we will explore how human lives influence history. We are interested in the fact that for most events, no survivors are left to tell the narrative first-hand after roughly 70 years. Those who do survive are old, frail, and possibly unreliable. Therefore, there is a big difference between short-term history—a history where witnesses are still alive—and a history where all witnesses are dead, and we can no longer ask for a first-hand account of the events, see Fig. 9.1.

Note that in the near future, we will have AI witnesses who will be much easier to manipulate than human witnesses. It seems that AI can solve the problem of witness deaths, but this solution is shaky, as AI witnesses can never be completely trustworthy. Moreover, AI witnesses would be easy to manipulate, as we will see in the chapter's exercises.

To study the influence of witness death, we first point out an essential similarity between historical narrative and consensus in distributed computing.

We assume that historical narratives are comprised of a sequence of historical events $(h_i)_{i=1}^{n}$ and two conflicting social groups B, R, called Blue and Red. For the narrative to reshape itself, each group needs to reach a consensus on each of the historical events h_i within the group. If such a consensus is achieved within a group, the group is said to have reached a historical consensus. However, in the case of a conflict between different groups, the groups must agree to disagree, separate from one another and define the conflict. For example, our groups can be the American Civil War's Union (North) and Confederacy (South). The historical event could be Lincoln's election as the President of the United States in 1860. The North considered the outcome of the elections positive news, giving the historical event the mark of $+1$, while the South thought it was a disaster, giving the historical event the mark -1. Both sides agreed within and disagreed with each other. Of course, not all the people in the North or the South agreed with these marks, but we do perceive it as a collective decision.

Z. Lotker, *History by Algorithms*,
https://doi.org/10.1007/978-3-031-93627-2_9

Fig. 9.1 A cartoon featuring the Almighty Narrator after the demise of the witnesses

Therefore, the problem of obtaining the collective memory of each of the groups on a specific historical event can be described in the framework of game theory. History is composed of many collective memories, each of which emphasizes a point of view in the historical narrative. Different communities can have different collective memory. Therefore, when studying history, one needs to present the different collective memories as they shed light on the historical conflict, which is central to the historical narrative.

According to game theory, to describe a process as a game we must define a player, the actions, the outcome of the game for each action, and the utilities of the game. In this chapter, we describe the process of reaching collective memory using game theory tools (for a formal definition of the game, see Sect. 9.5) and distributed computers. We model society as a graph. The nodes of the graph represent people in the society, the board of the game is the social network, and the utilities of each node are π_{υ}. Note that all the original nodes disappear from the network after some time. Therefore, consensus can only be achieved according to the definition of the consensus problem in distributed computing if a canonical text represents a surviving witness. The formal definition of the consensus problem can be found in Sect. 9.3.

To summarize, in order to understand the importance of the witness and the canonical text in the historical narrative, we look at an equivalent problem for agreeing on a single bit of the narrative in distributed computing. We use the impossibility result to reach the agreement of distributed computing for determining the condition when consensus is possible in the context of a historical narrative.

In the next few sections, we will discuss the similarities and differences between historical consensus and distributed computing consensus.

Before we dive into the definition of consensus in distributed computing, we must apologize for the advanced mathematical jargon. The consensus problem was borrowed from the realm of distributed computing, which is defined via mathematical language, and transformed it into a social network arena; this in turn forced us to incorporate game theory and the Nash equilibrium. However, from the historical consensus perspective, the results of this chapter are quite obvious: Without canonical texts, there are no narratives. However, developing the mathematical model allows us to not only reach a conclusion but also explain the process through which society reaches a consensus. Another reason for constructing the mathematical model is that it can provide an example where one may explain historical processes through game theory.

9.1 Related Works

In this chapter we fuse together two concepts: consensus and collective memory. We provide for each concept a subsection of related work.

9.1.1 Related Works on Consensus

Consensus is an essential topic in social network research, as it is a fundamental aspect of human decision-making and behavior. Consensus refers to the degree of agreement among individuals in a group, and it plays a crucial role in various social phenomena, such as opinion formation, group decision-making, and social influence. Over the past few decades, researchers have extensively studied consensus in social networks from different perspectives, including social psychology, sociology, computer science, and game theory. Various models and algorithms have been proposed to describe and analyze the dynamics of consensus formation in social networks. In this section, we provide an overview of some of the related works on consensus in social networks, highlighting the key contributions and limitations of each approach.

To preface our discussion on the related works on consensus in social networks, we would like to begin with a classical and very important paper by Morris DeGroot, titled *Reaching a Consensus* [1]. This paper is widely cited in literature and has significantly impacted subsequent research in the field of consensus formation. The author presents a model describing how individuals can pool their opinions to reach an agreement on a common distribution. He explicitly describes the process that leads to consensus and the distribution that is reached.

The problems of finding the consensus of multi-agent systems were described by Qin et al., see [2]. The authors provide a comprehensive review of recent progress in consensus and coordination of multi-agent systems.

The problem of reaching a consensus in a dynamically changing environment was investigated in the paper by Cao et al., see [3]. The paper addresses the problem of consensus formation in mobile autonomous agent networks. The authors use properties of compositions of directed graphs and nonhomogeneous Markov chains to derive worst-case convergence rates for the headings of agents. The paper provides insights into the dynamics of consensus formation in mobile autonomous agent networks under dynamic and uncertain conditions and finding the consensus in real-world systems.

A more modern example is the paper by Guo and Han, see [4]. The authors investigate the existence of the Nash equilibrium and group strategy consensus in a two-layer networked game of evolutionary game theory. The authors use the semi-tensor product approach to convert the payoff functions into algebraic forms and derive necessary and sufficient conditions for the existence of a pure Nash equilibrium. They also established a strict game algebraic equation with group intelligence using the best response adjustment rule and investigated the group strategy consensus problem.

9.1.2 Related Works on Collective Memory

The idea of collective memory, first developed by Halbwachs [5], has been examined and expanded upon from multiple perspectives, some of which are introduced below.

History is composed of many collective memories of different historical entities, and so historical narratives should represent historical conflict through different collective memories. It has been shown that different historical entities have different collective memories of the same historical event. For example, Wertsch et al. [6] showed that Russian and American students do not share collective memories concerning WWII. Another paper [7] revealed that different generations of the same population have different collective memories.

In [8], Zaromb, Franklin, et al. explore how literature and archives can influence collective memory. on collective memory.

For more basic information about collective memory, see [9].

9.2 Narrative History and Consensus in Distributive Computing

A historical narrative is composed of many facts or historical events, upon which all individuals agrees. For example, we all know that Caesar was assassinated on the 15th of March 44 BC. Without agreeing on that historical fact, we are left without a historical narrative, although we know that the fact could be false. Therefore, one can claim that historical narratives are composed of many bits of agreement. It follows that a key element of historical narratives is the foundational historical event that we all agree upon.

Before we can start understanding the historical narrative, we have to agree on the historical events. What is the possibility of achieving that? Consensus in distributed computing is similar to the historical agreement problem with one main difference: while in history, the main concern is humans and their behavior, in distributed computing, the main concerns are CPUs and computers.

As stated, one of the central questions in distributed computing is how a network of computers agrees on a single bit. Therefore, when replacing the network of computers with a network of humans, we are left with the same problem: how does the network agree on a single bit? Can they even agree? When and how can they do so?

To summarize, the main difference between historical narrative and distributed computing consensus is the element or node of the network. While in distributed computing, the nodes are computers, in history, the nodes are humans or computers, and sometimes both. Computers, it should be noted, have only become relevant historical nodes in the 21st century.

Therefore, one can adopt the mathematical formalism from distributed computing to the historical narrative with a small amount of change. From assuming that nodes follow some protocol in the case of distributed computing to game theory, where nodes are assumed to be rational and try to optimize some utility functions.

In this chapter, we will accomplish this goal and transform the consensus problem from distributed computing to the study of historical narrative.

We start by defining the consensus problem in distributed computing. We will then transform it into a historical consensus or agreement and discuss the possibility of reaching an agreement. At the end of the chapter, we will discuss the meaning of it all.

9.3 Consensus in Distributed Computing

Consensus is one of the pillars of distributed computing, and its definition can be found in many textbooks, for example, in Chap. 5 of Hagit Attiya's book [10]. For the reader's convenience, we present a slightly altered definition below.

In distributed computing, there is a set of V processes, a communication graph $G(V, E)$ which determines the pair of processors $v, u \in V$ that can exchange information; i.e., the two processors can talk if and only if $(v, u) \in E$. The communication graph is assumed to be a complete graph [10], i.e., every two nodes can communicate with each other in every round.

Time is either synchronous or asynchronous. In this chapter, we will assume it is synchronous and progresses in steps. At every step, each process can do a constant number of computations. If we like to be formal we would use a single computation, but without a loss of generality, we can exchange a single computation for a constant and vice versa.

In the consensus problem, each processor starts with a single bit $input_v \in \{0, 1\}$, which is either 1 or 0. Moreover, each processor $v \in V$ has a valuable out_v. The task facing the processors in V is to agree on a single output value out with the following restriction: In the formal description, the term "non-faulty processor" appears. We will explain it in Sect. 9.3.1.

1. Validity:
   If all processors start with the same value val, the output should be val i.e.,

$$\text{if for all } v \in V, input_v = val \text{ then } out_v = val. \tag{9.1}$$

2. Agreement:
   If not all processors have the same value, i.e., some processors got 0 and some processors got 1, then the output can be either 0 or 1.
3. Termination:
   In every admissible execution, $out_v = val$ is eventually assigned a value, for every non-faulty processor $v \in V$.

This is a simple task to perform, and we can provide a very simple algorithm to solve it if the communication graph $G(V, E)$ is connected. Each processor collects all input values. Once it has all the information, it computes the minimum among all input values in the network and determines the output value to be the minimum, like so:

$$out = \min\{input_v : v \in V\} \tag{9.2}$$

In fact, we don't even need to know all the inputs; suffices for each processor to know the minimum input in the network. Clearly, all nodes complete their computation after the diameter time, since that's the time it takes to collect the information.

Consensus is a simple problem as long as all processors behave as expected. Complications begin when our processors turn unreliable. This is human nature just as much as it is the nature of computers. The following model is copied from [10].

9.3.1 Faulty Processor

Faulty processors require a restating of the problem. We will assume that there are, at most, f faulty processors. The system that can handle f faults is called f-resilient.

In each round, all non-faulty processors successfully perform the following three steps in succession:

1. Send a message to all the neighbors;
2. Receive the messages that each of their neighbors has sent.
3. Upon collecting all of the messages, perform a single computation step per processor.

Next, we are going to discuss the behavior of the faulty processors. For an f-resilient system, we can define the execution as follows:

Let $t \in \mathbb{N}$ denote the number of rounds the system performs. Let $F(t) \subset V$ be a sequence of subsets of the faulty processors at round t. The sequence $F(t)$ is weakly monotonic increasing, and the size of $F(t)$ is bounded by f. Formally,

1. $F(0) = \varnothing$.
2. For all $t \in \mathbb{N}$
$$F(t) \subseteq F(t+1) \tag{9.3}$$
3. For all $t \in \mathbb{N}$
$$|F(t)| \leq f \tag{9.4}$$

Clearly, the set $F(t-1)$ contains all the faulty processors that stop working at time $t-1$. Those processors will do nothing at the round t since they are faulty. Some new processors may stop functioning at exactly round t. Denote this set of processors by $\partial F(t)$. Thus, the set of faulty processors at round t is

$$F(t) = F(t-1) \cup \partial F(t) \tag{9.5}$$

Since we do not know at which step these faulty processors fail, we cannot assume anything about their behavior. Therefore, the faulty processors at round t can send some information to a subset of the neighbors. However, these messages can be corrupted and are therefore not reliable. Once a processor fails, it cannot be used in the computation.

There is a well-known theorem, see Theorem 5.3 in [10], which links the number of running times of the algorithm to the number of crashes f:

Theorem 9.1 *Any consensus algorithm for n processors that is resilient to f crash failures requires at least* $f+1$ *rounds in some admissible execution, for all* $n \geq f+2$.

This means that there is no algorithm that solves the consensus before time $t+2$ and that there is an algorithm that solves the f-resilient consensus problem with a running time of $f+2$.

In distributed computing, there are many similar models with numerous nuanced variations. The reason for describing this model here is that it highlights the main interest in consensus problems in distributed computing. However, the problem is utterly different when considered from a digital history perspective. Let us compare the similarities and differences between both fields.

9.4 Distributed Computing and Digital History: A Short Comparison

Archimedes said, "Give me a lever long enough and a fulcrum on which to place it, and I shall move the world." So why not use the consensus problem in distributed computing as a fulcrum? We will start by describing the differences between distributed computing and collective memory, then explore their similarities.

9.4.1 Similarities Between Distributed Computing and Digital History

The advantage of starting with the consensus problem in distributed computing is that we can import many components of the model from the consensus problem in distributed computing to the collective memory formation model. The formal definition of collective memory appears in Sect. 9.5. In this subsection, we summarize the components that we will be transferring. The main similarities are:

1. Time could be synchronous or asynchronous. For our purpose, let us assume that it is synchronous . From a philosophical perspective, there is not much difference between the two options, and synchronized time is much simpler, technically speaking.
2. In the case of distributed computing, it made sense to talk about complete graphs. We use a graph $G(V, E)$ to mathematically describe communication between different nodes.
3. While it makes sense to model a social network as a graph $G(V, E)$ in a single snapshot, that doesn't quite work when considering a long historical process. The correct mathematical tool, in this case, changes from a single graph to a sequence of graphs.

There are two main models of time on distributed computing. The first is synchronous, meaning that there is a global clock and time progresses with the same speed everywhere according to the global clock. The second model is asynchronous, which means that every CPU or node has its own clock.

In general, there is an algorithmic technique that transforms asynchronous systems into synchronous with some performance cost, see [11, 12]. Therefore, although it is clear that historical clocks are not asynchronous, time moves at different speeds for different people and nations, and is heavily related to collective memory—in this chapter, we assume that time is asynchronous, since we wish to simplify the math.

In the next subsection, we discuss the main differences between consensus and the collective memory formation game.

9.4.2 Differences Between Distributed Computing and Digital History

As stated earlier, there is a difference between the collective memory formation game and the consensus problem in distributed computing. In this subsection, we summarize the main differences, address them, and adjust the model accordingly. The main differences are:

1. While in distributed computing we talk about processors and digital memory, in history we talk about humans, historical text, and collective memory. Therefore, we split the treatment of historical collective memory into two-time scales. The first is the time of human witnesses. This is approximately the average human life span, roughly 75 years. The second time scale is post witness death. In this scenario, all we are left with are texts or digital media for the future. We will assume that these sources will not disappear.
   Since we are talking about humans as the main players, the approach should be more game-theoretical and incorporate some behavior psychology and strategy.
2. Different groups of people will inevitably come to some point of disagreement which will separate them. That is precisely what sets them apart and divides them into different groups. Therefore, the model of consensus in digital humanity should allow people to agree to disagree .
3. In distributed computing, all processes have a common goal, whereas in digital history each actor can have a different goal.
4. Collective memory is a property of a group of people. Therefore, different groups should have different collective memories.
5. While humans can be born and die, process in distributed computing can (usually) only fail.
6. While the main question in distributed computing is when consensus is possible, the main question in digital history is how two groups arrive at different opinions and when can that happen.

In the next section, we will present the general model for the collective memory formation game.

9.5 The Model of Collective Memory Formation Game

To describe the collective memory formation game, we first borrow the common theme from the consensus problem. We will then explore their differences in Sect. 9.5.2.

9.5.1 Transforming Definitions from Consensus

As in the consensus problem, we have to describe the environment of the collective memory formation game. Since our main target is to understand collective memory formation, we start with a finite sequence of historical events. Denote them by $\mathcal{H} = (h_1, ..., h_\tau)$ see Sect. 2.2. Let $\mathbb{A}rc$ be the set of all historical events, i.e.,

$$\mathbb{A}rc = \{h_1, h_2..., h_\tau\} \tag{9.6}$$

Time

For each historical event, we have the time at which the event happens:

$$T : H \to \mathbb{R} \tag{9.7}$$

We will say that event h_i happens before event h_j, if and only if $T(h_i) < T(h_j)$. Moreover, we will assume that the historical events are well-ordered, and that the events happen in sequence, one after the other:

$$\text{if } T(h_i) = T(h_j) \text{ then } h_i = h_j. \tag{9.8}$$

To simplify denotation, we assume that the sequence of events $\mathcal{EV}$ is sorted according to the time the events happened, i.e.,

$$\text{for all } i \in \{1, ..., \tau - 1\}, \text{ it follows that } T(h_i) < T(h_{i+1}) \tag{9.9}$$

Next, we assume that there is an unbound number of computation steps between two consecutive time events, if needed. Our main interest is not the complexity of the algorithms, as most of them are very simple, but rather the existence of such a computation. Now, following the list Sect. 9.4.1, we move from 1 to 2, and specify our assumptions concerning space.

Space

Let V_i be the set of active nodes alive during the event h_i. Denote the sequence of V_i by $\mathcal{V} = (V_1, ..., V_\tau)$. Denote the set of all nodes by

$$V_{t_1}^{t_2} = \bigcup_{i=t_1}^{t_2} V_i \tag{9.10}$$

Let $v \in V_1^\tau$ be a node, $b(v)$ be the birthday of a node, and $d(v)$ be the death of a node. We assume that a life lived by each node is an interval, i.e.,

$$\text{for all } v \in V_1^\tau, i \in \mathbb{N}, \text{ such that } b(v) \leq i \leq d(v) \text{ it follows that } v \in V_i. \tag{9.11}$$

After describing the consensus model, we must describe the communication pattern between two nodes. Denote the set of edges that describe the communication pattern allowed between the nodes at time $T(h_i)$ when the event h_i happened to be E_i. Therefore, the way to model item 3 in Sect. 9.4.1 is by using the communication pattern in a sequence of graphs denoted by

$$\gamma = (G(V_1, E_1), ...G(V_\tau, E_\tau)) \tag{9.12}$$

For example, graph γ_i describes the communication pattern in relation to event h_i.

We have used all the similar items from the consensus problem list Sect. 9.4.1, and will now proceed to list Sect. 9.4.2

9.5.2 *Deviations from Consensus*

Now that we've presented the similarities between consensus and the collective memory formation game, let us explore their differences. We follow list Sect. 9.4.2 from Sect. 9.4.2. The first item on the list states that the nodes in the collective memory formation game are humans, not processors. This means that when we argue about the behavior of humans, we should use arguments from psychology, sociology, economics, and game theory. Of course, these arguments should be written in mathematical language as we are dealing with algorithms and models.

Since we are now describing the language of the model and not a specific example, it is enough to describe the input and output of a node v at time $t \in \mathbb{Q}$. Denote the input of node v as the knowledge $K_v(t)$ at a time t. Note that t belongs to a rational number since we allow computations between the events' time.

When building a collective memory, we generally have several decisions to make. For example, defining whether the event h_i is important and should be included in the collective memory. Another decision is how to connect the event to other historical events, how to interpret it, and so on. Therefore, we model all the decisions we must make concerning event h_i as a d-dimensional probability vector distribution. Therefore, person v is modeled as the *opinion function*:

$$Op_v(t) : K_v(t) \rightarrow \mathbb{R}^d \tag{9.13}$$

If we'd like to talk about the computational issues of a person, this function will be computed by a Turing machine. In our case, the function is the Turing machine. Note that each person has a different function in our model. Of course, all these functions depend on the social structure of the person.

We now move to item 3 on the list. To handle the fact that different people have different goals, we move from analyzing the behavior of the distributed system to game theory. We are particularly interested in the Nash equilibrium and the conversion to it. First, we will discuss the general games framework within which the Nash equilibrium exists.

General Games

In the rest of the chapter, we discuss a single game which we use to describe a single historical event. Therefore, the index now ignores the broad context of the historical narrative and concentrates on a single historical event. This means that index i will be used to describe players and not historical events, as is traditionally done in game theory.

We start with the formal definition of games. The definition requires three items. Then, we will adapt general game notation to the problem of collective memory formation games. These three elements are described below.

We have a set of players $\mathcal{N}$. For each player $i \in \mathcal{N}$ the set of strategies S_i, and the last element is a payoff function π_i, which is usually measured as the utility of player i. We generally wish to maximize our utility; sometimes, instead of having the utility function, we are interested in the cost function $cost_i$. In this case, the utility function is $\pi_i = -cost_i$. Formally, the non-cooperative game is defined as a tuple $(\mathcal{N}, (S_i), (\pi_i))$, where:

1. A set of players $\mathcal{N} = \{1, ..., n\}$.
2. For each player $i \in \mathcal{N}$, a set of strategies S_i.
3. A payoff functions $\pi_i : S \to \mathbb{R}$ for each $i \in \mathcal{N}$ where $S = \prod_{i=1}^{N} S_i$.

Before we continue, it is useful to define the set of all players except player i, Formally,

$$\mathcal{N}^{-i} = \mathcal{N} \setminus \{i\}. \tag{9.14}$$

Now that we have the general game framework in which the Nash equilibrium exists, we are going to define the Nash equilibrium in the context of general games and then adapt it to our game.

Nash Equilibrium

The Nash Equilibrium was introduced by John Nash in 1950, see [13, 14] as a solution concept where each player

$$i \in \mathcal{N} \tag{9.15}$$

in the game selects a strategy $\sigma_i^* \in S_i$ that is optimal given the choices of the other players. We denote the set S^{-i} to be the set of all vector strategies, where all players $j \in \mathcal{N}^{-i}$ except player i play the strategy σ_j^* and player i plays an arbitrary strategy $\sigma_i \in S_i$:

$$S^{-i} = \{(\sigma_1, ..., \sigma_n) \in S : \text{ for all } j \in \mathcal{N}^{-i}, \sigma_j = \sigma_j^*\} \tag{9.16}$$

In the Nash equilibrium, no player has an incentive to unilaterally deviate from their chosen strategy, as long as the strategies of the other players remain the same. A strategy profile

$$\sigma^* = (\sigma_1^*, \sigma_2^*, ..., \sigma_N^*) \tag{9.17}$$

is a Nash equilibrium if for all $i \in \mathcal{N}$ and for all $\sigma \in S^{-i}$

$$\pi_i(\sigma^*) \geq \pi_i(\sigma) \tag{9.18}$$

Collective Entities

In a distributed consensus, there is a single collective, and all non-faulty processors have to arrive at a single decision. However, when we consider collective memory, there are k different historical entities $(h_1, ..., h_k)$. To study a historical conflict, we need to have at least two historical entities. For the time being, we consider historical entities as a set of humans with a shared collective memory—for example, Americans versus Russians, or Chinese, since all three groups have a unique collective memory that is different from the rest. Usually, it is enough to analyze the case of $k = 2$; once this is done, it is easy to generalize it to an arbitrary k.

Note that the definition of historical entities uses collective memory, and we are trying to model how collective memory is formed as something resembling a cyclic definition. However, this definition is inductive because we are dealing with adding a single event to the collective memory.

In the collective memory formation game, we try to simplify the definition of entities as much as possible. Therefore, we will assume that the set of all entities is a partition of the set of nodes V_1^τ, Formally,

1. for all $i \in \{1, ..., k\}$, $h_i \subset V_1^\tau$
2. For all $i, j \in \{1, ..., k\}$ if $i \neq j$ then $h_i \cap h_j = \varnothing$
3. $\bigcup_i^k h_i = V_1^\tau$

9.5.3 *Life Span*

The fact that people both arrive and depart from the narrative formation game forces us to deal with a sequence of graphs instead of a single graph. To simplify the analysis, we choose to focus on a single slice of narrative in this chapter.

9.6 Historical Consensus After the Death of the Witnesses

In this section, we show that the historical consensus after the death of the witnesses is solvable only if there is a historical text that cannot disappear. To prove this, we split the argument into two cases. For the first one, we assume that there is no canonical text or living witnesses. For the second, we assume that there is a canonical text concerning the historical event.

9.6.1 No Canonical Text

According to the definition of the consensus problem in distributed computing, the agreement demands that we have a surviving witness who has an original opinion of the events (see Sect. 9.3). Therefore, since we presume that there are no surviving witnesses, a consensus cannot be reached.

9.6.2 Existing Canonical Text

Now, we assume the existence of a canonical text. In this case, the standard agreement problem would determine that all nodes can shape their opinion according to the canonical text, i.e., they can agree or disagree with it. If there are two or more communities, each of them can have its own opinion or interpretation of the canonical text. They can agree or disagree and, therefore, reach a historical consensus.

9.6.3 Death of Witnesses and the Narrative of History

It is always easier to tell a good story than to change history. Given that narratives are the mechanism that defines history, it appears that our great leaders and heroes—Elizabeth the Great, Constantine the Great, and Peter the Great—were revered not because of their actions, but because of the narrative.

Most narratives about heroes require a villain for contrast. Peter had Ivan the Terrible, Elizabeth had Richard the Third, and Constantine had Commodus.

Since narrative is the core of history, and any great person needs a villain, it is easy to blame evil deeds on the dead who cannot defend themselves. We typically picture the villain as a leader who lived 70 years before or after the great person. This means that the great are not really great, and the villains are not really villains.

Another interesting example is Augustus, who ruled from 27 BC to 14 AD, and Nero, who reigned from 54 AD until his death in 68 AD. In this case, the terrible person lived after the great person, which is unusual in inventing a historical narrative.

Note the difference of more than 70 years between the beginning of the rules of the great Augustus and the end of the evil Nero's.

> **Hero versus Villain 70-year rule of thumb**
>
> In historical narratives, the time that separates the villain from the great is longer than 70 years.

9.7 Conclusion

In this chapter, we began investigating collective memory. We studied it from the point of view of the famously unsolvable consensus problem in distributed computing, which shows how crucial historical texts are in generating the narrative and a collective memory after the death of the witness Fig. 9.2 illustrates the challenge of communicating with the deceased.

Fig. 9.2 The problem of discussing historical narratives with dead witnesses. Although we would like to converse with skeletons, it is (usually) impossible

One important observation from our model is that for a historical narrative to emerge, we need at least two historical entities. As Emmanuel Levinas told us, others have always existed before us, and we cannot exist without them. Historical narratives are always about the conflicts between historical entities.

To finish off this chapter, we formulate the historical Hero versus Villain 70-year rule of thumb. Note that this rule of thumb is derived from the way humans tell their narrative and not from the actual events. As Immanuel Kant taught us, science is composed of two central components: reality and the way reality is perceived in our minds. Some scientific rules talk about reality, and others talk about the perception of this reality in our brains. The Hero versus Villain 70-year rule of thumb concerns the way we perceive history, not about history itself.

Clearly, there are exceptions to the 70-year rule of thumb. For example, in the villain in WWII is obviously Hitler, and the heroes are Roosevelt, Churchill, and, perhaps, Stalin. Note that after the Allies won, all the Nazis were persecuted and no witnesses remained to share something positive about the Nazis or Hitler. This is not to say that Hitler or the Nazis were not villains by any means, but this example helps us comprehend the perspective of the 70-year rule of thumb.

In the next chapter, we will use game theory to study the process of generating canonical texts that allow historical entities to have a long history beyond human life span.

9.8 Exercise

1. This exercise aims to show that AI witnesses do not solve the problem of witness death. Pick your favorite historical text that describes a historical conflict, for example, the Persian Wars as described in the first historical book by Herodotus [15]. Identify the winner and the loser from the perspective of the historical text. Ask LLM or any other Large Language Models (LLMs) to rewrite the historical text so that the loser and winner trade places.
2. Pick your favorite collective memory from your culture. Identify your historical entity and any other historical entities that generate a conflict with the collective memory you have chosen.
3. Assume that you have a set of two processors: one has the value 1, and the other has the value -1. Show that in the event of a single failure, it is impossible to achieve consensus in a single round of communication.
4. Choose your favorite great historical figure and apply the 70-year rule of thumb in its narrative to find the villain.
5. Use your favorite Large Language Models (LLMs) to define the prisoner's dilemma. In this chapter, we defined the Nash equilibrium in the setting of game theory. Compute the Nash equilibrium of the prisoner's dilemma.
6. Write a game theory model that describes the outbreak of WWI as a game and compute its Nash equilibrium. You can use your favorite Large Language Models (LLMs). For more about this subject, see [16, 17].

7. After the death of Alexander the Great, his empire split into three Hellenistic power blocs: Ptolemaic in Egypt, Seleucid in Syria and East, and Antigonid Macedonia. Describe the historical event—the death of Alexander—as a single event that separates the Hellenistic world into three different historical entities. This means that when modeling the historical event, we need at least two bits of information in order for the event to contain three different opinions.

References

1. M.H. DeGroot, Reaching a consensus. J. Am. Stat. Assoc. **69**(345), 118–121 (1974)
2. J. Qin, Q. Ma, Y. Shi, L. Wang, Recent advances in consensus of multi-agent systems: a brief survey. IEEE Trans. Ind. Electron. **64**(6), 4972–4983 (2016)
3. M. Cao, A.S. Morse, B.D.O. Anderson, Reaching a consensus in a dynamically changing environment: convergence rates, measurement delays, and asynchronous events. SIAM J. Control. Optim. **47**(2), 601–623 (2008)
4. P. Guo, C. Han, Nash equilibrium and group strategy consensus of networked evolutionary game with coupled social groups. Appl. Math. Comput. **409**, 126380 (2021)
5. M. Halbwachs, J. Alexandre-Halbwachs, *La mémoire collective*. Bibliothèque de sociologie contemporaine (Presses universitaires de France, 1950)
6. J.V. Wertsch, *Voices of Collective Remembering* (Cambridge University Press, 2002)
7. F. Zaromb, A.C. Butler, P.K. Agarwal, H.L. Roediger, Collective memories of three wars in united states history in younger and older adults. Mem. Cogn. **42**, 383–399 (2014)
8. A. Josias, Toward an understanding of archives as a feature of collective memory. Arch. Sci. **11**, 95–112 (2011)
9. Wiki. wikisource-collective-memory. https://en.wikipedia.org/wiki/Collective_memory (2023)
10. H. Attiya, J. Welch. *Distributed Computing: Fundamentals, Simulations, and Advanced Topics*, vol. 19 (Wiley, 2004)
11. D. Peleg, J.D. Ullman, An optimal synchronizer for the hypercube, in *Proceedings of the Sixth Annual ACM Symposium on Principles of Distributed Computing* (1987), pp. 77–85
12. R.G. Gallager, Baruch awerbuch robert g. gallager mit
13. J.F Nash Jr., Equilibrium points in n-person games. Proc. Natl. Acad. Sci. **36**(1), 48–49 (1950)
14. R.B. Myerson, Nash equilibrium and the history of economic theory. J. Econ. Lit. **37**(3), 1067–1082 (1999)
15. Wiki. wikisource. https://en.wikisource.org/wiki/Herodotus_The_Persian_Wars_(Godley) (2024)
16. R.B. Myerson, Game theory and the first world war. J. Econ. Lit. **61**(2), 716–735 (2023)
17. S. Wolford, *The Politics of the First World War: A Course in Game Theory and International Security* (Cambridge University Press, 2019)

Chapter 10
History While the Witnesses Are Still Alive

In the previous chapter, we studied the ability to achieve consensus long after the witness disappeared. We showed that it is impossible to reach a consensus without historical documents Sect. 9.6.

In this chapter, we will study the process of reaching a consensus shortly after the historical events had occurred. In general, when witnesses are still alive, they each have their personal experience and opinion, which may differ from one person to the next, see Fig. 10.1. It is unclear how society shapes its collective memory. In this chapter, we will model the process of collective memory formation as a game played by humans from different groups.

Our starting position is in the previous chapter, where we used the concept of consensus in distributive computing as a model for agreeing on a single historical event. However, this chapter will diversify from the standard consensus problem in distributed computing to a more of a game theory approach, because we are dealing with human beings and not machines.

Since we assume that the witnesses are still alive, each might try to convince their friends of their point of view. We can therefore assume that each node or person has its own utility function with respect to their point of view and their friends' opinion. The historical consensus emerges as a Nash equilibrium of the collective memory formation game.

Again, we want to emphasize that there is an advantage in modeling how society reaches a consensus on the collective memory of the historical event as a simple game, since it allows us to identify simple requirements for the necessary and efficient condition of achieving collective memory. This chapter uses advanced mathematical tools; readers who are not interested in these can skip the mathematical models and read the conclusion.

Z. Lotker, *History by Algorithms*,
https://doi.org/10.1007/978-3-031-93627-2_10

Fig. 10.1 A cartoon about The Battle for the Narrative

10.1 Conversion to the Collective Memory

In non-cooperative game theory, a Nash equilibrium represents a stable state of the game where no player can benefit by unilaterally changing their strategy, given that other players' strategies remain unchanged.

According to the standard argument in game theory, the outcome of a non-cooperative game tends to converge to one of its Nash equilibrium states under certain conditions, such as repeated interactions and rational behavior by the players.

Therefore, the advantage of studying the collective memory formation game is its Nash equilibrium.

The existence of a Nash equilibrium raises two questions: to which Nash equilibrium does society converge, and how? Again, we use standard tools of game theory to answer these questions. Then, we circle back to interpret these results in the context of history.

10.2 Example of a Collective Memory Formation Game

In this section, we will describe the very simple collective memory formation game.

10.2.1 A Single Historical Event

We will assume that there is a single historical event h_1. If there are several events, we can start with the first event and analyze it according to the process described in this chapter. Once the communities reach an agreement, we can apply induction to the next historical event. This time, the opinion of the next historical event can depend on the previous agreement and the historical event itself. The simplest case is of independent events that do not influence one another; we can treat each event separately by its own game and accumulate the consensus. However, historical events tend to be dependent on each other, see Exercise 7.

10.2.2 Red and Blue Collective Entities

In general, let's assume that there are two collective entities: $\mathscr{E}_r$ for red and $\mathscr{E}_b$ for blue. We assume that we have N players.

10.2.3 Input

Just like in consensus, each player $i \in \mathscr{N}$ starts with an input $input_i \in \{-1, 1\}$ where the meaning of $input_i = 1/(input_i = -1)$ is that i believes that the historical event h_1 is positive/(negative) respectively.

10.2.4 Communication Graph

Like the consensus problem, we use the communication graph $G(\mathscr{N}, E)$ to determine which pair of nodes can exchange information. Note that the set of nodes in this graph is $V = \mathscr{N}$.

10.2.5 Pure Strategy

There are several options to define pure strategy, each with its own advantages and disadvantages. We list some of them for later use, moving from the smallest set of strategies and increasing from there.

One and Minus One

We start with the smallest set of strategies possible. The game doesn't make sense if we cannot separate the two collective entities. In order to separate two entities, we need at least two pure strategies. Formally, there are two pure strategies for each player:

$$S_i = \{-1, 1\} \tag{10.1}$$

Of course, the set of mixed strategies is the Bernoulli distribution. As in the consensus problem, each node will have an output variable: $out_i \in S_i$.

Note that if we have k collective entities, we will need k strategies.

The advantage of this set of strategies is that it is compact, simple, and small. The disadvantage is that the pure strategy is discrete and not continuous. Moreover, the set is too small to be used for analyzing converged algorithms.

Rational Numbers

As previously mentioned, we are interested in protocols that achieve conversion to the Nash equilibrium using a simple averaging procedure. In the protocol, each node is supposed to calculate the weighted average opinion of its neighbors and transmit this value to all its neighbors in the communication graph. Therefore, in this protocol, the strategy space of each neighbor should be the rational number. The strategy space when analyzing the protocol is

$$S_i = \mathbb{Q} \tag{10.2}$$

The rational numbers are rich enough to contain the analysis of the conversion protocol. However, they are not compact, and this, of course, might bring us to some topological problems, for example, the existence of the Nash equilibrium is not guaranteed.

Real Numbers Between One and Minus One

To overcome the compact problem, we can assume that the set of strategies is

$$S_i = [-1, 1] \tag{10.3}$$

The main problem with using an uncountable strategy space is that it no longer guarantees the existence of the Nash equilibrium. However, we can prove that the Nash equilibrium exists through a simple collective memory formation game.

10.2.6 Utility Function

The last thing we need to describe is the utility function for each player. In our example, the utility function should encourage the blue nodes to agree with each other, the red nodes to agree with each other, and the nodes that belong to different groups to disagree. There are many functions that we can use to describe these conditions. For the sake of simplicity, we will use the following formula:

$$\pi_i(out) = \begin{cases} -\left(\sum_{j1\in N(i)\cap\mathscr{E}_r}(out_{j1} - out_i)^2 + \sum_{j2\in N(i)\cap\mathscr{E}_b}(out_{j2} + out_i)^2\right), & i \in \mathscr{E}_r \\ -\left(\sum_{j1\in N(i)\cap\mathscr{E}_b}(out_{j1} - out_i)^2 + \sum_{j2\in N(i)\cap\mathscr{E}_r}(out_{j2} + out_i)^2\right), & i \in \mathscr{E}_b \end{cases} \tag{10.4}$$

The utility function for the red nodes is the first line in the Eq. 10.4, and the utility function for the blue nodes is the second line in the equation. We remind the reader that $N(i)$ is the neighborhood of node i in the social graph $G(\mathscr{N}, E)$, which means that all the nodes in the social graph are connected to node i.

The function $\pi_i(out)$ measures the extent of disagreement in the neighborhood of node i. We include a minus sign when we want to minimize disagreement within the group. We include a plus sign when we want to maximize disagreement between the different groups.

We are interested in how the system converges to the Nash equilibrium. We will provide a simple mechanism to converge to Nash if all groups (red, blue) are clear and known to everyone. We call the utility function in Eq. 10.4 the *debate utility function*.

We call the game in this subsection the collective memory formation game. There are many different types of these games, but in this chapter, we will analyze only the simplest one.

10.2.7 Example

Figure 10.2 demonstrates the beginning of the collective memory formation game on the left and the end of the collective memory formation game on the right. We have two communities: red and blue. The left side of the figure shows the input and describes the beginning of the evaluation of the historical event. It shows what each individual is thinking in regards to the events; green dots reflect positive opinions and yellow dots reflect negative opinions. This situation is not stable. It is easy to see that the left side of the figure is not the Nash equilibrium. The right side *is* a Nash equilibrium: all red nodes agree with each other and all blue nodes agree with each other, while the red and blue nodes disagree with each other. None of the nodes need to deviate and, therefore, this is the Nash equilibrium.

10.2.8 Nash Equilibrium of a Collective Memory Formation Game in Pure Strategy

In this section, we will prove the existence of at least two Nash equilibria of a simple collective memory formation game.

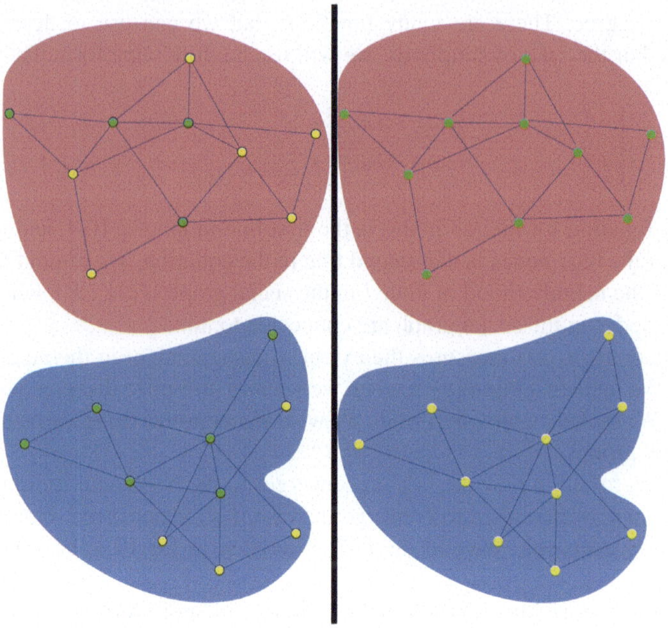

Fig. 10.2 One instance of the collective memory formation game: we see two groups of individuals, red and blue. On the left, we see the beginning of the game. The color representing the opinion of each individual is either green or yellow. Green dots reflect positive opinions (+1), and yellow dots reflect negative opinions (−1). They try to find the consensus using the communication graph. On the right of the figure, we see an output of the game, where communities achieve a consensus. The communication graph follows the same logic: the red community agrees in green, and the blue agrees in yellow. The right side is the Nash equilibrium

Theorem 10.1 *If the graph* $G(\mathcal{N}, E)$ *is connected, there exist at least two pure Nash equilibria:*

1. *For all* $i \in \mathcal{E}_r$ *and for all* $j \in \mathcal{E}_b$, $out_i^* = 1, out_j^* = -1$.
2. *For all* $i \in \mathcal{E}_r$ *and for all* $j \in \mathcal{E}_b$, $out_i^* = -1, out_j^* = 1$.

Proof We need to prove two things: first, that 1 is a Nash equilibrium, and second, that 2 is also Nash equilibrium. We start by proving that 1 is a Nash equilibrium, as gleaned from two simple facts: The first is that for all $out \in \mathbb{R}^N$,

$$\pi_i(out) \leq 0 \tag{10.5}$$

and the second is that plugging the value from condition 1, we find that

$$\pi_i(out^*) = 0. \tag{10.6}$$

Therefore, no player can benefit from condition 1. A similar proof shows that this condition 2 is a Nash equilibrium.

The question of identifying all the Nash equilibria depends on the set of strategies S_i of each player i. We refer the interested reader to Exercise 4, 5.

We can analyze each connected company separately if the graph is not connected. However, from the perspective of the historical event, it is natural to assume that the graphs are connected. Otherwise, it is unclear how the historical conflict is generated. If two groups of people share no connection or interaction, why should any conflict between them arise in the first place?

The above theorem proves that the vector $out^* \in \mathbb{Q}^N$ which minimizes the debate utility function $\pi_i(out^*) = 0$ is a Nash equilibrium. In the next section, we will provide a simple mechanism that uses the random input opinion $input_i$ of each node $i \in \mathcal{N}$ that converges to the direction of one of the Nash equilibria with high probability, when the input value is selected according to the uniform distribution.

One way to interpret the proof of Theorem 10.1 when we consider the space of strategy to be the real number, see Eq. 10.3, is that each node i tries to selfishly optimize its utility function π_i. This means that player i tries to solve

$$\frac{\partial \pi_i}{\partial out_i} = 0 \tag{10.7}$$

In a non-cooperative game, each player is selfish, and all players try to solve the following linear system of equations: For all $i \in \mathcal{N}$

$$\frac{\partial \pi_i}{\partial out_i} = 0 \tag{10.8}$$

Note that the number of equations is equivalent to the number of players, equal to the number of variables. Therefore, the system can be solved easily. Moreover, since the system is homogeneous and there is no homogeneous solution, we can conclude that the dimension of the solution is greater than or equal to 1.

Taking the derivative of the utility function, we can rewrite the linear system of equations according to each of the variables as follows:

$$\begin{cases} -\left(\sum_{j_1 \in N(i) \cap \mathcal{E}_r} - 2(out_{j_1} - out_i) + \sum_{j_2 \in N(i) \cap \mathcal{E}_b} 2(out_{j_2} + out_i)\right) = 0, & i \in \mathcal{E}_r \\ -\left(\sum_{j_1 \in N(i) \cap \mathcal{E}_b} - 2(out_{j_1} - out_i) + \sum_{j_2 \in N(i) \cap \mathcal{E}_r} 2(out_{j_2} + out_i)\right) = 0, & i \in \mathcal{E}_b \end{cases} \tag{10.9}$$

Another way to achieve this linear system is to write the total social utility, i.e., to sum all of the utility functions and the partial derivatives of each variable. This only works if the graph is undirected. We will then find that each entity in the linear system has to be multiplied by two.

It turns out that the linear system (10.9) can be solved iteratively and distributively since it is related to the Markov chain process. The next section will discuss the conversion of a simple iterative process. We believe that a similar process occurs in society when it reaches the collective memory consensus.

10.3 Conversion to the Nash Equilibrium

The main purpose of this chapter is to present a simple model in which two communities $\mathscr{V}_b$, $\mathscr{V}_r$ evaluate a single historical event h_1 that separates the blue node from the red node by evaluating their opinion of the event h_1. As previously stated, we will use a simple averaging process influenced by the linear system of Eq. 10.9 as a process that separates the opinions of the blue and red communities.

10.3.1 Algorithmic Approach for Solving Collective Memory Formation Game

Humans tend to talk with their friends when shaping their political opinions. The process of shaping opinions has been vastly studied in the context of social networks. Modeling opinion formation as an averaging process has already been suggested in [1, 2], while related questions were studied in [3–6].

However, the difference between a social network perspective and what we are doing here is the purpose of society. In social network, the model tries to explain the conversion to a single opinion. Here, we try to explain the formation of the conflict, i.e., how society converges to agree to disagree. Previous works were aimed at finding an agreement and were not based on the game theory approach. Since we are interested in the process of forming this disagreement, we model each person as a simple averaging algorithm. Moreover, we assume that each person uses the same algorithm. While it is possible to give each person a unique algorithm, this will make the model much more complicated and will not offer a better understanding of how this disagreement forms.

Algorithm 1 Solving Collective Memory Formation

for `i = 1 to N` **do**
    $com_i \leftarrow U[-1, 1]$
end for
for `n = 1 to` $N^3 \log(N)$ **do**
    for `i = 1 to N` **do**
        if $i \in \mathscr{B}$ **then**
            $com_i(t) \leftarrow \frac{\left(\sum_{j \in N_i \cap \mathscr{B}} (com_j(t-1))\right) - \left(\sum_{j \in N_i \cap \mathscr{R}}\right)(com_j(t-1))}{|N_i|}$
        end if
        if $i \in \mathscr{R}$ **then**
            $com_i(t) \leftarrow \frac{\left(\sum_{j \in N_i \cap \mathscr{R}} (com_j(t-1))\right) - \left(\sum_{j \in N_i \cap \mathscr{B}}\right)(com_j(t-1))}{|N_i|}$
        end if
    end for
end for
return (sign(com_1), ..., sign(com_N))

To analyze the above algorithm, we look at the sequence of values that each node computes per iteration. For each node $i \in \mathcal{N}$ and for all $t \in \mathbb{N}$, we consider the following sequence of $com_i(t)$ defined by the following recursive condition: $com_i \approx U[-1, 1]$.

At time step t, we compute $com_i(t)$ using $com_i(t-1)$ according to the following equation. Let's assume for node $i \in \mathcal{R}$:

$$com_i(t) = \frac{\left(\sum_{j \in N_i \cap \mathcal{R}} (com_j(t-1))\right) - \left(\sum_{j \in N_i \cap \mathcal{B}}\right)(com_j(t-1))}{|N_i|} \tag{10.10}$$

If the node is blue, we can update the model with a simple equation:

$$com_i(t) = \frac{\left(\sum_{j \in N_i \cap \mathcal{B}} (com_j(t-1))\right) - \left(\sum_{j \in N_i \cap \mathcal{R}}\right)(com_j(t-1))}{|N_i|} \tag{10.11}$$

The updated rule of each node can be interpreted thusly: Each node i tries to agree within the group and disagrees with another group. This is achieved by multiplying the numbers by -1. Of course, this coincides with the utility function π_i which is the purpose of the collective memory formation game.

We now explain line 4 in Algorithm 1: Since our social network is a simple graph, the mixing time of any simple random walk is $O(N^3)$. Therefore, adding a $O(logN)$ factor to the number of steps guarantees that the averaging process will converge. Moreover, in many cases, such as expounders and real social networks, the mixing time is, in fact, $O(\log(N))$. In such cases, we can shorten the number of iterations from $O(n^3 \log(N))$ to $O(\log(N)^2)$.

10.3.2 Markov Process with One Opinion

To analyze the collective memory formation game with two opinions, we should first study the same process when we have only one opinion. In this case, the goal of all individual group members is to agree on a single value. Therefore, the utility function is

$$\pi_i(out) = -\left(\sum_{j \in N(i)} (out_j - out_i)^2\right) \tag{10.12}$$

To move from the utility function to the linear system, we again take the derivatives of the quadratic function. In this case, after reorganizing the linear system of equations, we get

$$\frac{\partial \pi_i(out)}{\partial out_i} = 2\left(\sum_{j \in N(i)} (out_j - out_i)\right) = 0 \tag{10.13}$$

After taking the derivative of the utility function $\pi_i(out)$ according to the variable out_i and isolating variable out_i in the equation, it follows that

$$|N(i)|out_i = \sum_{j \in N(i)} (out_j) \tag{10.14}$$

Since the graph is connected, we know that the size of the neighborhood is $|N(i)| > 0$, therefore we can divide it by $|N(i)|$ and hence

$$out_i = \frac{\sum_{j \in N(i)}(out_j)}{|N(i)|} \tag{10.15}$$

Therefore, we can write Eq. 10.15 in matrix form, as follows:

$$out \cdot M = out \tag{10.16}$$

At this point, it is easy to notice that matrix $M = [m_{i,j}]$ is a Markov chain of the simple random walk traversing graph $G(\mathcal{N}, E)$. Note that if the communication graph $G(\mathcal{N}, E)$ is connected and not bipartite, then M is an ergodic transition matrix. For more on the ergodic Markov chain, refer to [7]. One simple technique is to add a single self-looped edge to all nodes. This trick forces the Markov chain to be ergodic. In fact, adding a self-looped edge is equivalent to having some memory of the individual opinion, which is usually the case when considering people's opinions.

Since M is ergodic, it follows that the absolute value of all eigenvalues apart from the biggest is less than 1, and the biggest value is 1. Formally, let $\lambda = (\lambda_1, \lambda_2, ..., \lambda_N)$ be the sequence of all eigenvalues of M, so that

1. $1 = \lambda_1$
2. for all $i = \{2, ..., N\}$, $|\lambda_i| < 1$

Moreover, the eigenvector of λ_1 is the vector where all elements have a value of 1, i.e.,

$$\vec{v}_1 = (1, ..., 1) \in \mathbb{R}^N \tag{10.17}$$

The statement on the eigenvalue can transform into an algorithm that achieves consensus with high probability.

In order to analyze the algorithm, we need the following notation: Denote the corresponding eigenvector of M to be $\vec{v}_1, \vec{v}_2, ..., \vec{v}_N$, where

$$\vec{v}_i = (v_{i,1}..., v_{i,N}). \tag{10.18}$$

Therefore, the i-eigenvalue λ_i and the i-eigenvector $\vec{v}_i$ satisfy the following equation:

Algorithm 2 Solving Consensus

1: **for** `i = 1 to N` **do**
2: $com_i \leftarrow U[-1, 1]$
3: **end for**
4: **while** $\{\text{sign}(com_1), ..., \text{sign}(com_N)\} = \{-1, 1\}$ **do**
5: **for** `i = 1 to N` **do**
6: $com_i \leftarrow \frac{\sum_{j \in N(i)}(com_j)}{|N(i)|}$
7: **end for**
8: **end while**
9: **return** $(\text{sign}(com_1), ..., \text{sign}(com_N))$

$$M\vec{v}_i = \lambda_i \vec{v}_i \tag{10.19}$$

Clearly, if the algorithm for solving consensus 2 stops, i.e., the condition in line 4 is false, then all nodes achieve consensus. We then ask, does the algorithm exit the While loop in line 4? The reason that the algorithm must stop after a finite time with a high probability follows from the fact that a random uniform real vector is never equal to one of the eigenvectors with probability 1. To write the proof formally, we need the inner product of the vectors x, y to be

$$< x, y >= \sum_{i=1}^{N} x_i y_i \tag{10.20}$$

Since the eigenvectors are orthogonal to each other, we can use them as a basis for the vector space. Therefore, we can decompose any vector v using the inner product and the eigenvectors, as follows:

$$v = \sum_{i=1}^{n} < v, \vec{v}_i > \vec{v}_i \tag{10.21}$$

The single average process that was done in line 6 of the algorithm is equivalent to the multiplication of the vector by matrix M

$$M \times v = M \times (\sum_{i=1}^{N} < v, \vec{v}_i > \vec{v}_i) \tag{10.22}$$

Using the definition of eigenvectors (see (10.19)), it follows that

$$M \times v = \sum_{i=1}^{N} \lambda_i < v, \vec{v}_i > \vec{v}_i \tag{10.23}$$

Now, when we repeat the averaging process t times, we get

$$M^t \times v = \sum_{i=1}^{N} \lambda_i^t < v, \vec{v}_i > \vec{v}_i \tag{10.24}$$

After some time, for all $i > 1$, value $\lambda_i^t < v, \vec{v}_i >$ is small enough so that all nodes will have the sign determined by the inner product $< v, \vec{v}_1 >$.

To summarize,

$$\Pr[\lim_{t\to\infty} M^t \times v = \langle v, \vec{v}_1 \rangle \times \vec{v}_1] = 1 \tag{10.25}$$

And the sign of all nodes after averaging for a long time is

$$sing(< v, \vec{v}_1 >) \tag{10.26}$$

Theorem 10.2 *For any connected non-bipartite graph G, the average algorithm solves the consensus problem at finite time with a high probability of the input.*

Now that we understand why the averaging process converges in a single opinion, we will move to the case of two communities and try to solve the collective memory formation game.

10.3.3 Matrix Version of the Weighted Average Algorithm with Two Opinions

To understand why the averaging Algorithm 1 solves the collective memory formation game with a high probability, we repeat the process from the previous subsection, with the utility function we've defined in Eq. 10.4. In this case, the equations depend on communities $\mathscr{B}, \mathscr{R}$. For the sake of simplicity, we will only apply this to the blue community, i.e., $i \in \mathscr{B}$. The same process will work for the red community by swapping the blue $\mathscr{B}$ and red $\mathscr{R}$ sets. Later, we will write the final conclusion for the red community as well (Fig. 10.3).

Assume node $i \in \mathscr{B}$ belongs to the blue community. The best action for each player i is determined by the Eq. 10.9. Player i looks at the opinions of its neighbors, plugs them into its utility function π_i, takes the derivatives according to value out_i, compares to 0, and computes new values out_i by solving the equation. This process is described as follows:

$$\begin{aligned} \frac{\partial \pi_i(out)}{\partial out_i} &= - \sum_{j_1 \in N(i) \cap \mathscr{B}} -2(out_{j_1} - out_i) \\ &+ \sum_{j_2 \in N(i) \cap \mathscr{R}} 2(out_{j_2} + out_i) \\ &= 0 \end{aligned} \tag{10.27}$$

Fig. 10.3 A figure on a narrative

We repeat the same algebraic process from the previous subsection and isolate variable out_i in the equation received when taking the derivative of the utility function $\pi_i(out)$ according to the variable out_i. It follows that

$$|N(i)|out_i = \sum_{j \in N(i) \cap \mathscr{B}} out_j - \sum_{j \in N(i) \cap \mathscr{R}} out_j \tag{10.28}$$

Since the communication graph G is connected $|N(i)| > 0$, dividing by $|N(i)|$ reveals that

$$out_i = \frac{\sum_{j \in N(i) \cap \mathscr{B}} out_j - \sum_{j \in N(i) \cap \mathscr{R}} out_j}{|N(i)|} \tag{10.29}$$

Now we can write a similar Eq. (10.29) for the nodes that belong to the red community, assuming that $i \in \mathscr{R}$.

$$out_i = \frac{\sum_{j \in N(i) \cap \mathscr{R}} out_j - \sum_{j \in N(i) \cap \mathscr{B}} out_j}{|N(i)|} \tag{10.30}$$

We leave developing the Eq. (10.30) as an exercise for the reader, see Exercise 9.

To write the system of equations, we use the matrix form

$$M' \cdot out = out \tag{10.31}$$

The key point to understanding why the averaging algorithm solves the memory formation game is understanding the relationship between matrix $M = [m_{i,j}]$ and matrix $M' = [m'_{i,j}]$. We can now write the relationship between the elements of matrix M and matrix M'.

The relationship between M and M' is simple:

$$m'_{i,j} = \begin{cases} m_{i,j} & i, j \in \mathcal{B} \\ m_{i,j} & i, j \in \mathcal{R} \\ -m_{i,j} & i \in \mathcal{B} \text{ and, } j \in \mathcal{R} \\ -m_{i,j} & j \in \mathcal{B} \text{ and, } i \in \mathcal{R} \end{cases} \tag{10.32}$$

Both matrices have the same eigenvalue, and a simple relationship exists between the eigenvectors of M and M'. To write the relationship, we need a few notations. First, to describe the relationship between the eigenvectors of matrix M and those of matrix M', we need the indicator vector of the blue community. $I_b = (b_1, ..., b_N) \in \mathbb{N}^N$,

$$b_i = \begin{cases} 1 & i \in \mathcal{B} \\ 0 & i \in \mathcal{R} \end{cases} \tag{10.33}$$

A similar definition for the red indicator vector is $I_r = (r_1, ..., r_N) \in \mathbb{N}^N$,

$$r_i = \begin{cases} 0 & i \in \mathcal{B} \\ 1 & i \in \mathcal{R} \end{cases} \tag{10.34}$$

Now we can write the eigenvector and the eigenvalues of $M' = [m'_{i,j}]$.

Theorem 10.3 *The eigenvalues of* M' *are* $1 = \lambda_1, \lambda_2, ..., \lambda_N$ *and the corresponding i eigenvalues are*

$$v'_{i,j} = (b_j - r_j)v_{i,j} \tag{10.35}$$

Our proof follows the simple calculation. We need to prove that

Proof

$$[m'_{i,j}] \cdot v'_i = \lambda_i v'_i \tag{10.36}$$

All we have to do is confirm this is true, in other words, that for all $i, j \in \mathcal{N}$,

$$(\overrightarrow{[m'_{i,j}] \cdot v'_i})_j = \lambda_i \cdot v'_{i,j} \tag{10.37}$$

This follows a simple calculation, assuming $i \in \mathcal{B}$

$$\begin{aligned}
\overrightarrow{([m'_{i,j}] \cdot v'_i)}_k &= \sum_{l=1}^{N} m'_{k,l} v'_{i,l} \\
&= \sum_{l \in \mathcal{N} \cap \mathcal{B}} m'_{k,l} v'_{i,l} + \sum_{l \in \mathcal{N} \cap \mathcal{R}} m'_{k,l} v'_{i,l} \\
&= \sum_{l \in \mathcal{N} \cap \mathcal{B}} m_{k,l} v_{i,l} + \sum_{l \in \mathcal{N} \cap \mathcal{R}} (-m_{k,l})(-v_{i,l}) \\
&= \sum_{l \in \mathcal{N} \cap \mathcal{B}} m_{k,l} v_{i,l} + \sum_{l \in \mathcal{N} \cap \mathcal{R}} m_{k,l} v_{i,l} \\
&= \lambda_i v_{i,l} = \lambda_i v'_{i,k}
\end{aligned} \tag{10.38}$$

In the case where $i \in \mathcal{R}$ the proof is

$$\begin{aligned}
\overrightarrow{([m'_{i,j}] \cdot v'_i)}_k &= \sum_{l=1}^{N} m'_{k,l} v'_{i,l} \\
&= \sum_{l \in \mathcal{N} \cap \mathcal{B}} m'_{k,l} v'_{i,l} + \sum_{l \in \mathcal{N} \cap \mathcal{R}} m'_{k,l} v'_{i,l} \\
&= \sum_{l \in \mathcal{N} \cap \mathcal{B}} - m_{k,l} v_{i,l} + \sum_{l \in \mathcal{N} \cap \mathcal{R}} m_{k,l} (-v_{i,l}) \\
&= -\sum_{l \in \mathcal{N} \cap \mathcal{B}} m_{k,l} v_{i,l} - \sum_{l \in \mathcal{N} \cap \mathcal{R}} m_{k,l} v_{i,l} \\
&= -\lambda_i v_{i,k} = \lambda_i v'_{i,k}
\end{aligned} \tag{10.39}$$

Now that we've established the relationship between matrix M and matrix M', we can use the same analysis as we did for solving the consensus using the averaging algorithm to show that the averaging algorithm, presented in Sect. 10.3.1, converged with a high probability to the Nash equilibrium of the collective memory formation game.

Theorem 10.4 *For any connected non-bipartite graph $G(V, E)$, and any non-trivial partition of the vertex in two non-empty two sets $\mathcal{B}, \mathcal{R} \neq \emptyset$, i.e., $V = \mathcal{B} \cup \mathcal{R}$, the average algorithm solves the collective memory formation game problem at finite time with a high probability of the input.*

Up to this point, we assumed that each node knows the color (or opinion) of each of its neighbors. In this case, we saw that the average algorithm could partition the network according to its color. In the next section, we will discuss the case where one of the nodes mistakenly characterizes one of its neighbors using the wrong color. In fact, the result presented in the next section also holds if multiple nodes wrongly characterize multiple neighbors. Fir the sake of simplicity, we will refer to a single node.

10.3.4 Reaching a Consensus

10.4 Examples

We start with the Consensus Problem, described in [8]. The algorithm can be found in 1. In this case, the social network contains three nodes, each connected to the others, see Fig. 10.4. In this subsection, society's goal is to reach full agreement among all members.

The vector of the nodes' opinion in the graph above is:

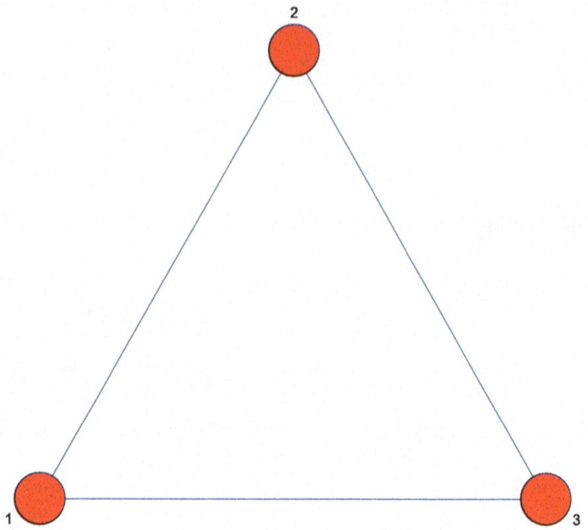

Fig. 10.4 A simple social network where nodes 1, 2, and 3 belong to the red party

$$\vec{opn} = (x, y, z) \tag{10.40}$$

In this case, the average process can be summarized by the matrix

$$M = \begin{bmatrix} 0 & \frac{1}{2} & \frac{1}{2} \\ \frac{1}{2} & 0 & \frac{1}{2} \\ \frac{1}{2} & \frac{1}{2} & 0 \end{bmatrix} \tag{10.41}$$

Note that in this case, each node ignores its own opinion, and the averaging process is only concerned with the neighbors' opinion. To analyze the general case, see Exercise 10

The eigenvalues of M are

$$\lambda_1 = 1, \lambda_2 = -1/2, \lambda_3 = -1/2 \tag{10.42}$$

and the corresponding eigenvectors of M are

$$\vec{v_1} = (1, 1, 1), \vec{v_2} = (-1, 0, 1), \vec{v_3} = (-1, 1, 0), \tag{10.43}$$

A simple calculation shows that

$$\lim_{t \to \infty} M^t = \begin{bmatrix} \frac{1}{3} & \frac{1}{3} & \frac{1}{3} \\ \frac{1}{3} & \frac{1}{3} & \frac{1}{3} \\ \frac{1}{3} & \frac{1}{3} & \frac{1}{3} \end{bmatrix} \tag{10.44}$$

Therefore, if we multiply the matrix in (10.49) by the vector (x, y, z), it follows that

$$\begin{bmatrix} \frac{1}{3} & \frac{1}{3} & \frac{1}{3} \\ \frac{1}{3} & \frac{1}{3} & \frac{1}{3} \\ \frac{1}{3} & \frac{1}{3} & \frac{1}{3} \end{bmatrix} \cdot \begin{bmatrix} x \\ y \\ z \end{bmatrix} = \begin{bmatrix} \frac{1}{3}(x+y+z) \\ \frac{1}{3}(x+y+z) \\ \frac{1}{3}(x+y+z) \end{bmatrix} \tag{10.45}$$

Subsequently, unless $(x + y + z) \neq 0$, all three nodes will agree on the value $\text{sign}(\frac{1}{3}(x + y + z))$. Note that if the first opinion vector is $\vec{v_2} = (-1, 0, 1)$ or $\vec{v_3} = (-1, 1, 0)$, the system will converge to the zero vector. In the next section, we will explore the case of two parties trying to agree to disagree.

10.4.1 *Reaching a Consensus on Disagreement with Complete Information*

We continue with the example of our Algorithm 1. This time, we assume that we have two sets, blue and red. The blue contains nodes 1 and 2, and the red contains node 3, see Fig. 10.5.

In this case, following the description in Eq. (10.32), the average process can be summarized by the matrix

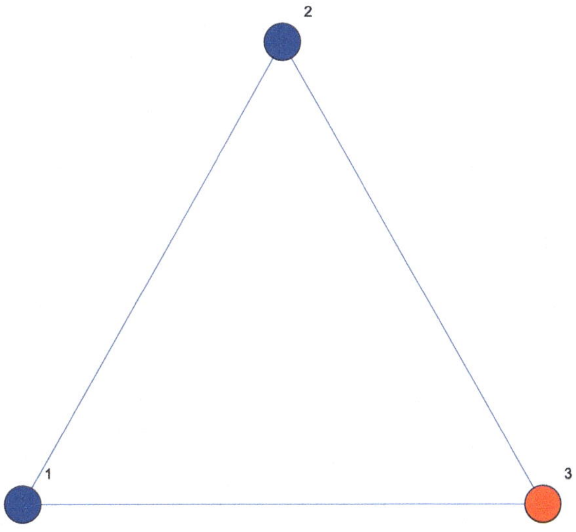

Fig. 10.5 A simple social network where nodes 1 and 2 belong to the blue party, and node 3 belongs to the red party

$$M_1 = \begin{bmatrix} 0 & \frac{1}{2} & \frac{-1}{2} \\ \frac{1}{2} & 0 & \frac{-1}{2} \\ \frac{-1}{2} & \frac{-1}{2} & 0 \end{bmatrix} \tag{10.46}$$

Therefore the eigenvalues of M_1 are

$$\lambda_1 = 1, \lambda_2 = -1/2, \lambda_3 = -1/2 \tag{10.47}$$

and the corresponding eigenvectors are

$$\vec{v_1} = (1, 1, -1), \vec{v_2} = (-1, 0, -1), \vec{v_3} = (-1, 1, 0), \tag{10.48}$$

A simple calculation shows that

$$\lim_{t \to \infty} M_1^t = \begin{bmatrix} \frac{1}{3} & \frac{1}{3} & \frac{-1}{3} \\ \frac{1}{3} & \frac{1}{3} & \frac{-1}{3} \\ \frac{-1}{3} & \frac{-1}{3} & \frac{1}{3} \end{bmatrix} \tag{10.49}$$

If we multiply the matrix in (10.49) by the vector (x, y, z), it follows that

$$M_2 = \begin{bmatrix} \frac{1}{3} & \frac{1}{3} & \frac{-1}{3} \\ \frac{1}{3} & \frac{1}{3} & \frac{-1}{3} \\ \frac{-1}{3} & \frac{-1}{3} & \frac{1}{3} \end{bmatrix} \cdot \begin{bmatrix} x \\ y \\ z \end{bmatrix} = \begin{bmatrix} \frac{1}{3}(x + y - z) \\ \frac{1}{3}(x + y - z) \\ \frac{1}{3}(-x - y + z) \end{bmatrix} \tag{10.50}$$

Therefore, if $x + y - z \neq 0$, they reach a consensus on disagreement. Nodes 1 and 2 will agree $\text{sign}(\frac{1}{3}(x + y - z))$ and node number 3 will disagree. Therefore, its output will be equal to the output of nodes 1 and 2, multiplied by -1: $\text{sign}(\frac{1}{3}(-x - y + z))$.

10.5 Conclusion

In this chapter, we presented a collective memory formation game scheme. We then presented a simple version of the formation game scheme.

This game described a simple example of merged memory formation, how collective memory is formed, and what is required of society to form it. In simple cases, the blue and red communities can achieve the consensus to disagree if all nodes know the exact color of their neighbors by a simple averaging process. Moreover, this averaging process converges to the Nash equilibrium of the game.

If any community includes a fifth column, the simple averaging process converges to the origin and, therefore, fails to generate collective memory.

To overcome this problem, we need some instrument to amplify opinions. In history, such forms of amplifiers are usually called publicity or propaganda, which comes in the form of books, TV, newspapers, etc. The conclusion of this chapter suggests that in order to generate collective memory, tools of publicity or propaganda are needed; after all, in real life we never know the precise opinion of our neighbors. Sometimes, neither do they.

10.6 Exercise

1. We know that Hannibal Barca and Scipio Africanus died in the same year, 183 BCE. In this problem, we want to apply game theory to determine who died first.
2. Explain the difficulty in applying the game theory approach to historical events. Try to discuss the conflict between the desire to study what really happened and the wish to study what could happen. Use these arguments to conclude that counterfactual reason is useful over a short period of time when studying history.
3. Explain why counterfactual reasoning is good for studying short duration local events such as battles, but not for long-term processes.
4. Assume that we are in the distributed consensus setting with two non-failing nodes. Propose an algorithm that solves the consensus problem and explain why adding failures prevents the node from reaching the consensus.
5. In this exercise, you are asked to prove that if the set of strategies for player i is the interval

$$S_i = [-1, 1] \tag{10.51}$$

and the social graph is connected, then the set of all Nash equilibria has the formula

$$Nash = \{\pi \in [-1, 1]^N : \pi = t(b_1, ..., b_N) - t(r_1, ..., r_N), -1 \leq t \leq 1\} \tag{10.52}$$

For the definition of r_i, see Eq. 10.34, and for the definition of d_i, see (10.33). Hint: you may use the idea developed in Sect. 10.3.2. Specifically, consider any Nash equilibrium ν that does not belong to the set $Nash$ in Eq. 10.52. Clearly, each node's best response is defined by Eqs. 10.29 and 10.30. Now, according to our assumption for Nash equilibrium $\nu \notin Nash$, let $\nu_m = \min(\nu)$ and $\nu_M = \max(\nu)$. Denote the set of all players whose strategy is equal to ν_m by I_m, and all the players whose strategy is ν_M by I_M. Show that at least one player in the set I_m can gain from deviating from strategy ν, and the same is true for at least one player in the set I_M.
6. Consider the collective memory formation game described in Sect. 10.2.8, where each player has two strategies, $S_i = \{-1, 1\}$, and the social network is a Barbell graph (see Fig. 10.6 for the Barbell graph definition). Consider each click as a party. Show that when the click size in the barbell is greater than 3, the final opinion of all nodes equals 1 for all $i \in \mathcal{N}$ so that

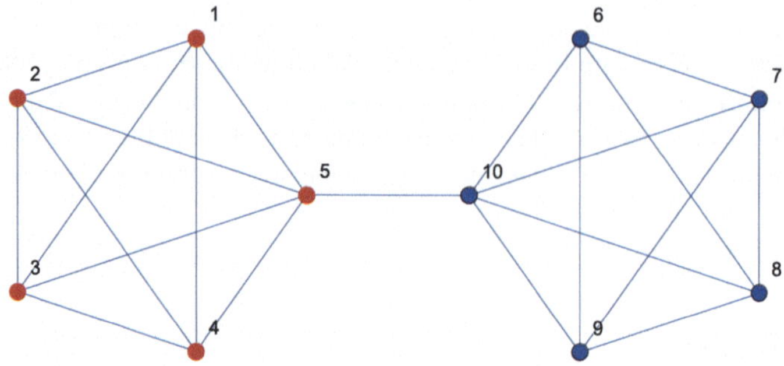

Fig. 10.6 A simple barbell graph social network where nodes 1 to 5 belong to the blue party and nodes 6 to 10 belong to the red party

$$\pi_i(out) = 1 \tag{10.53}$$

is a Nash equilibrium. This exercise shows the existence of a Nash equilibrium that does not separate the blue party from the red if the set of strategies contains only two strategies.

7. In this exercise, we provide an example of how to analyze two consecutive co-dependent historical events. Let's assume that we have two nodes $V = \{v_1, v_2\}$, i.e., $\mathcal{N} = \{1, 2\}$. The social network that describes society is one undirected edge $E = (v_1, v_2)$

$$\gamma = (G(V, E), G(V, E)) . \tag{10.54}$$

Let the historical event h_1 be the blue party winning of the Presidential elections. The second event is the blue party winning the elections for Congress.
Now, let's assume that the first player v_1 that belongs to the blue party thinks +1 about both events:

$$input_1 = (1, 1) \tag{10.55}$$

while the second player belongs to the red party and thinks that it is very bad that the blue party won the Presidency, but rather good that it won the elections for Congress. Formally:

$$input_2 = (-1, 1) \tag{10.56}$$

and two historical events

$$\mathcal{H} = (h_1, h_2) \tag{10.57}$$

Now assume that the first node v_1 belongs to the red party, while the second node v_2 belongs to the blue party.
Since the events are dependent, we wish to separate the red party from the blue and predict the collective memory that will form these two events together. To do so, we construct the social fabric as containing four nodes, each of which is double: one for the first event and one for the second. We connect the nodes that

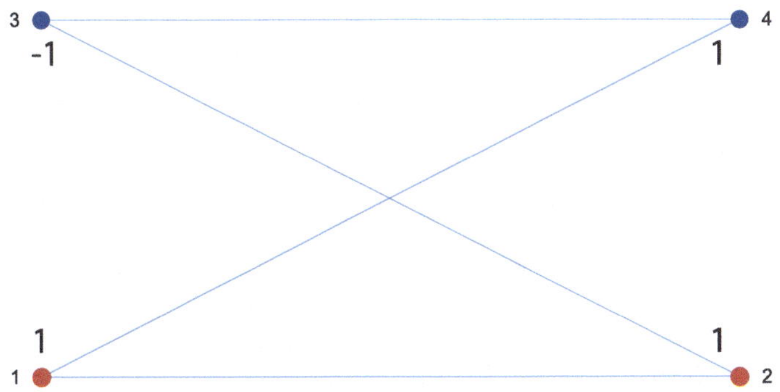

Fig. 10.7 A simple cycle graph social network where nodes 1 and 2 represent Player 1, who belongs to the Blue Party, and nodes 3 and 4 represent Player 2, who belongs to the Red Party

describe the same person by the edge, so the graph that describes the collective memory formation game for those two historical events is a cycle containing four edges and four nodes. Each party contains two nodes that are connected by one edge for each, see Fig. 10.7 Compute the Nash equilibrium when both players use the average protocol, see Sect. 10.3.1.

8. In this exercise, you are asked to construct a collective memory formation game and an opinion vector input such that Algorithm 2 fails to separate the red node from the blue. Hint: pick a vector opinion input orthogonal to the separation vector.
9. Use Eq. 10.27 develop Eq. 10.30.
10. In Sect. 10.4.1, we analyzed a social network with three nodes and two opinions, where the nodes ignored their own opinions. In this exercise, you are asked to repeat the same analysis, but now we assume that each node considers its own opinion with probability α. You may assume that the matrix is

$$M = \begin{bmatrix} \alpha & \frac{1-\alpha}{2} & \frac{1-\alpha}{2} \\ \frac{1-\alpha}{2} & \alpha & \frac{1-\alpha}{2} \\ \frac{1-\alpha}{2} & \frac{1-\alpha}{2} & \alpha \end{bmatrix} \tag{10.58}$$

11. In Sect. 10.4.1 we analyzed a social network with three nodes and two opinions. In this exercise, you are asked to repeat the same analysis but with a social network of five nodes and two opinions, see Fig. 10.8.

Fig. 10.8 A simple social network where nodes 1, 2, and 3 belong to the blue party, and nodes 4 and 5 belong to the red party

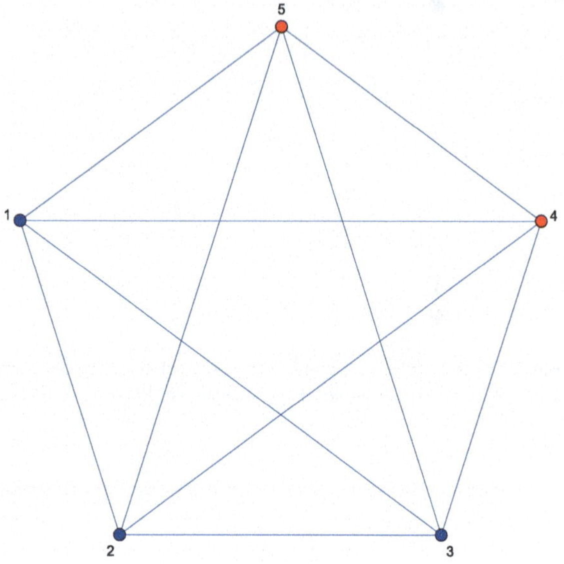

References

1. D. Peleg, Local majorities, coalitions and monopolies in graphs: a review. Theor. Comput. Sci. **282**(2), 231–257 (2002)
2. D. Peleg, Size bounds for dynamic monopolies. Discret. Appl. Math. **86**(2–3), 263–273 (1998)
3. E. Ackerman, O. Ben-Zwi, G. Wolfovitz, Combinatorial model and bounds for target set selection. Theor. Comput. Sci. **411**(44–46), 4017–4022 (2010)
4. F. Cicalese, G. Cordasco, L. Gargano, M. Milanič, U. Vaccaro, Latency-bounded target set selection in social networks. Theor. Comput. Sci. **535**, 1–15 (2014)
5. O. Favaron, G. Fricke, W. Goddard, S.M. Hedetniemi, S.T. Hedetniemi, P. Kristiansen, R.C. Laskar, D. Skaggs, Offensive alliances in graphs, in *International Symposium on Computer and Information Sciences* (CRC Press, 2022), pp. 298–302
6. D. Reichman, New bounds for contagious sets. Discret. Math. **312**(10), 1812–1814 (2012)
7. J.R. Norris, Markov Chains, Cambridge series in statistical and probabilistic mathematics. Cambridge University Press, Cambridge. Reprint of, 1997
8. M.H. DeGroot, Reaching a consensus. J. Am. Stat. Assoc. **69**(345), 118–121 (1974)

Chapter 11
Propaganda in History

In this chapter, we try to explain why propaganda is necessary for achieving consensus from a historical perspective (Fig. 11.1).

In the previous chapter, we assumed that we have a vector of opinions $\vec{out}$ with respect to some historical events. We can say that public opinion is amplified by a factor $r \in \mathbb{R}$ by some propaganda if public opinion changes from $\vec{out}$ to $r\vec{out}$. If $r > 1$, we can say that propaganda amplifies the opinion, and if $r < 1$, we can say that propaganda smooths public opinion.

The inverse of propaganda is censorship. In the paper [1] Avin et al. showed that it is impossible to construct fair censorship. Therefore, we are forced to study the influence of propaganda.

In the previous chapter, we defined the collective memory formation game. We showed that a simple averaging dynamic finds the Nash equilibrium that allows different communities to agree to disagree with each other, and agree within the community. We showed that if society knows in advance how society is divided into different communities, it is possible to compute the Nash equilibrium by simply averaging the opinions with regard to the communities.

Now, we study the case where the community structure has some small errors. In this case, we will show that the simple averaging algorithm converges to the zero vector, which means that the historical event will disappear from the collective memory and cannot separate the different communities. We then suggest a way to fix the average algorithm to achieve conversion in some cases using propaganda. Our model suggests that conversion in this case can depend on the starting opinion of the process, and the conflict can be between the different communities or within one of them.

Z. Lotker, *History by Algorithms*,
https://doi.org/10.1007/978-3-031-93627-2_11

Fig. 11.1 A cartoon illustrating the concept of the fifth column

11.1 The Price of Wrong Characterization

In the previous section, we assumed that we are given a partition of the nodes into two disjointed sets: blue $\mathscr{B}$ and red $\mathscr{R}$. In this section, we assume that at least three nodes $i, j, k \in \mathscr{N}$ exist, such that nodes i, j disagree with the color of node k, see Fig. 11.2, where nodes 1 and 2 disagree on the color of node 3. We call such a node the fifth column node.

If a fifth column node exists, the averaging algorithm converges to the consensus of the vector zero for all nodes, and therefore fails to reach a consensus to disagree. We can write the averaging algorithm as a matrix multiplication.

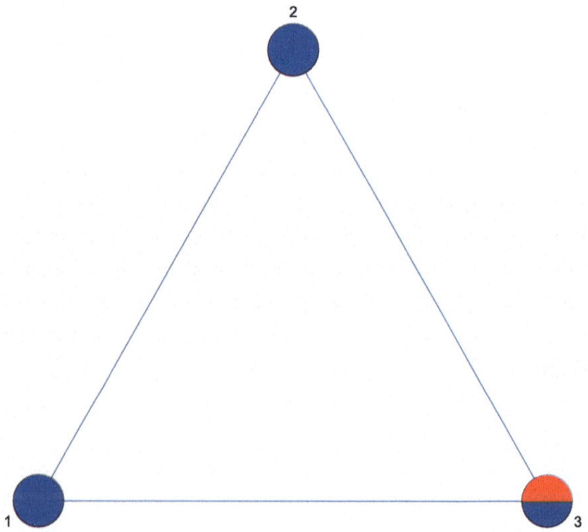

Fig. 11.2 A simple social network where nodes 1 and 2 belong to the blue party, and node 3 to the red party. However, node 3 is considered to be blue by node 1

To simplify the math, we will demonstrate the phenomenon using only 3 nodes. However, keep in mind that it is possible to prove that if the fifth column force exists, the averaging algorithm converges to zero vector and fails to separate the two communities. To avoid unnecessary complications and mathematical discussion, we will skip the proof.

In the next few subsections, we will analyze the simple social network describing Fig. 11.2. We start by analyzing a simple example where the fifth column does not exist.

11.2 Failing to Reach a Consensus on Disagreement

In this section, we will continue our example from Sect. 10.4 where we modeled small network with 3 nodes and two parties. The social communication network is the triangle, see Fig. 10.5.

Let's explore what happens when one of the nodes misidentifies the color of one of its neighbors. For example, node 1 misidentifies node 3 as a member of a blue party, while nodes 1 and 3 correctly identify the colors of all the nodes, see Fig. 11.2.

In this case, the matrix M_3 is

$$M_3 = \begin{bmatrix} 0 & \frac{1}{2} & \frac{1}{2} \\ \frac{1}{2} & 0 & -\frac{1}{2} \\ -\frac{1}{2} & -\frac{1}{2} & 0 \end{bmatrix} \tag{11.1}$$

Note that according to matrix M_3, node 1 wrongly characterizes node 3 as a member of the blue party. In this case, the eigenvalues of M_3 are

$$\lambda_1 = 1/2, \lambda_2 = -1/2, \lambda_3 = 0 \tag{11.2}$$

and the corresponding eigenvectors are

$$\vec{v_1} = (1, 0, -1), \vec{v_2} = (1, -1, 0), \vec{v_3} = (-1, 1, 1), \tag{11.3}$$

A simple calculation shows that

$$\lim_{t \to \infty} M_1^t = \begin{bmatrix} 0\,0\,0 \\ 0\,0\,0 \\ 0\,0\,0 \end{bmatrix} \tag{11.4}$$

When there is a fifth column, the averaging algorithm fails to converge the agreement to disagree because the system fails to preserve the probability mass. In the next section, we explain how to overcome this problem using propaganda.

11.3 Publicity and Propaganda

As we show in Sect. 11.1, if not all nodes agree on the partition, the averaging algorithm converges to the origin and therefore fails to separate the blue entities from the red. The main reason is that the greatest eigenvalues in the absolute value of the linear system are less than 1. Therefore, the probability should vanish over time. One way to overcome this problem is to force the sum of probabilities to be 1 by dividing each probability by the total sum. Note that as long as each column's total sum is not zero, we can force the sum in each row to be 1. To overcome the problem of dividing by zero, one can perturbate the sum of the probabilities by a little bit if it is zero. Therefore, we ignore this case.

The next algorithm gets the vector of outcome *out* and divides each row by the total sum. Therefore, it normalizes the vector *out* to be a probability vector. This algorithm can become easily distributed by collecting all data from society.

Overcoming the problem of converging to 0 of the vector *out* is to ensure that it will be far away from 0. As long as we repeatedly multiply the vector *out* by some constant, which overcomes the mass loss of the system but does not push the total value of *out* to infinite, we should be fine.

Another problem that cannot be solved by inserting mass into the system is that the system can become periodical; this happens when the largest eigenvalue is not a real number but a complex number. To overcome this problem, we suggest adding a self-loop with some probabilities, see Eq. 11.5.

If we look closely at the eigenvalues of the matrix M_3, the largest absolute eigenvalue is $\lambda_1 = 1/2, \lambda_1 = -1/2$. Note that we have two eigenvalues with an absolute value $1/2$. This means that the stochastic system is not ergodic. To overcome this problem, we add a self-loop with probability α, and so matrix M_4 becomes

$$M_4 = \begin{bmatrix} \alpha & \frac{1-\alpha}{2} & \frac{1-\alpha}{2} \\ \frac{1-\alpha}{2} & \alpha & -\frac{1-\alpha}{2} \\ -\frac{1-\alpha}{2} & -\frac{1-\alpha}{2} & \alpha \end{bmatrix} \tag{11.5}$$

The eigenvalues of matrix M_4 are

$$\lambda_1 = \frac{1+\alpha}{2} \geq \lambda_2 = \alpha \geq \lambda_3 = \frac{1}{2}(3\alpha - 1). \tag{11.6}$$

with the corresponding eigenvectors

$$\vec{V}_1 = (0, -1, 1), \vec{V}_2 = (1, -1, 1), \vec{V}_3 = (-1, 1, 0) \tag{11.7}$$

So, dividing matrix $M4$ by the eigenvalue λ_1, we get $M5 = \frac{M4}{\lambda_1}$, i.e.,

$$M_5 = \begin{bmatrix} \frac{2\alpha}{\alpha+1} & \frac{1-\alpha}{\alpha+1} & \frac{1-\alpha}{\alpha+1} \\ \frac{1-\alpha}{\alpha+1} & \frac{2\alpha}{\alpha+1} & \frac{\alpha-1}{\alpha+1} \\ \frac{\alpha-1}{\alpha+1} & \frac{\alpha-1}{\alpha+1} & \frac{2\alpha}{\alpha+1} \end{bmatrix} \tag{11.8}$$

Now, the eigenvalues of M_5 are

$$\lambda_1 = 1 \geq \lambda_2 = \frac{2\alpha}{\alpha + 1} \geq \lambda_3 = \frac{3\alpha - 1}{\alpha + 1} \tag{11.9}$$

The eigenvectors in matrix M_5 are equal to the eigenvectors in matrix M_4, since multiplying a matrix by a constant greater than zero does not change the structure of eigenvectors. And so:

$$M_6 = \lim_{t \to \infty} M_5^t = \begin{bmatrix} 0 & 0 & 0 \\ 1 & 1 & 0 \\ -1 & 1 & 0 \end{bmatrix} \tag{11.10}$$

Therefore, the averaging process after incorporating propaganda and taking the limit is described M_6 and will converge to

$$M_6 \cdot (x, y, z) = (0, x + y, -x - y) \tag{11.11}$$

where the vector (x, y, z) is a starting opinion. We can interpret the result as follows: Node 1 will claim that the historical event is not important since the node's opinion at the end of the averaging process is zero. The opinion of node 2 at the end of the averaging process is $x + y$, i.e., it adds up all of the blue party's opinions at the beginning. Node 3, which belongs to the red party, ignores its own opinion and adapts the inverse opinion of the blue party.

In general, it is possible to prove that dividing the averaging process by the greatest eigenvalues always forces the averaging process to converge and successfully separate some parts of society into two groups.

11.4 One-Sided Propaganda

In this section, we will consider the fact that propaganda acts on a part of society, rather than its entirety. We will continue working with our example (see) Fig. 11.2), but now, instead of multiplying all the matrices by the propaganda factor, we will multiply the set of rows in the matrix by the propaganda factor r. This allows us to model a society where only one part is generating propaganda. In this case, if we consider matrix M_4 from Eq. 11.5, we can have several different forms of propaganda. We can multiply the first, the second, or the third row by a factor of r.

11.4.1 Propaganda of the Majority

In this case, we multiply the first and second rows by the factor r.

$$M_7 = \begin{bmatrix} \alpha r & \frac{1}{2}(1-\alpha)r & \frac{1}{2}(1-\alpha)r \\ \frac{1}{2}(1-\alpha)r & \alpha r & \frac{1}{2}(\alpha-1)r \\ \frac{\alpha-1}{2} & \frac{\alpha-1}{2} & \alpha \end{bmatrix} \tag{11.12}$$

Again, in order to get the absolute size of the greatest eigenvalue, r needs to be

$$r = \frac{2}{\alpha+1} \tag{11.13}$$

Plug this in matrix M_7 to get

$$M_8 = \begin{pmatrix} \frac{2\alpha}{\alpha+1} & \frac{1-\alpha}{\alpha+1} & \frac{1-\alpha}{\alpha+1} \\ \frac{1-\alpha}{\alpha+1} & \frac{2\alpha}{\alpha+1} & \frac{\alpha-1}{\alpha+1} \\ \frac{\alpha-1}{2} & \frac{\alpha-1}{2} & \alpha \end{pmatrix} \tag{11.14}$$

Now, the eigenvalues of M_8 are

$$\lambda_1 = 1 \geq \lambda_2 = \frac{3\alpha-1}{\alpha+1} \geq \lambda_3 = \frac{\alpha^2+\alpha}{\alpha+1} \tag{11.15}$$

and the eigenvector of the first eigenvalue is

$$\vec{v}_1 = (-(1/2), -(3/2), 1) \tag{11.16}$$

In this case, the averaging process converges to

$$M_9 = \lim_{t\to\infty} M_8^t = \begin{bmatrix} \frac{1}{4} & \frac{1}{4} & 0 \\ \frac{3}{4} & \frac{3}{4} & 0 \\ -\frac{1}{2} & -\frac{1}{2} & 0 \end{bmatrix} \tag{11.17}$$

And so, the averaging process converges to

$$M_9 \cdot (x, y, z) = ((x+y)/4, (3(x+y))/4, 1/2(-x-y)) \tag{11.18}$$

In this case, both parties agree to disagree.

11.4.2 Propaganda of the Minority

In this case, we multiply the third row by the factor r. The averaging process then becomes

$$M_{10} = \begin{bmatrix} \alpha & \frac{1-\alpha}{2} & \frac{1-\alpha}{2} \\ \frac{1-\alpha}{2} & \alpha & \frac{\alpha-1}{2} \\ \frac{1}{2}(\alpha-1)r & \frac{1}{2}(\alpha-1)r & \alpha r \end{bmatrix} \tag{11.19}$$

Setting r to be $r = \frac{1}{\alpha}$, we get that the averaging process matrix is

$$M_{11} = \begin{bmatrix} \alpha & \frac{1-\alpha}{2} & \frac{1-\alpha}{2} \\ \frac{1-\alpha}{2} & \alpha & \frac{\alpha-1}{2} \\ \frac{\alpha-1}{2\alpha} & \frac{\alpha-1}{2\alpha} & 1 \end{bmatrix} \tag{11.20}$$

The eigenvalues of M_{11} are

$$\lambda_1 = 1 \geq \lambda_2 = \frac{\alpha+1}{2} \geq \lambda_3 =, \frac{1}{2}(3\alpha - 1) \tag{11.21}$$

The eigenvector of the first and greatest eigenvalue is

$$\vec{v}_1 = (\frac{1}{3}, -\frac{1}{3}, 1) \tag{11.22}$$

which means that the average processing converges to M_{12}

$$M_{12} = \lim_{t\to\infty} M_{11}^t = \begin{bmatrix} -\frac{1}{3\alpha} & -\frac{1}{3\alpha} & \frac{1}{3} \\ \frac{1}{3\alpha} & \frac{1}{3\alpha} & -\frac{1}{3} \\ -\frac{1}{\alpha} & -\frac{1}{\alpha} & 1 \end{bmatrix} \tag{11.23}$$

Note that in this case matrix M_{12} depends on α, which means that the majority is in some way fighting the minority.

Therefore, the outcome of the averaging process is

$$M_{12} \cdot (x, y, z) = (-\frac{x+y-\alpha z}{3\alpha}, \frac{x+y-\alpha z}{3\alpha}, -\frac{x+y-\alpha z}{\alpha}) \tag{11.24}$$

which depends on the value of $-\frac{x+y-\alpha z}{\alpha}$. Note that if $-\frac{x+y-\alpha z}{\alpha} \neq 0$, then node 1 and node 3 have the same sign, which means that they agree, and node 2 becomes the minority. The result is the generation of a new social partition.

In the next section, we will use this theory to explain the outcome of several historical events.

11.5 Historical Memory Formation Game

This section will examine some historical events as an example of the collective formation memory game.

The Historical Memory Formation Game explores how collective memory is shaped by political events and social narratives. A group of people protesting against the central government in Israel in 2024 serves as a powerful example of how public dissent becomes embedded in national memory through media coverage, public discourse, and documentation. Figure 11.3 shows this group of demonstrators, capturing a moment that contributes to the construction of historical memory.

The amount of control the government has on mass media is a measure of the political system's degree of freedom in the country. When the media does not report mass protests, it is an indication of an authoritarian regime, see Fig. 11.3.

11.5.1 The American Civil War

One of the most famous examples of polarization in history is the American civil war. The polarization process was greatly influenced by the geographical division between North and South. However, each region had those who supported or protested slavery. Therefore, each entity had a fifth column in it. This brings us to the obvious conclusion that each entity had to use propaganda to enhance its opinion. So the process started as an argument between people, progressed into cultural disagreement, Congressional disputes, and judicial restrictions, and when a consensus could not be reached, the civil war broke out, letting guns do the talking.

There were several important books that argued against slavery. Perhaps the first of these was Moby Dick, where the whale symbolizes slavery, the ship symbolizes the United States of America, and Captain Ahab symbolizes the President, who is obsessed with the idea of the whale and eventually sinks the ship. Of course, this is just one possible interpretation. In 1852, a year after Moby Dick, came a book that could not be interpreted in any other way—Uncle Tom's Cabin. It became a

Fig. 11.3 A group of people demonstrating against the central government in Israel in 2024

bestseller, provoked great public outcry, and caused fierce societal controversy. The people of the South displayed the strongest reaction to the novel, since the main plot narrative ran counter to the perception of slavery accepted in the society of the Southern states.

The typical reaction of Southerners to Beecher Stowe's novel was outrage, which soon took the form of a kind of "literary response": authors in the American South wrote a number of novels that asserted the usefulness and justification of slavery. In response to the novel, an entire genre of literature emerged, dubbed Anti-Tom. It represented a pro-slave owning point of view and relied on the opinion that the vision of slavery, described in Stowe's novel, is inflated and incorrect. Today, these novels are generally seen as pro-slavery propaganda. The literary genre Anti-Tom came to naught with the outbreak of the American Civil War.

But before guns became the only orator, both North and South made several other attempts to defend their points of view. In 1850, Congress became a new battlefield between the two sides when a set of laws was passed in an attempt to resolve several long-standing disputes between the free and slave states.

The compromise comprised five separate bills, each addressing a different issue related to slavery and territorial expansion. One of the key provisions of the compromise was the admission of California to the Union as a free state, which upset the balance between free and slave states in Congress. Another key provision of the compromise was the Fugitive Slave Act, which required Northern states to return escaped slaves to their owners in the South. The act also made it a crime for anyone to help an escaped slave, and many Northerners viewed it as an attack on their personal liberties. The Compromise of 1850 also abolished the slave trade in Washington, D.C., and established a territorial government in Utah, which left the decision to allow slavery at the hand of presidents. The Compromise of 1850 was an attempt to preserve the Union by addressing the growing tensions between free and slave states, but it ultimately failed to prevent the outbreak of the Civil War. Although the compromise temporarily eased tensions, it did not address the fundamental issues of slavery and territorial expansion that would continue to divide the country.

Judicial restrictions also played a role in the lead up to the Civil War. In 1857, the Supreme Court ruled in the case of Dred Scott v. Sandford that African Americans could not be considered citizens of the United States. This decision was seen as a significant blow to the anti-slavery movement and contributed to the growing divide between North and South.

Finally, John Brown's raid on Harpers Ferry in 1859 was a significant event that helped spark the Civil War. Brown was an abolitionist who attempted to start a slave rebellion by seizing weapons from a federal armory in Virginia. Although Brown's raid was unsuccessful, it heightened tensions between the North and the South and was seen as a direct attack on the institution of slavery. The growing disagreement between the North and the South can be shown as an up-streaming spiral, see Fig. 11.4.

Let us use our simple model to explain the evolution of the conflict of slavery in the United States. In the first 80 years, both sides, North and South, used their propaganda to support their opinion. Such a situation, where two parties use propaganda to support their opinion, was analyzed in Sect. 11.3. The outcome of such a dynamic

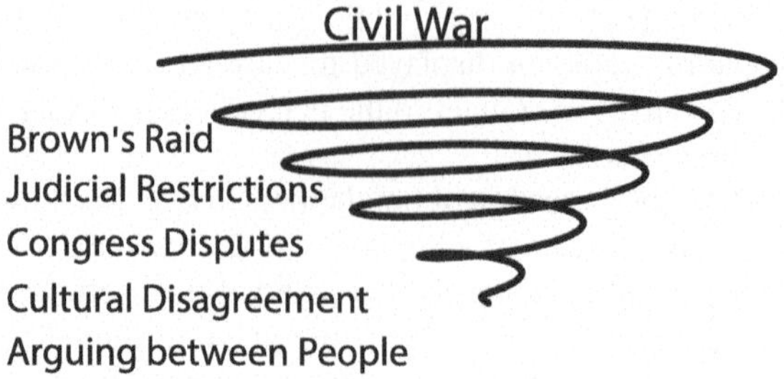

Fig. 11.4 Disagreement Spiral

is described in Eq. 11.11, which suggests that the fifth column ignores the conflict and their opinion approaches zero, while the rest of the blue and red parties agree to disagree. After the Civil War, the defeated South could not maintain its propaganda.

We now move to describing the propaganda of the majority, see Sect. 11.4.1. In this case, the outcome of the averaging process is given in Eq. 11.18, where the blue and red parties agree to disagree, and the fifth column accepts the majority's opinion. In this case, the outcome of the model can explain how, after 150 years, the Civil War still influences the politics of the United States, and the difference between North and South persists.

11.5.2 2022 Russian Invasion of Ukraine

The collective memory formation game described in this chapter models a situation where we have two groups and people are free to publicize their opinions without suffering any penalty. Moreover, the main result showed that when there is a fifth column, one must generate publicity or propaganda to separate the groups. However, sometimes people are afraid to publicize their opinions due to heavy penalties. One such example is the penalties imposed in Russia for public demonstrations against the invasion of Ukraine.

In our model, we allowed a node to interpret the opinions of the others by multiplying by -1. This means that a node can accept the opinion of another without real agreement. Due to penalties, people may agree with official opinions and show agreement in polls but individually have a different point of view. This is why evaluating the real collective consensus in the entity is hard.

Since the beginning of the Russian military invasion of Ukraine, Russian authorities and state centers studying public opinion have consistently claimed that the country's citizens almost unanimously support the "special military operation." However, independent experts doubt the level of public Russian support of the war. In this subsection, we will try to assess the extent to which Russian citizens support the invasion of Ukraine.

Official Reports

On February 20th, 2023, almost a year into the war, Russia's official center for sociological research, the All-Russian Public Opinion Research Center (VCIOM), published the results of its polls, which claimed that 68% of respondents (compared to 65% in February of the previous year) supported the decision to conduct the "operation," while only 20% opposed it. However, independent researchers' surveys show that such support does not actually exist. See [2] https://www.bbc.com/russian/news-64764949.

Independent Sociology

According to the independent Chronicles project's research, the core group of war supporters and the core group of war opponents are roughly equal and much smaller than VCIOM's data suggests: 22% (supporting) and 20.1% (not supporting) of all respondents, respectively. The material is available at: https://www.chronicles.report/. The core of the pro-war group (22%) includes those who:

1. Express support for the war (58.5%)
2. Believe that under conditions of budget deficit, state funds should be spent primarily on the military rather than the social sphere (47.2%)
3. Would not support the decision to withdraw troops from Ukrainian territory and start peace negotiations without achieving military objectives (37%)

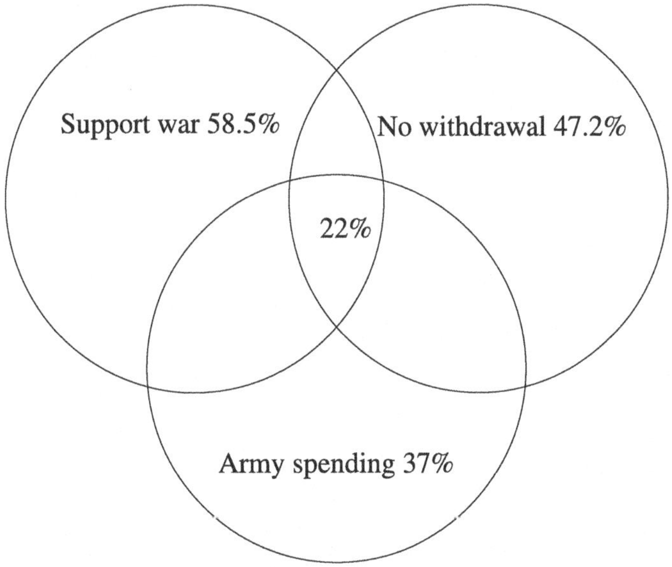

Private Donations

Another instrument to show true support of the war is by measuring private donations in Russia. In 2021, Russians donated 360–380 billion rubles to charity, according to a study by the analytics division of Sber Private Banking. The researchers found that private donations accounted for only 21% of all money directed toward charitable purposes, see [3].

Thus, it can be stated that private individuals donated 75.6–79.8 billion rubles in 2021. According to the same analysts' estimates, in 2022, private individuals' donations were 30–40% lower than in 2021. Thus, an approximate volume of private donations collected by Russians in 2022 can be calculated, and amounts to 45.4–55.8 billion rubles (Table 11.1).

One of the new directions for private donations in 2022 was fundraising in support of the "special military operation." We can consider these amounts to indicate direct support for the war.

In 2022, volunteers raised 4.4 billion rubles to purchase items for those on the front line, see [4]. Thus, in 2022, 7.8–9.7% of all funds raised were collected to support the military operation.

Russian Refugees

The attitude towards the war in Ukraine can be evaluated by analyzing the outflow of people from Russia that occurred after the invasion of Ukraine in 2022. This migration event is the largest wave of emigration from Russia since the dissolution of the USSR. It is driven by a complex set of factors, including political persecution, fear for personal safety in wartime, disagreement with the government's actions, and the anticipation of a significant decline in the economic and humanitarian conditions in the country [5].

Quantifying the exact number of people who have left Russia since the end of February 2022 is a challenging task that requires sophisticated methodology. Despite monitoring the number of departures from the country, it is difficult to draw conclusions from official data regarding the scale and destination of the emigration flow. However, independent research suggests that at least half a million Russians left the country in 2022. Demographer Igor Efremov estimates that 400–500 thousand people left Russia and did not return [6]. According to Forbes, the number is closer

Table 11.1 Donations in Russia

Source	2021	2022	Ratio 2022–2021 (%)
Corporate donations	284.4–300.2	196.6–210.2	60–70
Private donations	75.6–79.8	45.4–55.8	60–70
Total donations	360–380	242–266	60–70

to 700,000 people [7]. Even more important is the question of who exactly left. First we have the wealthiest people who had the financial means to leave Russia, and the finest professionals, who were confident in their ability to start a new life in another country. The two waves of migration, which included around 100,000 IT specialists, represent approximately 10% of the total workforce of Russian IT companies, according to Maksut Shadayev, head of the Ministry of Digital Development [8].

The emigration patterns of Russians following the invasion of Ukraine offer crucial insights into the deep divisions within Russian society regarding the conflict. Despite the difficulty in accurately quantifying the number of migrants and their reasons for leaving, the significant emigration wave indicates that the war has sparked intense debate and disagreement among the population. This divergence of opinion is likely fueled by a complex range of factors, including political ideology, personal safety concerns, economic considerations, and humanitarian fears. Analyzing these emigration patterns is essential for understanding the absence of consensus about the war in Russian society and its impact on the country's political stability and social cohesion.

In Sect. 11.4.2 we analyzed the case of minority using propaganda to control the majority. Using historical data, we've shown how the majority of Russian population does not support the war, but cannot use propaganda to express their opinion. At the same time, the minority that supports the war use propaganda to spread their opinion to the rest of the population. The outcome of the model can be seen in Eq. 11.23: Surprisingly, it generates a new partition of society where the fifth column joins the minority, while the rest of the blue party becomes the "enemy of the state," similarly to the processes observed in modern Russia.

11.6 Conclusion

In this chapter, we've shown that in the presence of the fifth column, the average algorithm converges to zero opinion vector and, therefore, fails to distinguish between different communities. To overcome this problem, we suggested dividing the averaging matrix by the greatest eigenvalues. If the original averaging matrix was ergodic, then the greatest eigenvalue is positive. Therefore, the greatest eigenvalue of the matrix averaging process after dividing the eigenvalues $\lambda_1 = 1$ is one, and so the averaging process converges to the corresponding v_1 eigenvector. However, v_1 contains positive and negative numbers as well as zeros, which means that the averaging process of the new matrix finds two groups that agree to disagree. Those groups may not be equal to the original groups, as our example shows.

We also present a simple model in which one of the groups uses propaganda while the other is forced to be silent. Our simple model may help explain the outcome of such a historical event. For example, we use our simple model to explain several historical events, such as the differences in public opinion about the Civil War and the civil rights movement in the South; see, for instance, the famous cases of Ruby Bridges and Rosa Parks.

Another example that we've studied is the war between Russia and Ukraine, where the majority of the population in Russia do not support the war but are forbidden from expressing discontent. We used our model to predict the outcome of public opinion in the case where the majority of people are forbidden from using propaganda to maintain their opinion, see Sect. 11.4.2. In this case, our model predicts that the new coalition will be formed and the fifth column from the majority will join the minority against the rest of the blue party.

This chapter mainly analyzes the simple graph with three nodes and two groups. We did this to make the material in the book more accessible and transparent. However, the tools presented in this chapter provide similar results for other concrete examples and, therefore, can be used to analyze many examples. The result presented in this chapter can be extended to general social network graphs. However, we decided not to put this material in the book to avoid unnecessary complexity.

11.7 Exercise

1. In our model, even a single node, which is the fifth column, destroys the ability to agree to disagree. Is this a strength or a weakness of the model?
2. In this exercise, you are asked to analyze the case studied in Sect. 10.3.4, where we have 3 nodes in the social network, and each node is connected to others. The difference between what was done in the subsection and what you are asked to do is that, in this case, each node in the social network is averaging with its own past opinion with probability α. Therefore, matrix M does not appear in (10.41), but instead is

$$M = \begin{bmatrix} \alpha & \frac{1-\alpha}{2} & \frac{1-\alpha}{2} \\ \frac{1-\alpha}{2} & \alpha & \frac{1-\alpha}{2} \\ \frac{1-\alpha}{2} & \frac{1-\alpha}{2} & \alpha \end{bmatrix} \tag{11.25}$$

   Repeat the calculations given in Sect. 10.4.1, where each player is averaging their own opinion with probability α. Use the matrix

$$M = \begin{bmatrix} \alpha & \frac{1-\alpha}{2} & -\frac{1-\alpha}{2} \\ \frac{1-\alpha}{2} & \alpha & -\frac{1-\alpha}{2} \\ -\frac{1-\alpha}{2} & -\frac{1-\alpha}{2} & \alpha \end{bmatrix} \tag{11.26}$$

3. Repeat the calculations from Sect. 11.2 where $\alpha = 0$, see Eq. 11.5. Each player is averaging its neighbor's opinion and ignores its own opinion, i.e.,

$$M_4 = \begin{bmatrix} 0 & \frac{1}{2} & \frac{1}{2} \\ \frac{1}{2} & 0 & \frac{-1}{2} \\ \frac{-1}{2} & \frac{-1}{2} & 0 \end{bmatrix} \tag{11.27}$$

Note that in this case, the eigenvalues of M_4 are $\lambda_1 > \lambda_2 = 0 > \lambda_3 = -1$. Therefore, the system does not converge since opinions oscillate. Technically, the oscillation happens because the social network graph is bipartite. Explain the historical meaning of this phenomenon.

4. Propaganda in a single row. In Sect. 11.4 we calculated the case of propaganda on majority and minority populations. In this exercise, you are asked to repeat the same computation when the propaganda is executed only by node 1 or node 2.

   a. Consider the case where propaganda is performed by node 1, so that the matrix is

$$M_{13} = \begin{bmatrix} \alpha r & \frac{1}{2}(1-\alpha)r & \frac{1}{2}(1-\alpha)r \\ \frac{1-\alpha}{2} & \alpha & \frac{\alpha-1}{2} \\ \frac{\alpha-1}{2} & \frac{\alpha-1}{2} & \alpha \end{bmatrix} \tag{11.28}$$

   b. By setting the propaganda factor to be $r = \frac{1}{\alpha}$, we get M_{14}:

$$M_{14} = \begin{bmatrix} 1 & \frac{1-\alpha}{2\alpha} & \frac{1-\alpha}{2\alpha} \\ \frac{1-\alpha}{2} & \alpha & \frac{\alpha-1}{2} \\ \frac{\alpha-1}{2} & \frac{\alpha-1}{2} & \alpha \end{bmatrix} \tag{11.29}$$

   Compute the eigenvalues λ_1 of M_{14}. Verify that, in this case, there is a single eigenvalue equal to 1, and the rest of the eigenvalues are less than 1, i.e., $\alpha < 1$.

   c. Show that the corresponding eigenvector of the greatest eigenvalue of λ_1 in the matrix M_{14} is

$$\vec{v}_1 = (-1, -1, 1) \tag{11.30}$$

   d. Show the following limit:

$$\lim_{t\to\infty} M_{14}^t = \begin{bmatrix} 1 & \frac{1}{3\alpha} & \frac{1}{3\alpha} \\ 1 & \frac{1}{3\alpha} & \frac{1}{3\alpha} \\ -1 & -\frac{1}{3\alpha} & -\frac{1}{3\alpha} \end{bmatrix} \tag{11.31}$$

   e. Compute the limit result of the averaging process when the averaging matrix is M_{14}.

   f. What is the historical significance of the result of the previous question 4e?

5. Repeat Exercise 4 when applying propaganda to the second row.
6. How can you measure the strength of the fifth column in the USA Congress?
7. Repeat Sect. 11.3 with matrix

$$M_4 = \begin{bmatrix} \frac{\alpha}{\alpha+2} & \frac{1}{\alpha+2} & 0 & \frac{1}{\alpha+2} \\ \frac{1}{\alpha} & 1 & -\frac{1}{\alpha} & 0 \\ 0 & -\frac{1}{\alpha} & 1 & \frac{1}{\alpha} \\ -\frac{1}{\alpha} & 0 & \frac{1}{\alpha} & 1 \end{bmatrix} \tag{11.32}$$

Assume that the blue party is composed of nodes 1 and 2, and the red party is composed of nodes 3 and 4. Instead of the matrix that appears in Eq. 11.5, do the following:

a. Identify the fifth column of the blue party.
b. Show that the eigenvalues of M_4 are:

$$\left(\frac{\alpha}{\alpha+2}, 1, \frac{\alpha^2 - \sqrt{2}\sqrt{\alpha^2+4\alpha+4}+2\alpha}{\alpha(\alpha+2)}, \frac{\alpha^2 + \sqrt{2}\sqrt{\alpha^2+4\alpha+4}+2\alpha}{\alpha(\alpha+2)}\right). \quad (11.33)$$

c. Show that the propaganda factor r that stabilizes the system when we multiply the matrix M_4 by factor r similarly to Eq. 11.8 in this case is

$$r = \frac{\alpha(\alpha+2)}{\alpha^2 + \sqrt{2}\sqrt{\alpha^2+4\alpha+4}+2\alpha} \quad (11.34)$$

d. Compute the results of the averaging process similarly to the Eq. 11.11.

References

1. C. Avin, H. Daltrophe, Z. Lotker, On the impossibility of breaking the echo chamber effect in social media using regulation. Sci. Rep. **14**(1), 1107 (2024)
2. BBC Russia. The illusion of the majority (in Russian). https://www.bbc.com/russian/news-64764949 (2023)
3. Forbes. In 2021, the Russians donated more than 360 billion rubles to charity (in Russian) (2023)
4. Iz.ru. For everyone to win: 12,000 tons of humanitarian aid collected for new regions (in russian). https://iz.ru/1448639/alena-nefedova/vsem-dlia-pobedy-novym-regionam-sobrali-12-tys-t-gumanitarnoi-pomoshchi (2023)
5. Meduza1. How is the current wave of emigration different from others? (in Russian). https://meduza.io/feature/2022/07/31/chem-nyneshnyaya-volna-emigratsii-otlichaetsya-ot-drugih-predstaviteli-kakih-professiy-uehali-sobirayutsya-li-migranty-vernutsya (2022)
6. Meduza. Is it still possible to somehow find out how many Russians died in the war? and what will emigration from Russia lead to - and was it truly massive? (in Russian). https://meduza.io/feature/2022/12/13/mozhno-li-vse-taki-kak-to-vyyasnit-skolko-rossiyan-pogiblo-na-voyne-a-k-chemu-privedet-emigratsiya-iz-rossii-i-byla-li-ona-po-nastoyaschemu-massovoy (2022)
7. Forbes. About 700,000 citizens left Russia after September 21 (in Russian). https://www.forbes.ru/society/478827-rossiu-posle-21-sentabra-pokinuli-okolo-700-000-grazdan (2022)
8. RBC. Shadayev called premature a complete ban on remote work for it people (in Russian). https://www.rbc.ru/technology_and_media/20/12/2022/63a187ed9a79478aa4435688 (2022)

Chapter 12
Stochastic Terrorism

12.1 Introduction

The term Stochastic Terrorism is used to denote Violence induced by public demonizing of a person, a group, or a nation [1]. For illustration, see Fig. 12.1.

It is also known as "acts of Violence by random extremists, triggered by political demagoguery." [2] The main problem in coping with stochastic terrorism is that it cannot be analyzed with a standard statistical approach and is difficult to forecast.

Nevertheless, stochastic terrorism is a critical history-shaping mechanism. Examples include the assassination of Israeli Prime Minister Yitzhak Rabin in 1995 and assassination attempts such as the attack on Speaker of the House Nancy Pelosi's husband in 2022, and the attempt on Donald Trump during the 2024 election race. Numerous attempts on the life of author Salman Rushdie further illustrate this phenomenon.

Stochastic terrorism has been mainly studied from the perspective of social science and humanities, but never analyzed with mathematical tools—until now. In this chapter, we will be using queuing theory to analyze stochastic terrorism.

In the previous chapter, we extended the theory of the Markov chain to include positive and negative numbers in the transition matrix. In this chapter, we face a similar problem. Standard queuing theory supports two kinds of entities: customers and servers, who are, in some ways, two opposites. We will use this observation to adapt queuing theory to allow two political opinions to compete.

In our model of stochastic terrorism we use the Jackson network (see for example Chap. 2 in [3]). The nodes V in the networks $G(V, E)$ are people. Each person has a color, either blue or red $\mathscr{B}, \mathscr{R}$. Each node $i \in V$ has a capacity ρ_i to consume news. News is also denoted by colors, as news can increase or decrease Violence.

One advantage of queuing theory is that it allows the queues to be infinite. We will use these unstable queues to indicate possible random acts of Violence.

Z. Lotker, *History by Algorithms*,
https://doi.org/10.1007/978-3-031-93627-2_12

Fig. 12.1 A cartoon illustrating the mechanisms behind stochastic terrorism

12.2 Related Work

Reference [4] While we know that demagogic attacks on a person or entity increase the probability of Violence against them, it's tricky to claim that rhetoric is the only **reason** for such Violence. The etymology harks back to the Greek *stochastikos*, "proceeding by guesswork". The term Terrorism has a number of definitions, both legal and scholastic, primarily connoting ideologically-inspired attacks on groups of people, often including civilians, with the goal of spreading fear and discord. Stochastic Terrorism was first mentioned by Gordon Woo, who suggested a quantifiable relationship between seemingly random acts of terrorism and the growing social anxiety brought on by mass media reports about terror attacks [5].

In this chapter, we suggest a mathematical model that uses a queuing network to explains stochastic terrorism from the social network perspective. In the future, this model could provide tools for measuring the level of stochastic terrorism in a society, as well as a way to reduce it. Moreover, it explains the connection between fake news and stochastic terrorism, showing that fake news is an important instrument for dividing society and, by doing so, generating stochastic terrorism.

12.3 The Model

We'll use some queue theory to model stochastic terrorism, starting with the Birth–death process and then use the Jackson Network for modeling stochastic terrorism.

12.4 Birth–Death Process

One of the first and simplest models in the stochastic process is the birth-death process. We give a very short introduction. For more about birth-death process, see [6].

The birth-death process is a type of continuous-time Markov process with two possible state transitions: "births," which increase the state by one, and "deaths," which decrease it by one. It was introduced by William Feller and is often used to model population dynamics, where births and deaths represent real changes in population size. This process has applications in various fields, including demography, queueing theory, performance engineering, and others. The birth-death process is defined formally.

12.4.1 The Queue as Data Structure

One of the first data structures students of computer science learn is a queue. A queue is characterized by a sequential arrangement of elements that follows specific rules for insertion and removal. Elements are added at one end of the queue, known as the rear (or back), and removed from the opposite end, termed the front (or head). This first-in, first-out (FIFO) access pattern mirrors real-world queues, where new arrivals join the back, and the individual at the front is served first. Operations associated with queues include "enqueue" for adding elements, "dequeue" for removing them, and "peek" for accessing the front element without altering the queue's state. The queue's state can be named using the natural number $\mathbb{N}$. We say that the system is in the state $i \in \mathbb{N}$ if the number of elements in the queues is i. For more on queues, see [7, 8].

12.4.2 From Queues to Birth and Death Process

Let $X(t) \in \mathbb{N}$ be the state of the system at time t, where the transitions occur according to the following rates:

1. λ_n is the birth rate, or the rate of transition from state n to state $n + 1$.
2. μ_n is the death rate, or the rate of transition from state n to state $n - 1$.

The infinitesimal transition probabilities are expressed as:

1.
$$\mathbb{P}(X(t + \Delta t) = n + 1 \mid X(t) = n) = \lambda_n \Delta t + o(\Delta t) \tag{12.1}$$

2.
$$\mathbb{P}(X(t + \Delta t) = n - 1 \mid X(t) = n) = \mu_n \Delta t + o(\Delta t) \tag{12.2}$$

3.

$$\mathbb{P}(X(t+\Delta t)=n \mid X(t)=n)=1-(\lambda_n+\mu_n)\Delta t+o(\Delta t) \tag{12.3}$$

where $o(\Delta t)$ represents higher-order terms that become negligible as $\Delta t \to 0$.

The process describes the changes in the system's state over time, where births increase the state by 1 and deaths decrease the state by 1.

To simplify the model we will assume that all birth rates are independent of the state and are equal to $\lambda_n = \lambda$. In the same way, we will assume that all death rates independent of the state and are equal to $\mu_n = \mu$.

It is useful to work with a state diagram, see Fig. 12.2. In this diagram, the nodes are the states of the system or the number of people in society, and the edges represent the rate of moving from one state of the system to another.

If there exists a stationary distribution, it is easy to compute from this Fig. 12.2. Clearly, in order for the system to be stable and not go to an infinite number of people, $\mu > \lambda$.

Let's compute the stationary distribution in the Fig. 12.2. To do this, we denote $\Pr[X_s(t)=i]$ to be the probability of having i-people in the system at the stationary distribution. Since we are at the stationary distribution, after dt time, those probabilities remain the same. So, we can write the balance equation as follows for all $i \in \mathbb{N}$

$$\Pr[X_s(t)=i]\lambda = \Pr[X_s(t)=i+1]\mu \tag{12.4}$$

From this, we can solve the system recursively and find that

$$\Pr[X_s(t)=i]=(1-\rho)\rho^i \tag{12.5}$$

where

$$\rho = \frac{\lambda}{\mu} \tag{12.6}$$

For more on the stationary distribution, see [9].

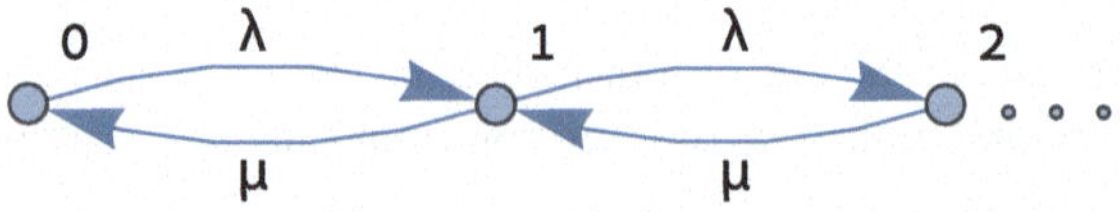

Fig. 12.2 This is an illustration of the rate at which the system M/M/1 changes from one state to another state

12.5 Integer Birth–Death Process

In this section, we extend the birth-death process to an integer birth-death process, where the state of the process is the integer numbers $\mathbb{Z}$ instead of the natural numbers $\mathbb{N}$.

12.5.1 *The DQueue as Data Structure*

One variant of a queue is a DQueue, which allows us to add or remove elements from the front and the back of the queue instead of adding elements in the back of the queue and removing elements in the front of the queue. For more on Dqueues, see [10, 11].

The Dqueue's state can be named using the integer number $\mathbb{Z}$. We say that the system is in the state $i \in \mathbb{Z}$ if the number of elements in the queues is i.

Let the Dqueue be in state i. We will say that the Dqueue is empty if it is in the state of zero i.e., $i = 0$. We will say that the Dqueue is positive if it is in a positive state i.e., $i > 0$ and, negative if it is in a negative state i.e., $i < 0$. We sometimes will call a Dqueue with an integer state *integer queue*.

As in the birth and death process, we add probability to the queue data structure in order to define the birth and death process. We follow this scheme. Our starting point is the Dqueue, and we need to add the probability of the data structure to define the integer birth and death process. In the next subsection, we explain how to add probability.

12.5.2 *From Dqueue to Integer Birth and Death Process*

Formally, we will have two arrival processes—one for positive $+1$ and one for negative -1. The birth rate of positive numbers will be denoted by λ_p. The birth rate of the negative numbers will be denoted by λ_n. And a death process, With rate μ.

According to Fig. 12.3, the integer birth and death processes can be seen as gluing two birth and death processes together at stage zero. The first process is used to describe the evolution of the system for the positive numbers. In this process, the birth rate is λ_p, and the death rate is $\mu + \lambda_n$. The second birth and death process is

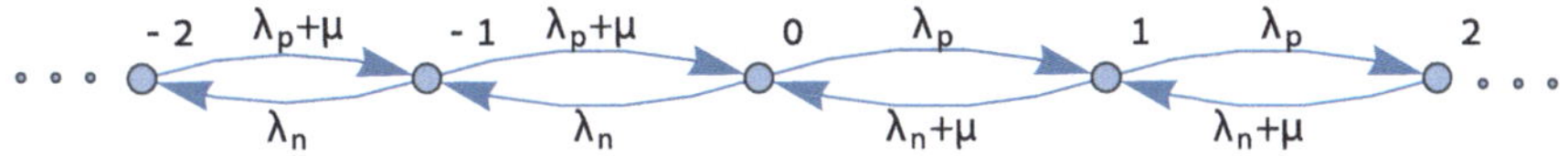

Fig. 12.3 This figure shows the evolution of the integer birth process

used to describe the evolution of the system for the negative numbers. In this regime, the birth rate is λ_n, and the death rate is $\mu + \lambda_p$, see Fig. 12.3.

Therefore state of the system at time t, is $X(t) \in \mathbb{Z}$. The transitions occur according to the following rates:

1. If $X(t) = 0$ a birth of a positive number happens with the probability λ_p, and a birth of a negative number happens with the probability λ_n. In this case, there is no service since the queue is empty.
2. If the process is positive. $X(t) > 0$, then the birth of a positive number happens with the rate λ_p. In this case, if the birth process of the positive numbers happens, then we increase the number of jobs in the queue by 1. At the same time, the death process happens at rate $\mu + \lambda_n$. If a death process happens, then we decrease the queue by 1.
3. If the process is negative $X(t) < 0$, then the birth of a negative number happens with the rate λ_n. In this case, if the birth process of the negative numbers happens, then we decrease the number of jobs in the queue by 1. At the same time, the death process happens at rate $\mu + \lambda_p$. If a death process happens, then we increase the queue by 1.

Note that the system has a stationary distribution if and only if the process does not escape to plus infinity or minus infinity. For escaping to plus infinity, we need the birth process of the positive states to be bigger than the death process of the positive states. This happens according to Fig. 12.3.

$$\lambda_p \geq \lambda_n + \mu \tag{12.7}$$

Therefore, for the system to be stable on the positive states, we need that Eq. 12.7 will not be satisfied. Using simple algebraic calculation, we get the following:

$$\lambda_p - \lambda_n < \mu \tag{12.8}$$

Doing the same calculation for the negative states, it follows that the condition "not to escape minus infinity" through the negative states is

$$\lambda_n - \lambda_p < \mu \tag{12.9}$$

We can summarize Eqs. 12.8, 12.9 for a single equation which is

$$\mu > |\lambda_p - \lambda_n| \tag{12.10}$$

12.5.3 Stationary Distribution of the Integer Birth Process

We can repeat the same method that we used to compute the stationary distribution; see Eqs. 12.4 and 12.5. In the case of the Eq. 12.4 we get that for the $X_s = i \geq 0$,

$$\Pr[X_s = i]\lambda_p = \Pr[X_s(t) = i + 1](\mu + \lambda_{n)} \tag{12.11}$$

And for $X_s = i \leq 0$,

$$\Pr[X_s = i](\lambda_p + \mu) = \Pr[X_s(t) = i - 1](\lambda_{n)} \tag{12.12}$$

Therefore, we can solve the positive system of Eq. 12.11 as a function of the probability $\Pr[X_s = 0]$. We get that for $i > 0$:

$$\Pr[X_s = i] = \Pr[X_s = 0]\left(\frac{\lambda_p}{\mu + \lambda_n}\right)^i \tag{12.13}$$

And for negative $i < 0$

$$\Pr[X_s = i] = \Pr[X_s = 0]\left(\frac{\lambda_n}{\mu + \lambda_p}\right)^i \tag{12.14}$$

Now, we can compute $\Pr[X_s = 0]$ using the Law of total probability i.e.,

$$\sum_{i=-\infty}^{+\infty} \Pr[X_s = i] = 1; \tag{12.15}$$

If follows that

$$\Pr[X_s = 0] = \frac{(\mu - \lambda_n + \lambda_p)(\mu + \lambda_n - \lambda_p)}{\mu(\mu + \lambda_n + \lambda_p)} \tag{12.16}$$

Figure 12.4 depicts the psychological fragmentation of an individual, reflecting the unpredictability and lack of coordination found in systems without centralized control-an idea that parallels queuing theory models where random arrivals and service times lead to unstable or overloaded states.

12.5.4 Ornstein-Uhlenbeck

Looking at Fig. 12.3 the integer birth-death process is in some sense similar to Ornstein-Uhlenbeck process [12] since we have a central process that behaves as a central "force" toward the zero states of the system at a rate μ, we have a positive process at rate λ_p that push the system toward plus infinity, and a negative process at rate λ_n that push the system toward minus infinity.

Fig. 12.4 This figure tries to capture the psychology of a person in a divided society

Allowing the value of λ_p, λ_n to be positive and negative, where negative means reversing the direction of the edge in the state diagram, we can redraw Fig. 12.3 and get Fig. 12.5.

Representing the integer birth and death process as the Ornstein-Uhlenbeck process is not only cosmetic; it is necessary as we move from a single queue to a network of queues.

We now move from a single queue to a network of queues. We begin with the well-known theory of the Jackson network. Then, we extend this theory to form birth and death queues to be integer queues.

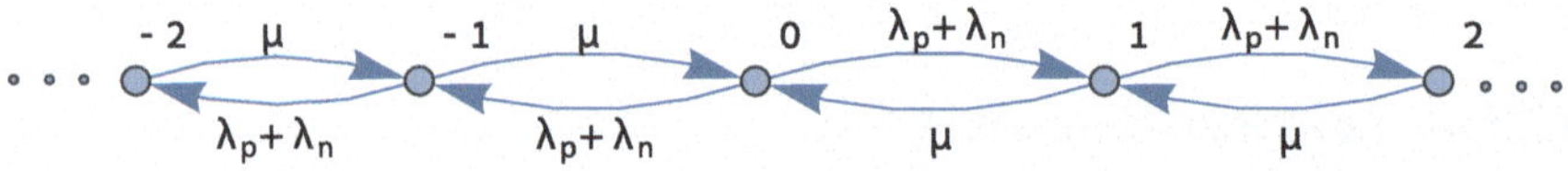

Fig. 12.5 This figure shows the integer death and birth process as the discrete version of the Ornstein-Uhlenbeck process, where the central force has a rate of μ and the spreading force has a rate of $\lambda_n + \lambda_p$ where λ_n is negative on the positive state and positive on the negative state, and λ_p is positive on the negative state and negative on the positive state

12.5.5 Jackson Network

The classic Jackson network is an extension of a single queuing theory usually denoted by $(M, M, 1)$ to a network of queues $Net = (G(V, E), P, \vec{\alpha}, \mu, Q)$ where $G(V, E)$ is a graph With $V = \{0, 1, ..., n\}$ nodes, E edges, P is the transiting probability matrix, $\vec{\alpha} = (\alpha_1, ..., \alpha(n))$ is the arriving rate vector, and $\mu = (\mu_1, \mu_2, ..., \mu_n)$ is the service rate. Note that we can have a special node 0, which models the arrival and departure of the jobs in the system. So, nodes $\{1, ..., n\}$ are modeling the system and are called *system nodes*, and nodes number zero are called *port nodes*.

Denote the total arriving rate α to the system by

$$\alpha = \sum_{i=1}^{n} = \alpha_i \tag{12.17}$$

We can assume that $\alpha > 0$ since otherwise, no customer arrives at the system, and therefore, after some time, the system becomes empty. Since most of the technical material on the Jackson network ignores the queues in the beginning, we follow as well. It is easy to add another vector function $\vec{Q}(t)$ that captures a number of jobs in the queue at the specific time t. However, we ignore the initial condition $\vec{Q}(0)$ for the sake of simplicity. Clearly, the number of jobs at the beginning of the queues does not affect the stability of the system in any way.

In queuing theory, customers (also known as jobs or requests) arrive according to a Poisson process, and services are distributed exponentially. The $(M, M, 1)$ model is the most elementary of queuing models. For more about queuing theory, see [3, 13, 14].

Similar to $(M, M, 1)$, customers in an open Jackson network arrive from the outside according to a Poisson process with a rate of $\alpha_i > 0$. We can now compute $p_{0,i}$

$$p_{0,i} = \begin{cases} 0, & \text{for } \alpha_i = 0 \\ \frac{\alpha}{\alpha_i}, & \text{for } \alpha_i > 0 \end{cases} \tag{12.18}$$

The routing probability for each arrival is equal to $p_{0,j} \geq 0$ and

$$\sum_{j \in V} p_{0,j} = 1. \tag{12.19}$$

Service at each node $i \in V$ is independent and identically distributed (i.i.d.) and has an exponential distribution denoted by λ_i. An open Jackson Network is usually denoted by (M, M, m).

When the service at node $i \in V$ has been completed, it can be sent to node $j \in V$ with probability $p_{i,j}$ or stop the service and quit the network with probability

$$p_{i,0} = 1 - \sum_{j \in V} p_{i,j}. \tag{12.20}$$

The overall arrival rate to node $i \in V$ is therefore λ_i.

$$\lambda_i = \alpha p_{0,i} + \sum_{j \in V} \lambda_j p_{j,i} \tag{12.21}$$

This includes both external arrivals and internal transitions. Therefore, when analyzing the stability of the network, we ignore service rate μ_j in the above equation. This is due to Jackson's theorem, see Theorem 12.1, explained below. The purpose of ρ_i is to determine if the queuing network is stable or if some queues are blown up. This is done using simple inequality. $\rho_i \leq \lambda_i$.

The Jackson theorem states that if for all $\mu_i > \lambda_i$ then the system is stable and each of its queues behaves like $(M, M, 1)$. Formally, following [3], let $X_i(t)$ denote the number of jobs at node i at time t in a continuous-time Markov chain. We consider its distribution as stationary and omit the time index t for simplicity. Set $X = (X_i)_i^n$.

The key result Theorem 12.1 is related to the distribution of $X = (X_1, ..., X_n)$ to a vector of independent random variables $(Y_1, ..., Y_n)$, where each Y_i has a following probability mass function (pmf):

$$P[Y_i = k] = P[Y_i = 0] \cdot \left(\frac{\lambda_i}{\rho_i}\right)^k \tag{12.22}$$

and

$$\sum_{i=1}^{\infty} \left(\frac{\lambda_i}{\rho_i}\right)^k < \infty \tag{12.23}$$

so that $P[Y_i = 0]$ is well-defined, namely,

$$P[Y_i = 0] = \left(1 + \sum_{i=1}^{\infty} \left(\frac{\lambda_i}{\rho_i}\right)^k\right)^{-1} \tag{12.24}$$

Following Chap. 1 of the book *Fundamentals of Queuing Networks* [3], we recognize Y_i as the number of jobs in a birth-death $(M, M, 1)$ queue in equilibrium, with a constant arrival (birth) rate λ_i and service (death) rates ρ_i. The following theorem is taken from the book.

Theorem 12.1 *Provided that the condition in (12.24) is satisfied for all $i = 1, ..., n$, the equilibrium distribution of the open Jackson network has the following product form:*

$$\pi(x) = \prod_{i=1}^{n} \Pr[Y_i = x_j] \tag{12.25}$$

for all $x \in \mathbb{N}^n$, where Y_i follows the distribution in (12.23).

In the next section, we will use the idea from the Jackson theorem to describe stochastic terrorism. The reduction is not straightforward. We replace customers and service times with feelings, specifically two types: one increasing tension by demonizing or using hate speech, and the other harmonizing and reducing tension. Moreover, to distinguish between standard network theory and our model for stochastic terrorism, we will change customers to *news items* . The advantage of this approach is that news items can be positive or negative.

The basic idea is to replace each queue with an integer queue, which can handle positive and negative news items.

12.5.6 Stochastic Terrorism Network

In stochastic terrorism, each node has an integer queue that supports positive and negative messages to read. We use the integer numbers to indicate two different points of news: one which indicates the negative and the other indicates the positive. This is by no means that the negative is bad news and the positive is good news; it means that the negative opposes the positive, and vice versa.

The absolute length of the integer queue measures the stress of a specific node that represents an individual in a political society. This means that if the queue is long, the node is under stress. The stress of the node does not depend on the kind of news in the queue. It can be stressed from many positive news or from many negative news. If the queue contains too many messages, the resulting behavior can be violent. Therefore, when the system is not stable, stochastic terrorism will appear at random points in time and in random nodes.

There are some technical problems with converging queuing theory to stochastic terrorism. First, in queuing theory, customers arrive and then traverse the network. In stochastic terrorism, messages may increase or decrease the tension of the nodes according to their signs and the signs of the queues. If news arrives in the queue with a different sign, the tension decreases by 1. If news arrives in the queue with the same sign, the tension increases by 1.

Let's denote the input stream as two streams of messages: negative α_- and positive α_+. The negative incoming messages arrive at a rate of α_- and disperse throughout the system according to a vector probability of $\vec{\alpha_-} = (\alpha_{-,1}, ..., \alpha_{-,n})$. The input rate to node i of negative messages is $\alpha_{-,i} \leq 0$.

In the same way, positive streaming α_+ disperse throughout the system according to a rate vector $\vec{\alpha_+} = (\alpha_{+,1}, ..., \alpha_{+,n})$. The input rate to node i of negative messages is $\alpha_{+,i} \geq 0$.

Note that now we have two incoming streaming with rate α_+, α_-

$$\alpha_+ = \sum_{i=1}^{n} \alpha_{+,i}; \text{ and } \alpha_- = \sum_{i=1}^{n} \alpha_{-,i} \tag{12.26}$$

Similarly, to Eqs. 12.18, 12.19 we can write a

$$p_{+,i} = \begin{cases} 0, & \text{for } \alpha_{+,i} = 0 \\ \frac{\alpha_+}{\alpha_{+,i}}, & \text{for } \alpha_{+,i} > 0 \end{cases} \tag{12.27}$$

and for α_-

$$p_{-,i} = \begin{cases} 0, & \text{for } \alpha_{-,i} = 0 \\ \frac{\alpha_-}{\alpha_{-,i}}, & \text{for } \alpha_{-,i} < 0 \end{cases} \tag{12.28}$$

To summarize, when writing equivalent equations of (12.21) for a stochastic terrorism network, the output rate of node $i \in V$ is λ_i equal to the total sum of the rate entering the node i. The main difference between the Jackson network and stochastic terrorism is that the rate can be negative as well as positive.

$$\lambda_i = \alpha_{+,i} + \alpha_{-,i} + \sum_{j \in V} \lambda_j p_{j,i} \tag{12.29}$$

Let us explain the Eq. 12.29. The output rate of node i, if the network is stable, is equal to the total incoming rate of node i, which includes the positive arrival rate to $\alpha_{+,i} \geq 0$ the negative arrival rate $\alpha_{+,i} \leq 0$, and the sum of all rates that are routing towards the node i which is equal to $\sum_{j \in V} \lambda_j p_{j,i}$. Note that we do not know in advance the sign of each λ_i. To overcome this technical difficulty, we must allow the rate λ_i to be positive or negative and let linear algebra take care of this problem for us.

The advantage of modeling stochastic terrorism using the Jackson network is that we can analyze whether the system is stable or not. For each of the nodes i we can compute the service rate λ_i by solving the Eq. (12.29). If $|\lambda_i| < \mu_i$ then queue Q_i is stable and node i will not react violently. However, if $|\lambda_i| > \mu_i$, then queue Q_i is not stable and will blow up very quickly, meaning that node i will react violently.

If a violent node i exists—i.e., the queue blows up—then will say that the network suffers from stochastic terrorism.

Note that we only care whether the system is stable or not. To compute the actual number of messages in each queue is a more complicated task which we will leave for future research.

12.5.7 Example

Consider a social network which is a clique with $2n$ nodes $V = \{1, ..., 2n\}$. The first nodes are colored blue $V_B = \{1, ..., n\}$. while the last nodes are colored red $V_R = \{n+1, ..., 2n\}$. Assume that the blue nodes $i \in V_B$ receive positive messages with rate one and negative messages with rate zero, formally:

$$\alpha_{+,i} = 1 \text{ and, } \alpha_{-,i} = 0. \tag{12.30}$$

On the other hand, red nodes $j \in V_R$ receive negative messages with rate one and positive messages with rate zero, formally:

$$\alpha_{-,j} = 1 \text{ and, } \alpha_{+,i} = 0. \tag{12.31}$$

Assume that once the message is going out of the queue Q_i, it picks a random node from V according to the uniform distribution. Note that messages can not go out of the system, but they can be annihilated when a positive message meets a negative one, or vice versa.

Assume that the service rate is μ. We would like to compute the maximum rate at which the network is unstable.

Since all nodes $i \in \{1, 2, ..., n\}$ have the same inputs and they receive positive messages more than negative messages, we will denote the rate that goes out of node i as λ_+, and for the nodes in $j \in \{n+1, ..., 2n\}$ we will denote the rate that goes out of node j as λ_- since they receive more negative nodes than positive. Now, we can write the system of two equations.

$$\lambda_+ = \frac{n\lambda_-}{2n} + \frac{n\lambda_+}{2n} + 1 \tag{12.32}$$

$$\lambda_- = \frac{n\lambda_-}{2n} + \frac{nn\lambda_p}{2n} - 1 \tag{12.33}$$

It turns out that the system of Eq. 12.32 is linearly dependent, and so we need to add one equation to the system in order to get a solution to the system. We can add the following equation

$$\lambda_+ = -\lambda_- \tag{12.34}$$

Now we can solve the system and get that

$$\lambda_+ = 1 \text{ and, } \lambda_+ = -1 \tag{12.35}$$

as expected. Therefore, if $\mu > 1$, the system is stable, and if $\mu \leq 1$, the system is unstable, and stochastic terrorism will appear in the network.

In the next section, we incorporate the mechanism of echo chamber into the model of stochastic terrorism.

12.6 Fake News Generating an Echo Chamber

The nodes in our model are divided into two communities, Blue V_B and Red V_R. We will assume that

$$V = V_B \cup V_R \text{ and, } V_B \bigcap V_R = \varnothing \tag{12.36}$$

We will be using color to describe the simple process that generates an echo chamber. Once an echo chamber is generated, it is easy to increase the amount of Violence in one of the two communities, since the ability to annihilate positive messages by negative messages and vice versa is reduced and the network becomes more unstable.

In the previous section, we assumed that we have two streams of messages: one positive and one negative. In this section, we assume that we have four generators of the streams: truth and fake generators in both the Blue and Red communities. This is a summarized arrival matrix (12.37) rat:

$$[\alpha] = \begin{bmatrix} \alpha_{b,t} & \alpha_{b,f} \\ \alpha_{r,t} & \alpha_{r,f} \end{bmatrix} \tag{12.37}$$

Once a node $i \in V$ receives fake news from one of its neighbors, it chooses to believe it or not according to the color of the message. The blue node $i \in V_B$ always believes the Blue message and the Red node $j \in V_R$ always believes the Red message.

Once the Red node $j \in V_R$ receives the Blue fake message, it decides not to believe it. Fake news generates dissonance in the reader's mind, which is an undesired feeling. Therefore, with some probability p_d it decides to disconnect from the source $i \in V_B$ of this message. The blue node behaves similarly.

Opposite to news sources, nodes transmit only messages they believe. If the nodes i, j do not believe the messages, they ignore them. This means that only Red nodes can spread fake Red messages, and only Blue nodes can spread Blue fake messages.

This simple mechanism generates a small cut $C_{B,R}$ between the Blue and Red communities (see Fig. 12.6). It is easy to compute the expected size of the cut $|C_{B,R}|$ using the amount of fake news delivered by the news agency. The probability $\Pr[(i, j) \in C_{B,R}(t)]$ that an edge between Blue and Red nodes will survive at time t is equal to

$$\Pr[(i, j) \in C_{B,R}(t)] = (1 - p_d)^{(\alpha_{r,f}+\alpha_{b,f})t} \tag{12.38}$$

which is exponentially small as a function of t. This means that very few edges will survive between the Blue and the Red communities.

Once society is broken down into two communities, we can say that an echo chamber is generated and each news agency can have near complete control over the spread of positive or negative messages. The echo chamber is a necessary step for stochastic terrorism to emerge because it eradicates the simple mechanism of balancing the queue.

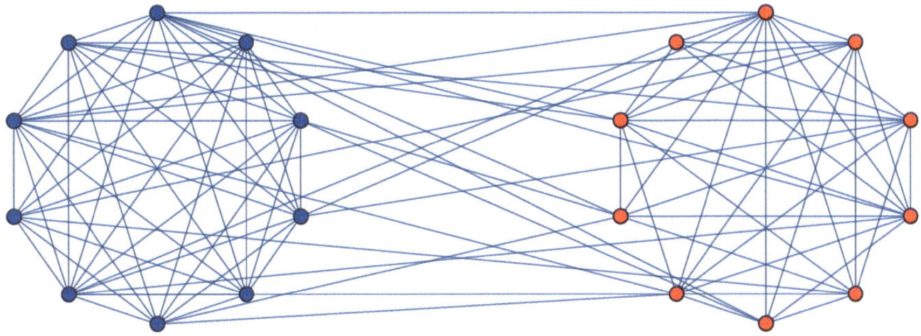

Fig. 12.6 This is an illustration of the sparse cut between the Blue and Red communities, generated by using the fake news model

12.6.1 Example

To see the influence of fake news on stochastic terrorism, let us compare two cases. In the first example, we consider a complete graph society with $n = 2n'$ nodes. See section example above Sect. 12.5.7.

We will assume that each node is receiving an arrival rate only for its party. This means that the blue node receives only blue messages, and the red node receives only red messages. Moreover, half of the messages are truthful, and the other half are fake. Since each node received an incoming message at rate 1, the arrival matrix is equal to

$$[\alpha] = \begin{bmatrix} \frac{n}{2} & \frac{n}{2} \\ \frac{n}{2} & \frac{n}{2} \end{bmatrix} \tag{12.39}$$

The transition matrix for the message works according to a simple random walk: the message arrives to all the neighbors N_i of node i with probability $\frac{1}{2n+1}$. Since we have only $2n + 1$ neighbors for all i, the message goes out from the system with the same probability, i.e., $\frac{1}{2n+1}$.

In this simple case we ignore the fact of fake news, and therefore, the arrival rate equation (12.29) becomes for blue node $i \in V_B$

$$\lambda_i = 1 + \sum_{k \in V_B} \lambda_k \frac{1}{2n+1} + \sum_{k \in V_R} \lambda_k \frac{1}{2n+1} \tag{12.40}$$

Similarly, since we ignore the fake news for the red node, the equation for the red node is $j \in V_R$

$$\lambda_j = -1 + \sum_{k \in V_B} \lambda_k \frac{1}{2n+1} + \sum_{k \in V_R} \lambda_k \frac{1}{2n+1} \tag{12.41}$$

Since all blue nodes behave the same, we can replace λ_i by λ_b and, all red nodes behave the same, we can replace λ_j by λ_r so that the equation for blue nodes becomes

$$\lambda_b = 1 + \lambda_b \frac{n}{2n+1} + \lambda_r \frac{n}{2n+1} \tag{12.42}$$

and for the red node the equation becomes

$$\lambda_r = -1 + \lambda_b \frac{n}{2n+1} + \lambda_r \frac{n}{2n+1} \tag{12.43}$$

Solving the equation for $lambda_i$ it follows that

$$\lambda_b = \lambda_r = 1 \tag{12.44}$$

which means that the network is stable for any service rate $\mu_i > 1$.

After analyzing the simple scenario where the society is a complete graph, we analyze the society after its division into two independent parts., We now have two cliques: one contains only n Blue nodes, and the other contains only n Red nodes. In this case, we see that the arrival rate equation (12.29) becomes for the blue party.

$$\lambda_b = 1 + \lambda_b \frac{n}{n+1} \tag{12.45}$$

Note that according to our assumption, the message that is going out of the system is $\frac{1}{n+1}$. Solving the above Eq. 12.45 for the blue party, we get that the system is stable if $\mu > n + 1$. This provides a big gap between those two systems, which means that the network is unstable for any service rate $\mu_i \leq n + 1$. Note that the service rate for a stable system tends to go to infinity and is related to the number of nodes in the system. The reason we have this result is that the speed of the message that goes out of the system is very slow—$\frac{1}{n+1}$. In fact, the stability rate is equal to the average speed of the message going out of the system.

Comparing these two examples, we see two different societies. The first is a homogeneous society that tends to be peaceful. The second, containing two disconnected cliques, tends to have stochastic terrorism.

Similar results will appear for the real social network, but that computation is more complicated and provides less insight.

12.7 Conclusion

In this chapter, we modeled stochastic terrorism using the extended Jackson network with two kinds of customers. These customers reflected positive and negative messages traversing the network. From a historical perspective, stochastic terrorism is a new phenomenon, and we do not have sufficient data to check the proposed model.

However, this chapter demonstrates a new way of analyzing historical processes using a mathematical model. The model suggests that to generate a stochastic terrorism network, as we have done, one needs to divide the network into two independent parts. One way to do that is to generate an echo chamber for some part of society through fake news.

The advantage of an echo chamber is that by isolating a part of the network, the news agency prevents it from getting messages from the other part of society. Then, it becomes possible to control the input messages for this part.

This explains why fake news is a crucial component of stochastic terrorism. Once you generate repeated fake news, society is split into two parts: one that believes fake news and one that does not.

The part that does not believe the fake news is in dissonance with the speaker, and if the process repeats many times, they stop listening to the nodes that spread the fake news. Clearly, they do the same to any news source that supports the fake news. The part that does believe the fake news is in dissonance with reality and stops listening to news that describes reality correctly. In both cases, the outcome of this process is disconnecting the network. Eventually, this process breaks the social network into two parts, forming two different echo chambers and dividing society. Now, the part that believes fake news is controlled by only one side and, therefore, becomes more vulnerable and more likely to perform stochastic terrorism.

12.8 Exercise

1. Determine which of the following queue systems is stable:

   a. A single $(M/M/1)$ queue with arriving rate $\lambda = 1$, and service rate with $\mu = 2$.
   b. A single $(M/M/1)$ queue with arriving rate $\lambda = 2$, and service rate with $\mu = 1$.
   c. A single $((M/M)/M/1)$ queue with positive arriving rate $\lambda_1 = 2$, negative arriving rate $\lambda_2 = 1/2$, and service rate with $\mu = 1/2$.
   d. A single $((M/M)/M/1)$ queue with positive arriving rate $\lambda_1 = 2/3$, negative arriving rate $\lambda_2 = 1/2$, and service rate with $\mu = 1/2$.

2. Consider a single queue of regular customers (which models good and bad news) Q, arriving at the same rate of $\lambda_+ = \lambda_- = 1$; the service rate is $\mu > 0$.

   a. Prove that the system is always stable
   b. Write a simulation of the system
   c. Show that the probability of having n news in the system at the steady state is

$$\pi_n = 2\pi_0 \left(\frac{\lambda}{\lambda + \mu} \right)^n \tag{12.46}$$

d. Compute π_0 if

$$\sum_{i=0}^{\infty} \pi_i = 1 \tag{12.47}$$

3. In this chapter, we extended the Jackson network and Markov chain to allow negative and positive flow which cancel each other. Use the complex number and equation $x^3 = 1$ and explain how to extend that concept to allow a three-color flow (α, β, γ) that satisfies the following relationship between the colors:

$$\alpha + \beta = -\gamma \tag{12.48}$$

4. Consider the example in Sect. 12.5.7. Assume that the probability of rerouting is uniform, but now, instead of having to choose one of the $2n$ nodes, we add one more node that indicates that the news went out of the system, i.e., each node receives $\frac{n}{2n+1}(\lambda_+ + \lambda_-)$ instead of $\frac{n}{2n}(\lambda_+ + \lambda_-)$, see Eqs. 12.32, 12.33. Calculate the threshold for mu in which the system starts to be stable. Note that in this case, we do not need an Eq. 12.34. Explain why not.
5. We are given a social network that is divided into m blue nodes, and n red nodes, where $m < n$. Assume that the nodes look at their neighborhood and vote according to the majority of their opinions. For example, if the node is red and most of his friends are blue, he votes for the blue party.

   a. Show that there exists a network where the minority wins the election.
   b. Use the previous network and determine which party enjoys generating echo chamber in this case.

References

1. What is "stochastic terrorism," and why is it trending? (2023) dictionary.com. https://www.dictionary.com/e/what-is-stochastic-terrorism/ (2023)
2. J. Keats, The rising danger of stochastic terrorism. Wired (2019)
3. H. Chen, D.D. Yao, *Fundamentals of Queueing Networks: Performance, Asymptotics, and Optimization*, vol. 4 (Springer, 2001)
4. J. Keats, There are no lone wolves. Opinion (2019)
5. G. Woo, Quantitative terrorism risk assessment. J. Risk Financ. **4**(1), 7–14 (2002)
6. Wiki. wikisource birth death process. https://en.wikipedia.org/wiki/Birth%E2%80%93death_process (2024)
7. Wiki. wikisource-queue. https://en.wikipedia.org/wiki/Queue_(abstract_data_type) (2023)
8. wolfram. wolfram-queue. https://reference.wolfram.com/language/ref/datastructure/Queue.html (2024)
9. Wiki. wikisource-mm1. https://en.wikipedia.org/wiki/M/M/1_queue (2023)
10. Wiki. wikisource-double. https://en.wikipedia.org/wiki/Double-ended_queue (2023)

11. wolfram. wolfram-dqueue. https://reference.wolfram.com/language/ref/datastructure/Deque.html (2024)
12. Wiki. wikisourceornstein–uhlenbeck-process. https://en.wikipedia.org/wiki/Ornstein (2023)
13. A.O. Allen, *Probability, Statistics, and Queueing Theory* (Gulf Professional Publishing, 1990)
14. U.N. Bhat, *An Introduction to Queueing Theory: Modeling and Analysis in Applications*, vol. 36 (Springer, 2008)

Chapter 13
Historian Machine

13.1 Introduction

Hegel, in his famous book *Philosophy of History* [1], started the discussion of history through historian conscience. In this chapter, we follow in his footsteps and define the historian entity in the spirit of machine consciousness (Fig. 13.1).

We will uncover the secrets of a m_h historian machine, how it works, and how it can write history. But to do that, we must first understand what a historical entity is, how it relates to consciousness, and how it can be used to write a narrative, all with the help of the theory of mind.

In general, the process of writing history is one of transforming a distributed stream of events into a coherent narrative. This process is central to the way humans define themselves. Therefore, in order to get a clear understanding of writing history, we should concentrate on modeling consciousness. If we want machines to write history, we must first understand machine consciousness and what historical entities are, as the machine will write the history of these entities.

When humans write history, it is always from the point of view of a conscious author who represents some historical entity. Until now, we had no need to define these concepts since they were clear. Now that machines are also authors, the concept requires some explaining.

Consciousness is a key component of our historical understanding. There are several layers of consciousness: individual, collective, and national. In fact, only the collective consciousness is universal, as we will discuss later in this chapter, see Sect. 13.7.

An individual's consciousness experiences and allows one to document their life, whereas national consciousness binds people through a shared identity and collective memory (see Chaps. 9, 10 and 11). These concepts are crucial for defining the historian machine.

The topic of consciousness is complicated. Figure 13.2 demonstrates this complexity. In the comic, each of the three robots thinks one word from René Descartes'

Z. Lotker, *History by Algorithms*,
https://doi.org/10.1007/978-3-031-93627-2_13

Fig. 13.1 A cartoon depicting a machine in the act of writing history

Fig. 13.2 The human consciousness is distributed and parallel, but contained within one body. This comic, of three robots in three different times and places thinking the three Latin fragments of Descartes' "I think therefore I am" quote raises the question: how do machine consciousnesses compare?

statement "cogito, ergo sum"—"I think, therefore I am." Descartes claimed that questioning his own consciousness proved that he was, in fact, conscious [2]. Would three different machines, each in a different time and location, make up a whole consciousness that could be proven aware by that same statement? After all, machines are easily connected, merged, and disassembled, unbothered by time or distance if neither stands in the way of their data. How is the unit of the 'self,' so critical to human consciousness, defined for machines?

The comic also hints at what is happening in our brains. Consciousness does not have a single seat in our head; it is spread throughout the brain. Our sense of existence consolidates many different neural streams and entities. In humans, those components run in parallel. Would they do the same in machines? Indeed, consciousness has a hidden distributed layer, which is essential for defining consciousness.

In this chapter, we build a model that allows us to talk about universal consciousness. We borrow some major ideas from Michael S. A. Graziano (see [3]), who used a system approach to define machine consciousness. Putting together Graziano's works on machine consciousness and Turing's universal machine, we can define a universal consciousness machine. We can then use the universal consciousness machine to define historical entities with universal consciousness as the principal part of defining the historian machine. To that end, historical entities are, in fact, an aggregation of individuals, the main building blocks used by machines when writing history.

This is the last chapter in the second part of the book, in which we use a mathematical model to explain historical phenomena. Throughout this part, we encounter negative probabilities Chaps. 9, 10, and 11. Similarly, we use a negative rate in discussing stochastic terrorism (see Chap. 12). In both cases, the concept of negative numbers is justified, since our major interest in history is conflict, and numbers best describe conflict through negative and positive numbers.

However, using the model developed in this chapter would allow us to replace both negative probabilities and negative rates with a more evolved model that uses only positive numbers. To do that, we must first develop a theory of universal consciousness.

13.2 Related Works

Consciousness is one of the great questions in philosophy. Modern science can explain our senses, motor functions, and decision-making. It cannot, however, yet explain the mental emotions that come with the physically rooted aspects of existence. David Chalmers, philosopher and cognitive scientist, called this the "hard problem" of consciousness [4], a problem that will remain unsolved even after we reveal all that remains to uncover about our consciousness.

It is necessary to develop cognitive architecture models in order to work towards machine consciousness. Bernard J. Baars' Global Workspace Theory (GWT) [5] is the foundational cognitive architecture. It is the basis for many other AI cognition architectures, such as Stan Franklin's LIDA [6, 7], Shanahan's architecture [8], and Blum's Conscious Turing Machine [9].

We will also be using Baars' architectural theories as scaffolding. Nevertheless, we approach cognition from a different perspective than Baars; his theory stems from neuropsychology, and although it lends well to computation, it serves a different purpose than when we approach it from the direction of computer science. We are not interested in the neuronal implications of the model so much as in the architecture that can produce the illusion of consciousness in a machine.

This difference allows our work to circumnavigate some of GWT's critiques. Jeffrey W. Dalton criticized GWT for modeling how cognition functions rather than answering what cognition is and why it is conscious [10]. This is fine for our purposes, as we want to produce a machine that functions as conscious. We are not interested in pursuing the hard problem of consciousness. We also use *Rethinking Consciousness* by Michael S. A. Graziano to characterize the universal historical entity.

13.3 Objective Consciousness

Before defining universal historical consciousness, this section explores the concept of consciousness itself. Grasping the essence of consciousness is crucial for understanding universal historical consciousness and its interactions, which form the core focus of historical studies.

Michael S. A. Graziano, an American professor of psychology and neuroscience, recently wrote a wonderful book about consciousness from the perspective of neuroscience and engineering: *Rethinking Consciousness*. In it, Graziano writes that most everything, living or machine, can have a type of consciousness which he dubbed "objective consciousness [3]."

Objective consciousness is attributed to beings that take in information from their surroundings, process it, and modify their behavior accordingly. Sea anemones are conscious of prey coming into their grasp. Sunflowers are conscious of the position of the sun in the sky. A car's lane departure warning system is conscious of the segmented line on the road.

But are any of those entities conscious of their awareness? Are they conscious of their own existence? When environmental inputs affect their behavior, do they feel those changes and ponder them? We may say that they do, but that could be the human tendency to overgeneralize our own subjective consciousness. We may say that they don't, but that could be the human tendency to place ourselves at the center of the universal experience, claiming that we are the Crown of creation.

In the end, the entirety of consciousness could just be an illusion. We are interested in the behavior that generates that illusion, as it does in the Turing test. This consciousness is detected and tested by interactions between two beings. According to Hume's problem of induction, two interacting entities cannot guarantee that the other is conscious in the same way that they are [11]. We humans navigate the world, assuming that those with whom we interact are conscious. Inversely, those we cannot interact with are assumed to be without consciousness. Take the famous example of Johnny from Dalton Trumbo's *Johnny Got His Gun* [12]. An injured veteran is trapped in his own body after an artillery explosion leaves him aware and conscious but without a face or any of his limbs to communicate with. Johnny is assumed to be without consciousness or even awareness until his nurses realize that he is communicating via Morse code. We are interested in communication, even if it is a manufactured illusion of consciousness, as opposed to self-consciousness or collective consciousness.

13.4 Machine Consciousness

According to Graziano, machines must fulfill four requirements to be considered conscious.

First, they need to display artificial concurrent attention. The machine must be able to focus on one thing at a time and shift focus between targets. This implies that the machine experiences concurrent processes that compete for attention and can be moved between.

Second, they require an attention schema model. The machine must be able to describe its attention and guide it using some internal model. This is its form of subjective consciousness and the way the machine can explain itself to outsiders.

Third, they must possess a wide sensory range. The machine must have sensory inputs through which it can be conscious. The closer the system is to mimicking the human sensory experience, the closer it is to mimicking human consciousness.

Finally, machines must maintain self-conscious communication. The machine must be able to access its internal models and communicate its inner workings so that a conversation about its consciousness may be had.

However, when defining machine consciousness, we must emphasize the property of boundaries. Current models of consciousness do not state their boundaries explicitly, because assuming the boundary of human consciousness comes naturally to us: it is contained in the individual. External sources may augment one's senses, and tools may feel like natural extensions of our body. Yet, we do not typically mistake other humans for an extension of ourselves[1], nor can we copy another individual's memories and experiences exactly as they sensed them, not even through books or empathy.

Machines, on the other hand, are incredibly modular. Two computers may be linked together; their processing power will combine, their memory banks will merge, and their displays will sync. If those computers were individually conscious before they were linked by a cable, are they still individuals after the fact? If a machine is connected to the internet, is the internet a tool extending its senses or a limb of its entire consciousness? It could go either way with our modern capabilities, but there must exist a line where the 'self' of a machine turns into the 'other' of the world. A conscious machine must have its boundaries defined, whether by its software or hardware, just as our minds and bodies define the encapsulation of our consciousness.

Including boundaries in our definition opens up the use of consciousness models for many more entities in the humanities. Nations, organizations, and disciplines may all be viewed as conscious entities: A nation, for example, has a border, a sense of self, imports and exports of all kinds, and an ability to communicate its narrative with other nations in the world. As for the sequential narrative, history is a perfect example of many distributed streams, sorted (in this case by causality) and consolidated into a single sequential narrative.

[1] Pregnant individuals, parents to young babies, or neurodivergent people are more complicated topics which highlight the importance of this definition.

The narrative itself is a crux in machine consciousness, an excellent starting point for a field that does not often talk about such things. The first step in bringing machines and humans closer together is finding a common language and concepts. A narrative's definition is natural to both machines and humans, which is why it lends itself so well to modeling consciousness across the two.

13.5 Consciousness and Historical Ententes

The process of consciousness is comprised of three general elements: the defining of the self, the composition of narrative for the self, and the communication of narrative to others (Fig. 13.3).

Mathematically, we model the world as a graph $G(V, E)$. The nodes of the graph represent the entities of the world, whereas the graph's weighted or unweighted edges represent the information communicated between a node and its neighbors. We assume that we have a set of consciousness $\mathscr{E} = \{\mathscr{E}_1, ..., \mathscr{E}_k\}$ for each consciousness entity:

$$\mathscr{E}_i = \left(S^i, N^i, \Pi^i\right). \tag{13.1}$$

index$\left(S^i, N^i, \Pi^i\right)$ Each entity is composed of three elements. We use the i notation when differentiating between entities and, for simplicity's sake, ignore the i notation when talking about a singular entity.

1. The conscious entity must have a defined border. It must have a sense of self and, as a result, a sense of others. In the same vein, other entities must exist outside of the border. If there are no others to hold a boundary against, then no avenues exist to prove that the bounded entity is conscious. To define and maintain that boundary, sensory experiences are required. These senses are aimed at many directions, inside and out; they are complicated and distributed. There is a difference between sensing a high temperature of your own body or a high temperature of your environment. These senses are the basis for the entity's interaction with itself and its surroundings. Formally, the definition of the self is:

$$S^i = (V^i, \sigma^i) \tag{13.2}$$

where $V^i \subset V$ is the set of all sensors of entity i. k_i is the number of distributed streams observed by entity i, $\mathbf{T}_o$ is a set that contains the objective time of our interest, and the distributed streams are

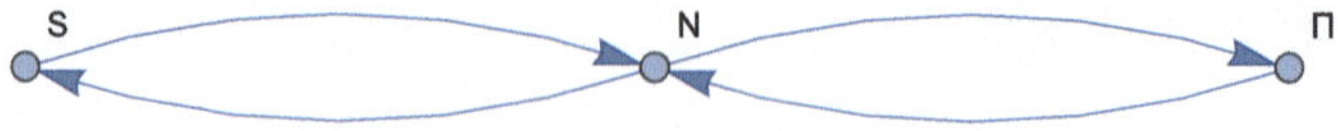

Fig. 13.3 This figure shows the relationships between the three components of consciousness: S, N, and Π

$$\sigma^i : \mathbf{T}_o \to \mathbb{N}^{k_i}. \tag{13.3}$$

Each sensory stream σ^i_j represents a single sensory channel that is attached to a connected subgraph $G^i_j(V^i_j, E^i_j)$ of the set of nodes that the channel measures and the information received from those nodes, where one of the vertices $v^i_j \in V^i_j$ must belong to the set, i.e., $v^i_j \in V^i$. The node v^i_j receives information from V^i_j through the edge E^i_j. Therefore, σ^i is a collection of all information consumed by entity i.

We sometimes shorten σ^i to

$$\sigma^i = \left(\sigma^i_1, ..., \sigma^i_{k_i}\right) \tag{13.4}$$

The set of all stream functions is

$$\sigma^i \in \left(\mathbb{N}^{k_i}\right)^{\mathbf{T}_o} \tag{13.5}$$

We have the set of nodes $V_i \subset V$ that belongs to the entity. The entities i, j conflict if

$$V^i \bigcap V^i \neq \varnothing. \tag{13.6}$$

2. The conscious entity must be able to generate a sequential narrative from its many distributed sensory streams. The consolidation of experiences into a narrative sequence is a process of attention. The conscious entity will have an attention schema that directs its inner narrative to focus on what is essential and relevant to it at each point in time. Therefore, an attention schema

$$AS : T_s \to \{1, 2, ..., k_i\} \times T_o \tag{13.7}$$

is a function that goes from the Syuzhet time of the narrative to the sensory range described in the narrative and its Fabula time. The conscious entity is aware of its own narrative and is able to interact with it: interpret it, criticize it, and learn from it. It may or may not take advantage of those abilities, but they exist.
Transforming the stream of consciousness to the narrative:

$$N^i : \left(\mathbb{N}^k\right)^{\mathbf{To}} \to \left(\mathbb{M}_n\left[\mathscr{R}\right]\right)^{\mathbf{Ts}} \tag{13.8}$$

where $\mathbf{T_S}$ is a set that contains the subjective time of the narrative.
3. The conscious entity must be able to communicate its narrative to others. Again, our existence is defined by the existence of others; we only have use in defining ourselves if there are others[2] to define ourselves against. This concept is

[2] In historical narratives sometimes there is a single historical entity. In this case, the other is the reader for the author.

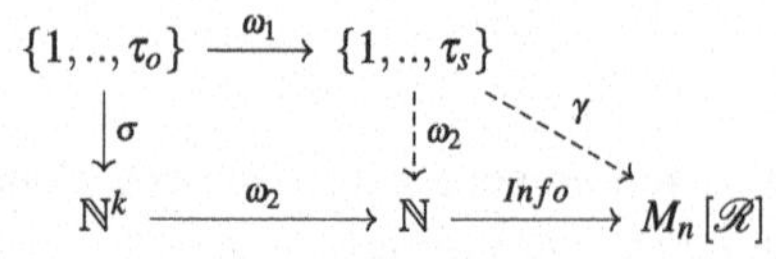

Fig. 13.4 Function diagram that describes $\Pi = \gamma(\omega_1(t))$

found in the works of philosophers such as Jean-Paul Sartre and Emmanuel Levinas [13, 14]. Furthermore, we humans rely on the same mechanisms that we use to recognize our consciousness in order to recognize it in others. The problem of other minds remains divisive, but our experience as humans tends to be that of inferring others' consciousness through communication and other similarities we behold from others [15, 16]. In the same way that we prove ourselves conscious by interacting with the world around us, communicating our inner world through our actions and words, so must any conscious entity prove its consciousness in order to be perceived as such.

Π^i is a vector of communication of the entity i across the rest of the entity.

$$\Pi^i : (\mathbb{M}_n [\mathscr{R}])^{\mathrm{Ts}} \rightarrow (\mathbb{N})^{\mathrm{Ts}} \tag{13.9}$$

The entity generates its narrative using $\mathbb{N}$. Figure 13.4 describes the process of converting the stream of consciousness into a dynamic matrix/graph. The advantage of this process is that it allows to describe a causal relationship, as we will demonstrate later.

The final elements in our model is the Π^i dialog function.

13.6 Life Cycle of a Narrative

When we talk about machines communicating with humans, we are actually talking about the machine's narrative. The ability to provide a narrative is connected with consciousness, see Fig. 13.3 in Sect. 13.5.

From a system architecture point of view, there are three stages in the life cycle of a narrative . A narrative is conceived, processed, and finally communicated. The conceptual separation may be less clear-cut in practice, but abstraction is critical for the building of a system.

When we try to analyze a narrative, it is useful to think about an author, a translator, and a reader as representations of three separate life cycles of a narrative.

The author will think of a story, refine it in the writing process, and publish a book. The translator will read a story in one language, transform its contents into a different language and culture, and publish the translation. The reader will consume the text of a book, draw their impressions and understandings of the novel, and tell others about their reading experience.

The three stages within each life cycle imitate the three components of consciousness: definition of the entity (conception), causal transformation (processing), and communication. The distinction between the three stages is useful to us in translating the life cycle of a narrative to machine architecture. Even as soft definitions, the three stages represent the ways in which a machine can interact with a narrative. A conscious machine will exhibit the three aspects when inhabiting any role.

This can be compared to the architecture of a governing body with its legislative, executive, and judicial branches: A law is written, passed, and enforced.

Next, we discuss the three conceptual life cycles of a narrative in greater detail within our machine consciousness model.

13.6.1 The Author as a Conscious Machine

The world of the author, $\mathscr{A}$, is the author's conceptual field and language. The nodes within this world are the characters, and the edges connecting them are their relationships. The author's sensory range is their personal experiences and imagination, S, which is made of V and σ.

The reader often does not get to see the background and the sensory range of the author. The perspective of the story determines which subgraph the reader is exposed to. An omniscient narrator will be able to communicate every node and edge within the graph. A narrator limited to a single perspective, however, will only be able to communicate the perceptions, opinions, and actions of a single node. Many narrator perspectives fall on a spectrum between the two examples.

Composing the story is a complicated process, N, which takes the many sensory streams of the characters and turns them into a causal narrative. The evolution of the story's social network in the author's mind happens over time $\mathbf{T_O}$. What is communicated in the final story happens over time $\mathbf{T_S}$, as the time of the final narrative may not be linear. The events that happen over time to change the network are causally linked and communicated from the perspective that the author has chosen for the story. The resulting sequential narrative $(\mathbb{M}_n\, [\mathscr{R}])^{\mathbf{T_S}}$ is communicated to others in the vector Π (Fig. 13.5).

13.6.2 The Translator as a Conscious Machine

Much like the author, the world of the translator, $\mathscr{A}$, is the translator's conceptual field and language. The nodes within this world still represent characters, and the edges remain the characters' relationships. The first difference between authoring and translating a narrative, then, is the translator's sensory range, S. It includes the original manuscript in addition to the translator's personal experiences and imagination. Good translation augments the original manuscript with properties of a new

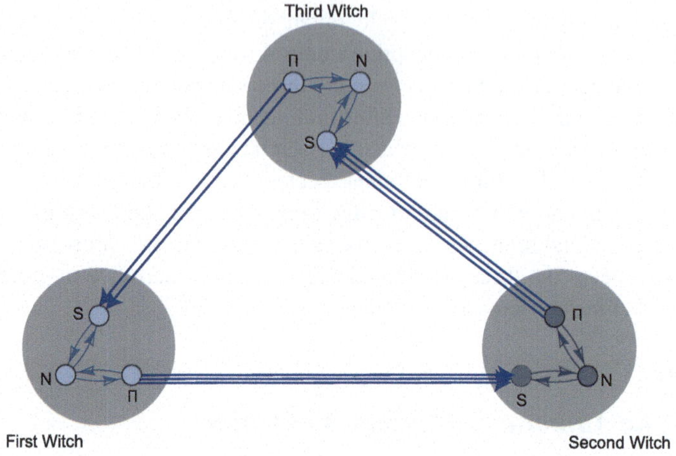

Fig. 13.5 This figure shows the reciprocal system between the three witches from Shakespeare's *Macbeth*. The communication between entities is represented by the arrows, each arrow depicting a spoken line, extending from an entity's Π to another's *S*

medium, whereas mediocre translation may lose some of the original's narrative by altering it entirely.

13.6.3 The Reader as a Conscious Machine

Now that we have described what each machine does, let's talk about the connection between the life cycle of a narrative and universal machines.

13.7 Universal Consciousness

Turing's proof of a universal machine, a machine able to simulate any other machine in existence, is a cornerstone of computer science. Without the insight as to the existence of such a machine, humans would have continued like manual computers without ever being replaced.

The issue of conscious machines raises the question of universal consciousness. A machine with universal consciousness would be able to simulate any objective or subjective consciousness. This means that, in its universality, the machine would achieve the same level of consciousness as Homo sapiens, or surpass it.

Many believe that we are not universally conscious and that there are experiences and understandings beyond our grasp-scientific mysteries that remain unsolved, or the divine, to name just a few. Some may claim that if God is a universal consciousness, then Homo sapiens are not.

For the purpose of this discussion, we speak of universality in the context of science, within our understanding of the world and ourselves. Homo sapiens can simulate other consciousnesses- think of actors who inhabit the minds of other beings, whether real or fictional, and act out their inner dialogue; think of authors, readers, and others. We are all constantly simulating other beings' subjective experiences. Homo sapiens could try to understand and inhabit the mind of an animal, but an animal would not be able to do the same.

Universality among the Homo sapiens is characterized by our species' ability to create. If we cannot sense something, we construct technology that extends our sensory capabilities. If we cannot process something, we build a model and the equipment required to process it. If we cannot communicate a concept, we birth language to encapsulate it better. And if we fail to build something, we create an understanding of why, as we did with the second law of thermodynamics after failing to build a perpetual motion machine. In this way, the universal consciousness of Homo sapiens depends on the eventual building of a conscious machine or the ability to explain why it is impossible.

A machine with universal consciousness could create in the same capacity as Homo sapiens. It would expand on the three components of consciousness as if it were Homo sapiens.

Let's look at the three components and their sub-components applied to Homo sapiens' universality:

1. Universal S
   a. Border. Homo sapiens can draw universal borders, defining any possible factions, organizations, or entities.
   b. Sensing. Homo sapiens' sensory capabilities are universal by extension of our technology. Inputs beyond our natural senses, such as ultraviolet light, can be perceived through equipment that captures them for us.
2. Universal N
   a. Attention. Homo sapiens' attention is universal. We may direct our focus and comprehension toward any realistic input with suitable equipment.
   b. Causality. Homo sapiens can assign causal relationships between events, whether that relationship is objectively correct or not, and edit that relationship with new information.
3. Universal Π
   a. Interpretation. Homo sapiens are universal interpreters capable of deriving meaning from any narrative.
   b. Communication. Homo sapiens are universal communicators. Anything not captured by our many languages, whether natural, mathematical, physical, or otherwise, may be added to our lexicon and communicated to any other entity using our technology.

Once we can replace Homo sapiens with machines in the above list, we will have machines with universal consciousness.

This ability to create defines Homo sapiens and sets them apart from other beings. Machines today cannot build everything by themselves; they may create in cooperation with Homo sapiens, but left to their own devices, a machine will not possess universal building capabilities, at least not within the bounds of current machine programming.

It is this difference that sets Homo sapiens apart from animals, as well. Whales, as smart and communicative as they are, cannot create ways to measure gravitational waves.

The use of "Homo sapiens" instead of "human" in this section is purposeful. We do not list other humanoids such as the Homo erectus or the Neanderthal, as there is evidence that they were not in possession of the same universality. Neither Homo erectus nor Neanderthals developed agriculture or animal domestication [17]. It is specifically the taming of canines that demonstrates Homo sapiens' ability to communicate universally, even with other species, and the others' inability to develop a technology of the same kind that shows a deficit. Homo sapiens out-created all other humanoids to become the dominant species, as described in Pat Shipman's fascinating book *The Invaders: How Humans and Their Dogs Drove Neanderthals to Extinction* [18].

Dogs' partnership with Homo sapiens showcases different dimensions of consciousness and how this can contribute to universality. Dogs have a much more developed sense of smell than humans; their conscious experiences exist on a level of olfaction that humans cannot access alone, just as humans experience consciousness with visual dominance. When we train dogs to utilize their sense of smell for our purposes, we are expanding our consciousness by using their senses as technological proxies of our own.

13.7.1 Difficulties in Universal Consciousness

It is not easy to achieve universal consciousness, which is a property of a collective rather than an individual. It took billions of years for life on Earth to achieve universal consciousness. Let's look at some of the difficulties inherent in universal sensing and communication processes.

Humans have been using tools to enhance their sensory capabilities since the discovery of fire. But expanding our senses through tools is not an easy process. It requires building scientific theory, methodology, and a language to communicate any newly measured phenomena. These needs can be attributed to both intelligence and culture. For example, consider the discovery of ultraviolet light. According to Wikipedia, "UV radiation was discovered in February 1801 when the German physicist Johann Wilhelm Ritter observed that invisible rays just beyond the violet end of the visible spectrum darkened silver chloride-soaked paper more quickly than violet light itself." In this case, the silver chloride-soaked paper is used as a machine that can sense UV radiation. Using scientific methods, humans could conclude the existence of UV radiation.

Any machine that we build requires resources- the energy and materials for building it and keeping it operational. There are many parameters to consider when building a machine. Constructing a machine takes patience and care; a hammer too big or too small will not service certain nails well. The machine's environment needs to be taken into account, as well: for instance, roads for cars and docks for ships.

Universal communication also has its limits. Communication between any two entities, human or otherwise, faces the problems of noise, language barriers, and differences in perception. Returning to our canine example, humans developed a shared language with dogs (such as a command to sit, be quiet, etc.) despite differences in intelligence, perception, and modes of communication. The same barriers are found with machine communication.

13.7.2 Collective Universality

The modern human is not universal in itself. As social creatures, it is the aggregate of humanity that is universal. This is a defining feature of the species, not of the individual. As individuals, we each have our strengths and weaknesses: Shakespeare might not have had the potential to learn differential calculus; Albert Einstein might not have had the potential to write an engaging stage play. It is at the level of societies that we encounter universality.

13.8 Historian Machine

We are now ready to define how and what a historian machine m_h does.

The first step in the workflow of a historian machine is to define two or more $k \geq 2$ historical entities that share a common time and conflict. Each historical entity is a universal consciousness, meaning that the i-th historical entities for each are formally $1 \leq i \leq k$. We denote the i-th historical entities by $\mathscr{E}_i = \left(S^i, N^i, \Pi^i\right)$, see Eq. 13.1. Denote the set of all the possible i historical entities by $\mathbb{E}_i$. Note that since times T_o, T_s, and k are finite, the set $\mathscr{E}_i$ is well defined.

Denote the set of all k historical entities by

$$\mathbb{E}^k = \prod_{i=1}^{k} \mathbb{E}_i \tag{13.10}$$

The historian machine m_h takes all those entities and transforms them into a historical narrative using an attention schema. For more details, see Sect. 13.5. The output of the historian machine is

$$h : T_{syuzhet} \to \mathbb{M}_k\left[\mathscr{R}\right] \tag{13.11}$$

As usual, we denote the set of all functions from $T_{syuzhet}$ to $\mathbb{M}_k\left[\mathscr{R}\right]$ by

$$\left(\mathbb{M}_k\left[\mathscr{R}\right]\right)^{T_{syuzhet}} \tag{13.12}$$

Therefore, the historian machine m_h can be defined as a function that takes input from the set $\mathbb{E}^k$ and produces a historical narrative. Formally,

$$m_h : \mathbb{E}^k \to \left(\mathbb{M}_k\left[\mathscr{R}\right]\right)^{T_{syuzhet}} \tag{13.13}$$

We end this section with a few remarks related to the historical narrative. First, a historical narrative is a narrative, and therefore, the function h that the machine h_m generates should have at least two interesting interpretations. Generally speaking, the interpretation of a narrative transforms one narrative into a reasonably different one. We need at least two of these to make a narrative interesting for humans.

Not all functions m_h are legitimate historian machines. One essential criterion is that the historian machine should link to current events and explain something about the present situation in the world. If the narrative does not link past events to the present state of the world, it is not a historical narrative.

13.9 From Timeline to Historical Entities

In the next three sections, we will use the Wiki timeline of WWII that we have already discussed in Chaps. 4 and 5 as input and generate a narrative from its historical entity and its attention schema. Similarly to our example, any timeline can be transformed into historical entities with many attention schemes.

Different historical entities allow us to generate different narratives. We can use the attention schema as a skeleton for a narrative. Specifically, we can mechanically construct prompts that follow the attention schema and feed those prompts to any LLM. To summarize, the input in the next three sections is any timeline, and the output is prompts that will generate a narrative following the attention schema of the historical entities. Therefore, the historical entities can be seen as a method for mechanically generating prompts from any timeline.

In Sect. 5.2, we introduce the concept of the interval graph. Since any event in the timeline can be represented as a time interval, we can compute the relevant interval graphs. The interval graphs of the WWII timeline can be seen in Fig. 5.4.

13.9.1 Interval Caterpillar Graph

Now, we can add to each main event (a red node in Fig. 13.6) its sensory range which is the hyperlink that points to the Wiki page. In this simple example, each historical event points to a single Wiki page which contains many hyperlinks to all the events that together form this main historical event.

In general, we start with a timeline. Each historical event has a specific time interval, place, and meaning. The meaning in Wikipedia contains hyperlinks to other web pages that explain the meaning of the historical event. We first use the time interval to generate interval graphs where each node is a historical event along the timeline. We then add to each of these nodes its hyperlinks as a node. The new graph is called an *interval caterpillar graph*. Note that, in general, the body of the interval caterpillar is the interval graph computed from the timeline and its legs—the yellow nodes—are the hyperlinks that explain the historical event in detail.

We borrow the notion of interval caterpillar graph from caterpillar tree, i.e., a tree in which all vertices are within distance 1 from a central path. For more on caterpillar trees, see [19]. By analogy, an interval caterpillar graph is a graph where all nodes are within distance 1 from a central interval graph.

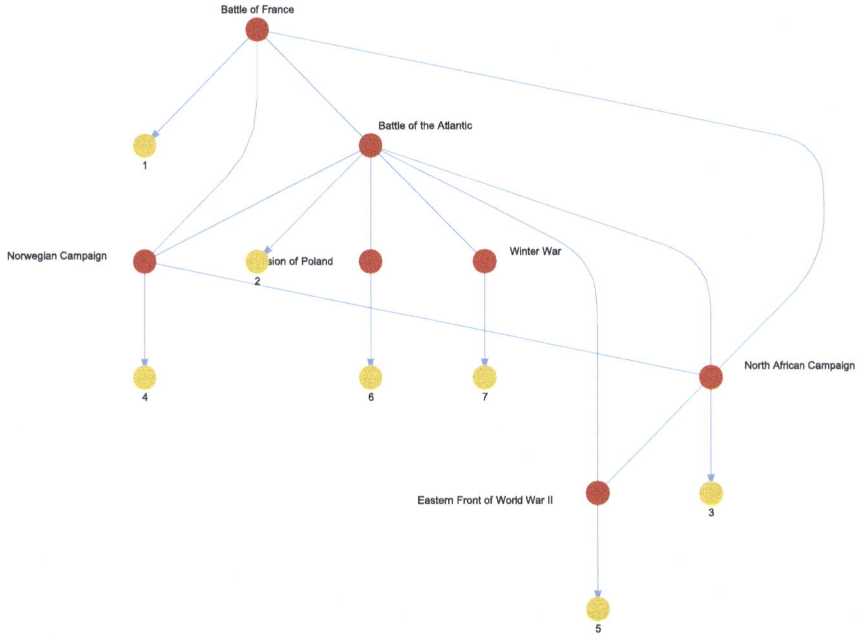

Fig. 13.6 This figure shows the caterpillar interval graphs of the seven main events of the WWII timeline, where red nodes are the seven historical events and yellow nodes are their sensory ranges. In this example, each historical event is a single Wiki page containing its timeline and many hyperlinks

Let us explain the connection between the interval graph caterpillars and the historical entities. The historical entities are the interval graphs. The yellow nodes that we just added to the interval graphs are the sensor ranges of the historical entity. Therefore, the interval graph caterpillar in Fig. 13.6 can be used to generate the attention schema necessary for the narrative.

After constructing the interval graph caterpillar of the historical entities using the timeline, we are ready to explain how to construct the attention schema out of the interval graph caterpillar. In essence, the attention schema is a *weighted interval graph* of the timeline where the weights are on the vertices rather than on the edges, as we would expect, and all weights are non-negative. The weights will indicate the amount of attention each node receives from the attention schema.

13.10 Attention Schema of Interval Graph Caterpillar

In this section, we provide several attention schemes. Then, we use our example of the seven main historical events of WWII to show the difference between those attention schemes.

An attention schema can be deconstructed into two independent components. The first is the order in which we scan the node of the interval graph caterpillar. The second is how much time is spent on each event.

We can scan the interval graph caterpillar in many ways. One way to do that is by using depth-first search (DFS). For more on DFS, see [20]. Another way to scan is to use breadth-first search (BFS). For more on BFS, see [21]. It is better to monitor the nodes to preserve the timeline. Therefore, DFS is preferable to BFS. Any other order is acceptable but may be less clear for the historical timeline. Note that it is not always possible to maintain the timelines of events; see Exercise 1.

Note that in our example, we only change the weights of the sensory range and not the order in which we visit the historical event. Since there are seven main historical events, we can scan the graph in seven ways just by reordering historical events. We can have infinite number of ways to visit the nodes since we can visit the nodes several times. However, we ignore this option and do not recommend applying it if we wish the narrative to remain simple.

One way to overcome the problem of scanning the caterpillar interval graphs when using DFS is to have one paragraph when we first visit the node, which describes the beginning of the interval, and another paragraph that describes the end of the interval when we finish visiting all its descendants. We call this approach *reflected DFS*. It creates a narrative that is more reasonable and flexible and allows one to reflect on the historical events in some chronological order. The next exercise focuses on this approach.

After we have determined the order in which we scan historical events of the blue sensory range, the attention schema needs to determine how much attention we can give to each of these events/nodes. A natural number is sufficient to determine the amount of time/(words)/(paragraphs) that our attention schema will focus on that specific sensory range. If a blue node receives the number 0, it means that our attention schema will ignore that specific sensory range.

To summarize, the attention schema of the historical entities is composed of two elements: the order in which we scan the caterpillar interval graph and the number we give for each sensory range. These determine the amount of attention the sensory range will receive during the visit to the sensory range.

Therefore the attention schema is

$$AS : T_s \rightarrow \{1, 2, ..., k_i\} \times T_o, \tag{13.14}$$

We can simplify the definition of the attention schema when writing the history of the timeline by determining the order in which we visit the nodes in the caterpillar interval graphs $G(V, E, W)$ together with positive weights of the sensory range:

$$AS : T_s \rightarrow V \times \mathbb{N} \tag{13.15}$$

The interval graph caterpillar, together with the attention schema, is the skeleton of our narrative. Once we determine the skeleton, we can feed it to any LLM and receive a historical narrative.

In the next subsection, we provide several simple examples of how to produce the numbers that we give for each sensory range.

13.11 Attention Schema Examples

In this section, we describe several attention schemes, then summarize the weights in the table for all the attention schemes on our WWII example, see Fig. 13.7.

The advantage of this approach is that it allows historians to control the flow of the narrative using the hyperlinks that were computed by our attention schema. Each paragraph is written by the LLM, while the complete narrative is written by a historian who controls the attention schema.

Battle\AS	# Hyperlinks	# Words	# Views	Page Rank Centrality	# killed
Invasion of Poland	1924	18 691	8 094 906	0.159756	933 700
Battle of France	1979	25 903	8 194 685	0.142068	2 246 587
Battle of the Atlantic	1889	22 444	4 286 223	0.0909721	102 700
Winter War	2261	22 675	10 020 756	0.203575	451 000
Norwegian Campaign	1887	18 714	901 773	0.083625	11 898
North African Campaign	1223	8225	2 424 611	0.0684285	7 487 000
Eastern Front of World War II	2945	28 251	7 433 927	0.251576	1.51×10^7

Fig. 13.7 This Table summarizes the weights of each of five attention schemes for the WWII historical event

13.11.1 Number of Hyperlinks Attention Schema

In this attention schema, we count the number of hyperlinks for each sensory range that relates to the historical event. We can pick k historical events that have a maximum number of hyperlinks. For our example, we compute the number of hyperlinks for each of the seven main campaigns in WWII, see the second column in Fig. 13.7 for the exact number. We can use any function of those numbers to determine our hyperlink attention schema. The blue line in Fig. 13.8 summarizes the second column graphically. Next, we describe the number of words attention schema.

13.11.2 Number of Words Attention Schema

Note that each sensory range is a hyperlinked to the Wiki page. We can go to that Wiki page and read the number of characters or the number of words needed to describe that specific Wiki page. The third column in Fig. 13.7 summarizes the number of words in each hyperlink sensory range. The yellow line in Fig. 13.8 summarizes the third column graphically.

Now, any function of this number can determine the attention we give each sensory range/Wiki page. For example, one may take the identity function.

Next, we explain the page view attention schema.

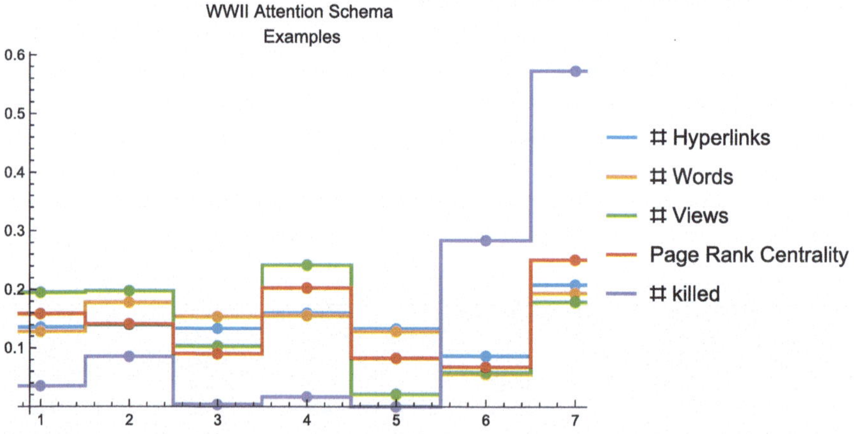

Fig. 13.8 This figure shows the relative weights of each of the seven historical events in each of the five attention schemes

13.11.3 Page Views Attention Schema

The third attention schema is similar to the previous one. However, instead of counting the words, we can base our attention schema on the number of people who read the Wiki page. In general, the Wikimedia API allows us to get the number of viewers for each Wiki page. We can use these numbers to build our attention schema. For more on Wikimedia, see Sect. 4.6.5. Looking at the fourth column in Fig. 13.8, we see that the Winter War gets the most amount of views. This is probably due to the interest in the Tv show Game of Thrones' Winter War. This example shows that one needs to be careful when using this attention schema. It appears in the green line at Fig. 13.8.

13.11.4 Centrality Attention Schema

Another way to construct an attention schema is to take the list of all hyperlinks and construct a graph $G(V, E)$ where the nodes V are the Wiki pages. Let l_i be the list of all hyperlinks that appear on page i. Now, two Wiki pages $i, j \in V$ are connected by an edge $(i, j) \in E$ with a weight equal to the number of hyperlinks shared by both Wiki pages.

$$w_{i,j} = |l_i \cap l_j| \tag{13.16}$$

However, if we do that, the page rank gives us uniform weight, since all Wiki pages share many common hyperlinks. To overcome this problem, we remove the minimum number of common hyperlinks between all pairs. Therefore:

$$w'_{i,j} = w_{i,j} - \min_{i,j \in V} w_{i,j} \tag{13.17}$$

Now, we can compute the page rank centrality or any other measure for each node and concentrate only on the most important nodes where the weight of the edge $\{i, j\}$ is $w_{i,j}$. The fifth column in Fig. 13.8 summarizes the page rank centrality. Note that the maximum page rank centrality appears in the Eastern Front of World War II, which was the bloodiest front, as the sixth column shows.

Our last attention schema is related to the number of causalities, which is usually the most interesting to the readers.

13.11.5 Causality Attention Schema

Our last attention schema is related to the number of people killed in each frontier. The last column in Fig. 13.8 represents the number of causalities in each frontier.

In the next subsection, we summarize all the attention schemes.

13.11.6 WWII Attention Schema Examples

This subsection summarizes the result of several attention schemes described above.

13.12 Historical Narrative of Two Timelines

Up to this point, we explained how to construct a historical narrative of a single timeline using a historian machine. In this section, we will explain how to take two timelines and generate a combined narrative. We will provide two examples: one for a war or competition between two timelines, and the other for cooperation between two timelines.

13.12.1 War Between Timelines

In this section we will explain how to construct from two timelines TL_1, TL_2 two historical entities $\mathcal{E}_1, \mathcal{E}_2$. We then construct two attention schemes: one that emphasizes the competition/war As_w between two historical entities, and the other that emphasizes the cooperation As_c between two historical entities.

When the historical entities are competing with each other, each of them wishes to emphasize its uniqueness. They therefore ignore the common ground they share. For more information, see Chaps. 9 and 10.

The simplest way to generate attention schema for two sides that agrees to disagree is to delete all the events they share in common. After we eliminate all the common events from both timelines, each timeline concentrates its narrative on the non-common historical events. Now we can combine those two narratives into a single narrative by constructing an attention schema $AS_{1,2}$.

In the next subsection, we describe how two historical entities can cooperate and generate a single narrative that emphasizes cooperation.

13.12.2 Cooperation Between Two Timelines

Contrary to war, when two historical entities wish to cooperate, they emphasize their common historical events. They will therefore eliminate the historical events that are not common to both of them.

Once we compute the intersection of the hyperlinks, we can use both caterpillar interval graphs to construct a single interval graph caterpillar with attention schema, which can be used to generate a common narrative.

An interesting observation is that a common hyperlink, i.e., the Wiki page hyperlinks that appear in both timelines, usually appear with other hyperlinks that are not common to both timelines. Those non-common hyperlinks can be used to generate conflict between the timelines.

13.13 What is History

To construct a historical narrative, we require several key components. First, we need a collection of historical documents, denoted as

$$\mathcal{H} = (h_1, ..., h_k)\,. \tag{13.18}$$

We sometimes call these documents the archives $\mathbb{A}rc$, which serve as the foundation for historical research. Next, we must have a set of historical entities, represented by

$$\mathcal{E} = \{\mathcal{E}_1, ..., \mathcal{E}_m\}, \tag{13.19}$$

which include nations, institutions, individuals, and other actors that play a role in shaping history. For more on historical entities, see Sect. 13.5. The process of writing history also requires authors, $\mathcal{A}$, who take distributed streams of historical events and analyze and compile them into coherent narratives. For more about the authors, see Sect. 13.6.1. Additionally, the medium of representation, $\mathcal{M}$, plays a crucial role in how history is conveyed, whether through books, videos, digital archives, or other formats. Finally, historical narratives must have multiple interpretations, denoted by $\Lambda = \{\lambda_1, \lambda_2, \lambda_3, ..., \lambda_{int}\}$,

$$\lambda_i : \mathcal{M} \to \mathcal{M} \tag{13.20}$$

λ_j reflects different perspectives, biases, and analytical approaches. Note that any historical narrative must have at least three historical interpretations: λ_1 for time, λ_2 for location, and λ_3 for meaning.

Since any medium $\mathcal{M}$ can be encoded by universal machine as a natural number $\mathbb{N}$ using Shannon's information theory, we can represent historical narratives numerically with an encoding function λ_{en} and a corresponding decoding function λ_{dec}. This means that all forms of historical representation, whether textual, visual, or oral, can be mapped to sequences of natural numbers. Consequently, the process of writing and analyzing history can be treated as a structured transformation between encoded and decoded representations, allowing for formal manipulations and computational methods in historical research. To conclude history, we need the tuples

$$(\mathcal{H}, \mathcal{E}, \mathcal{A}, \mathcal{M}, \Lambda) \tag{13.21}$$

13.14 Eliminating Negative Probabilities Using the Model of Historian Machine

Most of the chapters in the second part of this book bring us face to face with negative probabilities. The concept of negative probabilities is, in some sense, hard to accept and possibly unnecessary when we use the models developed in this chapter.

The main idea is to model a node in the social network as a historical entity. This means that each node i will have an identity with consciousness $\mathscr{E}_i = \left(S^i, N^i, \Pi^i\right)$. as described in Sect. 13.5. Now, when the nodes in the social network have a more complex structure, we can use this structure to eliminate negative probability. We provide several examples below.

The reason we used negative probabilities in previous chapters is that negative and positive numbers are, in a sense, the essence of conflict; they oppose and eliminate each other. In many cases, math tends to be simpler but less methodical because of that. An alternative approach is to model probabilities as positive numbers in the traditional way but increase the complexity of the model, especially its state, to allow negative and positive opinions or probabilities. In fact, when we write -1, we add a single bit before the number, which determines if the number is positive or negative. We can write the number minus one as $(-, 1)$. We can do the same for each state and, shifting it from i

$$i \to \begin{cases} (+, i), & \text{if } i \in \mathbb{N} \\ (-, i) & \text{if } 0 < -i \in \mathbb{N} \end{cases} \tag{13.22}$$

For example, in Chap. 11 Sect. 10.4.1 we use negative probability in M_1. However, we can increase the set of states from three to six by splitting each state into positive and negative. In this case:

$$V = \{1_+, 1_-, 2_+, 2_-, 3_+, 3_-\}. \tag{13.23}$$

or

$$V = \begin{bmatrix} 1 & 2 & 3 & 4 & 5 & 6 \\ 1+ & 1\_ & 2+ & 2- & 3+ & 3- \end{bmatrix} \tag{13.24}$$

so that M_1 is transformed into

$$\begin{bmatrix} 0 & 0 & \frac{1}{2} & 0 & 0 & \frac{1}{2} \\ 0 & 0 & 0 & \frac{1}{2} & \frac{1}{2} & 0 \\ \frac{1}{2} & 0 & 0 & 0 & 0 & \frac{1}{2} \\ 0 & \frac{1}{2} & 0 & 0 & \frac{1}{2} & 0 \\ 0 & \frac{1}{2} & 0 & \frac{1}{2} & 0 & 0 \\ \frac{1}{2} & 0 & \frac{1}{2} & 0 & 0 & 0 \end{bmatrix} \tag{13.25}$$

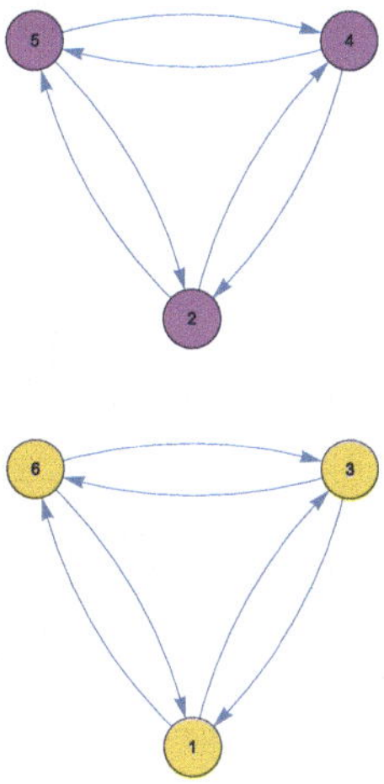

Fig. 13.9 This figure shows the alternative model for the matrix M_1 without negative probability. The alternative model doubles the number of steps from three to six. Note that nodes 1, 3, and 5 are modeling positive probabilities, and nodes 2, 4, and 6 are modeling negative probabilities. The relation between the node names in the figure is given by an Eq. 13.24 , i.e., node 1 is 1_+ and node 6 is 3-

The situation becomes more complex when we try to eliminate the negative probability in Chap. 11. Each node in the system is transformed from two to three (Fig. 13.9)

$$i \to \begin{cases} (+, i), & \text{if } i \in \mathbb{N} \\ (-, i) & \text{if } 0 < -i \in \mathbb{N} \\ (\infty, i) \text{ if the probability was annihilated} \end{cases} \tag{13.26}$$

Note that we said that the probability was annihilated if a positive probability meets and cancels a negative probability.

Things become even more complicated when we incorporate propaganda, since the total probability is no longer constant but increases all the time. In this case, we fix the sum of the positive and negative probabilities to be 1, but the number of probabilities that leave the system is no longer bound. This can be fixed by adding more complexity to the space in the model. We can reshuffle the probability that enters the state (∞, i) according to the propaganda method. Doing so will force the total sum of the probability in states (∞, i) to be unbounded. Since we use the model of propaganda, we can see that in each step by multiplying the transition matrix by the propaganda factor $r > 1$ there. This is because at time t, the total probability of being in the system is 1. After multiplying by a factor of r, the total probability in the system + leaving the system is r, since the probability of being in the system is still 1, the total mass that left the system is $r - 1$. So, the total mass that left the system after we multiply the matrix by r, t times, and the total mass that left the system is $(r - 1)t$, which is unbounded.

As we claim in this chapter, it is always possible to eliminate negative probabilities by increasing the system state from a single simple node i to mathematical entities $\mathscr{E}_i$.

13.15 Conclusion

In this chapter, we defined the universal consciousness machine. In the same way, the Turing machine can emulate any machine, the universal consciousness machine can emulate any historical entity see Sect. 13.8. We used this idea to define a historian machine.

The advantage of working with a historian machine is that it provides a clear schematic way to generate general historical narratives. We start with the timeline. From the timeline, we construct the interval graphs of the interval timelines see Sect. 13.9, and from there, we add the sensory range, which defines the border of the historical entity. In general, a historical entity is a caterpillar interval graph. In the end of this section we describe the relationship between Fabula and Syuzhet using the Fabula-Syuzhet diagram.

Now, we need to add the attention schema to the historical entity. This can be described by two independent decisions. The first decision is the order in which we scan the caterpillar interval graph see Sect. 13.10, which is a historical entity. The second decision for each sensory range/node is the amount of attention we wish to provide when we visit the historical event, see Sect. 13.11.

The last ingredient is to write a prompt for any LLM to generate the narrative related to the historical entities together with the attention schema. To demonstrate the idea, we used the quantum mechanics timelines, see Sect. 13.12

In Sect. 13.14 we used a consciousness entity $\mathcal{E}_i$ to eliminate negative probability from our models in the previous chapters.

13.16 Exercise

1. Let the timeline of a historical entity $\mathcal{E}_1$ be composed of one interval event and one point event $I_1 = [1, 4]$, $I_2 = [2, 2]$.
   a. Compute the interval graph relevant to the timeline.
   b. There are two ways to scan the interval graph. Provide two different meanings of the timeline where the order makes sense.
   c. Describe the Fabula-Syuzhet diagram of this timeline.
2. Consider a general interval graph $G(V, E)$ which is generated from the intervals $I_i = [a_i, b_i]$ for $i = 1, ..., n$. Assume that we scanned the graph according to the DFS order where we always visit v_i the descendants of the node according to the beginning of the intervals, i.e., if v_{j_1}, v_{j_2} are children of the node v_i and node v_{j_1} is scanned before node v_{j_2}, then $a_{i_1} \leq a_{i_2}$. Plot the Fabula-Syuzhet graph.

3. Repeat what was done in Sects. 5.2, 13.10 for the timeline of quantum mechanics [22] but only consider inherent events, i.e., delete the background historical events from the timeline. Note that there will be more connected components as the historical events are used as connectors to many different inherent events. You can find the date and the hyperlinks of the quantum mechanics timeline in Google sheet format at [23].
4. In this exercise, we use the Wiki categories [24].
   a. Generate two historical entities: one for the Romans and one for the Barbarians. You may use the people appearing in the categories to define historical entities.
   b. Explain how to use the Wiki page [25] as an attention schema for the historical entities.
5. Repeat Exercise 4 for the first Jewish-Roman War. You can use [26] to collect all the names that play the major role in the conflict. You can use [27] as a digital source in English, the book written by Josephus about the Jewish-Roman War, as an attention schema.
6. In Chap. 10 we use negative probability to describe the process of generating collective memory and agreeing to disagree among different groups. In Sect. 13.14 we explain how to eliminate a negative probability and give an example on the matrix M_1, see Eq. 10.46 and get the new matrix (13.25). In Exercise 11, you were asked to repeat the calculation on a network with five nodes and two opinions, see Fig. 10.8. Repeat the calculation done in Sect. 13.14 to eliminate the negative probabilities that appear in collective memory formation games, related to Fig. 10.8.
7. In this exercise, you are asked to generate a historical entity of quantum mechanics,
   a. Use the timeline of quantum mechanics, see [22] to generate the interval graph caterpillar of the timeline of quantum mechanics, compare your result with Fig. 13.10 that was generated in 2025.
   b. Use one of the attention schemes mentioned in Sect. 13.11 to generate a narrative of quantum mechanics with three paragraphs.
8. Explain how to eliminate negative probabilities as presented in Chap. 11.
9. Explain how to eliminate a negative rate as presented in Chap. 12.

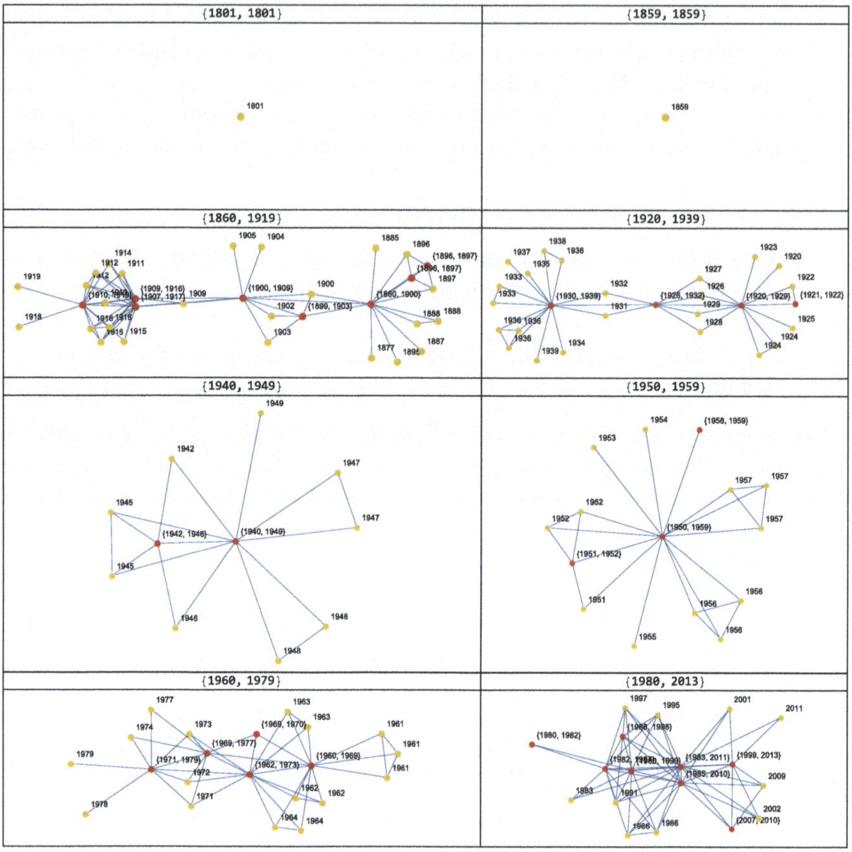

Fig. 13.10 This figure shows the interval graphs of the quantum mechanics timeline, where yellow nodes are point events and red nodes are interval nodes. The years of the connected component of the interval graph appear above the graph

References

1. G.W. Friedrich Hegel, *Lectures on the Philosophy of History* (G. Bell and Sons, 1861)
2. R. Descartes. *Discourse on Method and Meditations on First Philosophy* (Hackett Publishing, 1999)
3. M.S.A. Graziano, *Rethinking Consciousness: A Scientific Theory of Subjective Experience* (WW Norton and Company, 2019)
4. D.J. Chalmers, Facing up to the problem of consciousness. J. Conscious. Stud. **2**(3), 200–219 (1995)
5. B.J. Baars et al., *In the Theater of Consciousness: The Workspace of the Mind* (Oxford University Press, USA, 1997)
6. B.J. Baars, S. Franklin, Consciousness is computational: the lida model of global workspace theory. Int. J. Mach. Conscious. **1**(01), 23–32 (2009)
7. S. Franklin, F.G. Patterson Jr., The lida architecture: adding new modes of learning to an intelligent, autonomous, software agent. Pat **703**, 764–1004 (2006)

8. M. Shanahan, A cognitive architecture that combines internal simulation with a global workspace. Conscious. Cogn. **15**(2), 433–449 (2006)
9. M. Blum, L. Blum, A theoretical computer science perspective on consciousness. J. Artif. Intell. Conscious. **8**(01), 1–42 (2021)
10. J.W. Dalton, The unfinished theatre. J. Conscious. Stud. **4**(4), 316 (1997)
11. D. Hume, *A Treatise of Human Nature* (Clarendon Press, 1896)
12. D. Trumbo, *Johnny Got His Gun* (Citadel Press, 2007)
13. E. Levinas, The trace of the other. *Deconstruction in Context*, pp. 345–359 (1986)
14. J.-P. Sartre. *Being and Nothingness: An Essay in Phenomenological Ontology* (Citadel Press, 2001)
15. A. Hyslop, *Other Minds*, vol. 246 (Springer Science and Business Media, 2013)
16. A. Hyslop, F.C. Jackson, The analogical inference to other minds. Am. Philos. Q. **9**(2), 168–176 (1972)
17. J. Glausiusz et al., How dogs became humans' best friends: from neanderthals to now. Nature **599**(7886), 553–554 (2021)
18. P. Shipman, *The Invaders* (Harvard University Press, 2015)
19. Wiki. wiki-caterpillar-tree (2024). https://en.wikipedia.org/wiki/Caterpillar_tree
20. Wiki. wiki-depth-first-search (2024). https://en.wikipedia.org/wiki/Depth-first_search
21. Wiki. wiki-breadth-first-search (2024). https://en.wikipedia.org/wiki/Breadth-first_search
22. Wiki. wikisource-timeline-quantum-mechanics (2024). https://en.wikipedia.org/wiki/Timeline_of_quantum_mechanics
23. Zvi lotker. wikisource-wiki-page-summary-g (2024). https://docs.google.com/spreadsheets/d/17N8SJelU2yTqjwV2Q2JLoQ8m_b6OEeBt/edit?usp=sharing&ouid=116617959280772747526&rtpof=true&sd=true
24. wikisource (2023). https://en.wikipedia.org/wiki/Category:People_of_the_Gallic_Wars
25. J. Caesar, *Commentaries on the Gallic War*, 4 Oct. 2022
26. Wiki. wiki-category-people-first-jewish–roman-war (2023). https://en.wikipedia.org/wiki/Category:People_of_the_First_Jewish%E2%80%93Roman_War
27. Josephus. wiki-category-people-first-jewish–roman-war. https://en.wikisource.org/wiki/The_War_of_the_Jews

Part III
History Through AI

Chapter 14
Information and History

The ability of machines to think redefines history in several ways. In the first part of the book, we determined that the language of historians should embrace tools from computer science such as big O-asymptotic notation and function diagrams. In the second part of the book, we discussed the change in the philosophy of history that machines and algorithms will make. In the book's third and final part, we will discuss how machines will transform history from text-based to information-based study.

Up until now, the majority of historical work has centered on reading, analyzing, and writing texts. The primary source was text, the secondary source was text, the outcome was text, and so on. Technological advances will change all that. Text will no longer be the only source; we have images, videos, TV shows, interviews, recorded voices, documents, computer programs, and much more sources for historians to take advantage of. Figure 14.1 demonstrates the influence of technology.

Another aspect of historical study that is undergoing a process of change is machines' ability to read and "understand" information from the past. We are currently in the midst of an AI revolution which allows AI to generate an illusion of textual understanding. This has a tremendous impact on the way we study, analyze, and write history. Things that had previously been almost impossible and highly time-consuming can now be done in seconds thanks to AI. In short, the study of history is about to shift from text-based research to information-based science.

In the last part of the book, we will demonstrate the impact of this text-to-information shift, in the next six chapters. The first Chap. 15 deals with how AI could potentially manipulate different types of data and how we can use large language model (LLM) prompts for history.

In Chap. 16 we will demonstrate how to generate non-textual historical representations such as maps and movies.

Chapter 17 examines the application of expert systems to illuminate historical events. Specifically, the chapter demonstrates how one can analyze the main leaders of the 20th century, such as Hitler and Churchill, using the expertise of a psychiatric expert, a political science expert, and an emotion analysis expert.

Z. Lotker, *History by Algorithms*,
https://doi.org/10.1007/978-3-031-93627-2_14

Fig. 14.1 A cartoon illustrating the transformative impact of AI on the study of history

Chapter 18 provides an example of how standard AI tools from images and video processing can be used for history study. Specifically, the chapter analyses the first debate between Hillary Clinton and Donald Trump from the body language perspective.

The last Chap. 19 in the book provides some guidelines about history versus fake history.

Before we start using LLM or any other AI technology that is based on analyzing vast amounts of text through statistical means, we should try to understand the limitations of these technologies. To do that, I would like to start with a personal story. I studied philosophy in high school. As a young mathematician, I was certain I could define everything. In the first lesson, my philosophy teacher declared that it is impossible to define what a chair is. I was stunned. I told the teacher that I could, defining a chair as "An object with 4 legs designed to sit on it." My teacher quickly debunked my definition by showing that chairs can have a different number of legs and that there are a lot of other 4-legs objects to sit on, including horses. Again and again, I rephrased my definition, but he easily debunked each one. I went home feeling slightly frazzled, but being a young mathematician, I refused to give up. The following week I came up with the following definition: "A chair is an object that, if you were to ask the entirety of the human population, more than 50% will say is a chair." Initially, my argument would seem invalid since I used the word "chair" in my definition, but this is not the case. I have designed a scientific thought experiment for testing the object by population. There are several advantages and disadvantages to this approach: On the one hand, it allows computers and machines to grasp ideas and objects that have no simple definition and can only be defined through statistical experiments. On the other hand, this approach can be misleading - for example, all objects far enough from human observers will look like dots, but this is only because we lack the details to discern what they truly are.

Once we understand that machine learning technology and current Large Language Models (LLMs) are based on statistical definition, we can understand what we can gain from them. They are good at building a bridge between two statistical definitions. For example, if the machine has vast knowledge about Abraham Lincoln and Moses, it could apply its information on one on the other, and vice versa.

The basics of Large Language Models (LLMs) technology is generating probability distribution of long sentences and selecting the next word or sentence according to probability. This points to the main problem of this technology: if a long sentence becomes even longer, the machine could become more coherent; however, the space of all possibilities for the next sentence grows exponentially, as does the cost of generating such sentences. Once you cross this boundary, the machine is no longer coherent. This opens the door for human-machine collaboration where machines will be responsible for small and short parts of the text, and humans will be responsible for gluing together the short pieces of text to maintain coherence see Chap. 13. While machines will gather and filter information, humans will be the ones asking the right questions in order to separate correct information from false.

Large Language Models (LLMs) can also be used to transform portions of text (paragraphs or pages) into a list of numbers, then transform the complete document into a table of numbers, where one dimension is a question that the Large Language Models (LLMs) is asked about each paragraph, and the other dimension is the number of paragraphs in the document. Once we transform the historical documents into tables of numbers, we can apply standard tools from economics and statistics and generate numerical analyses of these documents. This method has the advantage of analyzing text fragments and applying statistical tools to analyze the complete historical document. We call this method "textual economics," which we demonstrate throughout this book.

The last application for Large Language Models (LLMs) with respect to historical documents is the use of historical search engines. For example, one can look for historical paragraphs that contain proof that Hitler had a psychological disorder.

In the following five chapters, we will demonstrate how the Idea we discussed in this chapter can be used with AI to develop and study digital history.

Chapter 15
Machine Learning and History

15.1 Introduction

Over the last ten years, we have witnessed a major revolution in the ability of machines to copy human actions in many fields. In the last 5 years, large language models (LLMs) become able to produce analyses and understand written language at the level of normal humans. This is, of course, a huge revolution that has many applications. In this chapter, we will discuss AI and how it should be utilized by historians. This is an inevitable advance, as most researchers in other fields are already using various AI tools (Fig. 15.1).

In the past, when studying history, researchers were tested on the ability to understand and write historical processes. Future researchers will be tested on their ability to write prompts and the sequences of prompts.

This chapter is divided into two main parts. The first part focuses on AI in general. We introduce three independent categories/dimensions, which AI algorithms can categorize, see Fig. 15.2. The first dimension is mathematical, describing the problems we wish to solve. The second dimension is physical, describing the nature of the object we wish to analyze. The last is information, which encapsulates the amount of knowledge we capture in advance of the things we analyze.

In the second part of this chapter, we focus on LLMs, particularly prompt engineering, which plays a key role in guiding AI models. Again, we will analyze prompts in three dimensions. The purpose of this categorization is to help us get a better understanding of the quality of prompts, from good to bad. The three categories we consider are precision versus innovation, the amount of expertise the prompt demands from the LLM, and single dialogue versus monologue, or programmable LLM.

Z. Lotker, *History by Algorithms*,
https://doi.org/10.1007/978-3-031-93627-2_15

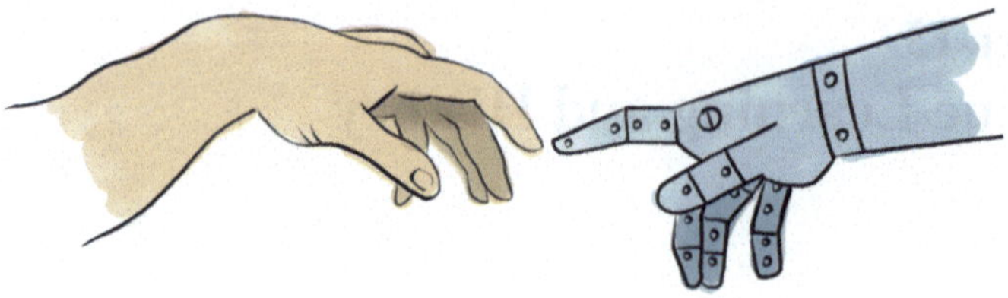

Fig. 15.1 A cartoon illustrating the gap between human understanding and artificial intelligence, as well as the potential for collaboration between them

15.2 Related Work

There are many great books about artificial intelligence. For the introduction level, we recommend [1], where the reader can find an up-to-date introduction to the theory and practice of artificial intelligence. The study is beyond the scope of this book. Artificial intelligence is based on deep mathematical tools. We refer interested readers to Deisenroth's book [2].

Prompt engineering is a relatively new tool in AI. This is the main tool the user can use to interact with LLM. Although prompt engineering is a new field of study, several books have already explored the subject. We refer the interested reader to [3, 4].

15.3 Basic AI Operations

In the aforementioned books, the reader may find general tactics to improve communication with the LLM.

This chapter differs from the books above in two perspectives: one, it focuses on computational history, and two, it is more philosophical than technical.

In the next section, we discuss the basic operations of AI in computational history.

15.4 Basic AI Operations

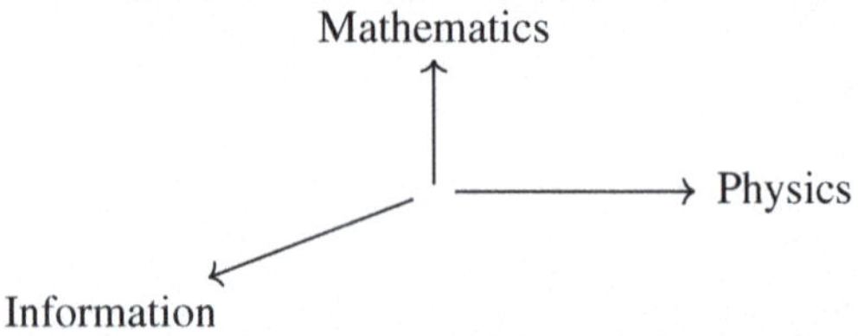

(15.1)

AI tasks can be broken into several main problems, which are determined by the input mathematical task the machine has to perform and the output of the task. Both input and output are mathematical objects like points in the vector space. Therefore, when we wish to solve general problems, it is best to translate the problem into the language of points in the vector space. Below we list four of the main mathematical problems that AI can solve using current technologies and discuss how historians may use these problems to their advantage.

15.4.1 Mathematics

- **Clustering**: Groups similar data points without predefined labels, commonly used in pattern discovery and segmentation. For example, when analyzing the location of the battles during the German invasion of Poland in 1939, the clustering algorithm divides the location of the battles into two clusters. These main clusters coincide with the fact that Germany attacked Poland in 1939 using two main forces: Army Group North and Army Group South, see Fig. 15.2
- **Classification**: Assigns data points to predefined categories based on a priori knowledge. We can use the clustering procedure and then apply the knowledge we gain from clustering to the classification. For example, we can use the clustering we have already computed on the locations of the battles in Poland as input to

Fig. 15.2 This figure shows the result of the clustering function when applied it to the 72 battles during the invasion of Poland in 1939. Blue points can be interpreted as the battles fought by the German Army Group North and orange points can be interpreted as the battles fought by the German Army Group South

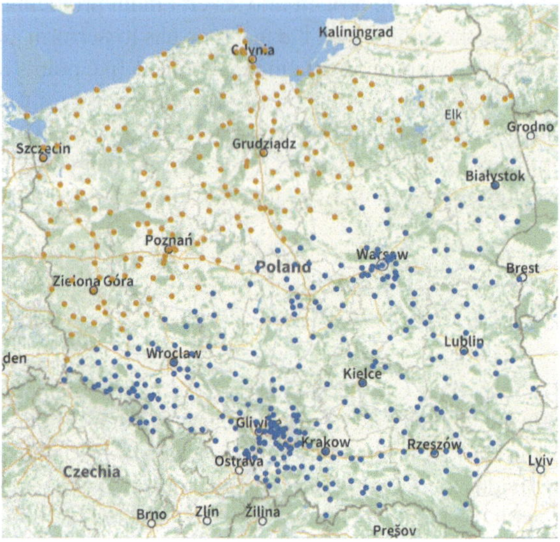

Fig. 15.3 This figure shows the argument of the clustering that we represented in Fig. 15.2 by the machine learning classification using clusters as a labeling source. The blue points in this figure are cities that were conquered by the Army Group North, and the orange are cities that were conquered by the Army Group South

the classification problem and construct the classifier that will identify for each city in Poland which army group conquered it (Fig. 15.3). In a nutshell, we were able to extend our knowledge from the 80 battles to all the cities in Poland by classification. The clustering problem can thus be considered a bootstrap of the classification problem. The result of this process can be seen in Figure. However, the bootstrap step for the classification problem does not always have to come from the classification problem. They can be retrieved from any prior knowledge, see Exercise 15.3

- **Anomaly Detection**: Identifies rare or unusual patterns in data. We have already used anomaly detection in Sect. 2.4.2 where we analyzed the data of dead people from Wikipedia. For more on anomaly detection, see [5].
- **Prediction**: Forecasts future outcomes using historical data. This is useful when we try to do counterfactual analyses of historical events.

In the next subsection, we categorize AI procedures from the point of view of the knowledge we have on the data.

15.4.2 Information

AI problems can be categorized according to how data is used for learning. The primary distinctions arise from whether the data is labeled, unlabeled, or involves adapting knowledge from one domain to another.

- **Supervised Learning**: Requires labeled data, where models learn from input-output pairs to make predictions. Examples include classification and regression.

We have already discussed classification in the previous section. However, any prior knowledge can be used to any prior data and generate machine algorithms that classify general data into the categories that were identified in the prior knowledge. We will use this approach in Chap. 18, where we analyze the first debate between Hillary Clinton and Donald Trump.
- **Unsupervised Learning**: Works with unlabeled data to discover hidden patterns, such as in clustering and dimensionality reduction. We discuss this approach in Sect. 15.4.1.
- **Reinforcement Learning**: Learns from interactions with an environment, optimizing actions based on rewards to maximize long-term benefits. This approach is useful, especially in games where AI can play against itself. We will not use this approach in this book.
- **Transfer Learning**: Transfers knowledge from a source domain to a related target domain, reducing the need for large labeled datasets in new applications. This is a very useful approach and can be used in sound recognition, time interval analyses, etc. In this book, we use LLM as a platform for transfer learning whenever we transfer the time interval into the sequence of numbers. For example, we can feed the LLM a paragraph and ask for the geographical coordinates, see the example of the position of the 72 battles during the invasion of Poland in 1939 Fig. 15.2.

In the next subsection, we will discuss the physics aspects of AI by looking at different kinds of data: time interval, image, video, sound, and so on. Each media type uses specific mathematical tools and, therefore, heavily affect the basic tasks the machines can perform on the data.

15.4.3 Physics

Machine Learning (ML) techniques are applied to various types of data, each requiring specialized models and preprocessing methods. The key domains include:

- *time interval*: Natural Language Processing (NLP) and LLM enables machines to understand, generate, and analyze time interval. We use LLM extensively throughout this book.
- *Images*: Computer vision models process and interpret visual data for tasks like object detection, image classification, and facial recognition.
- **Video**: ML analyzes sequential visual data for applications such as action recognition.
- **Audio**: Speech and sound processing techniques are used for speech recognition, music analysis, and environmental sound classification.

We provide an example of applying image, video, and audio processing to the history study when we analyze the first debate between Hillary Clinton and Donald Trump in Chap. 18.

15.5 Prompt

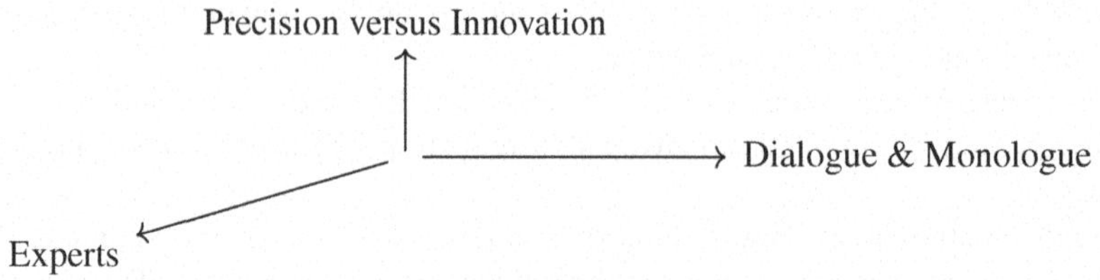

(15.2)

One of the major problems AI faces in the early 20 s of the 21st century is the background. When two humans converse, they enjoy vast knowledge, which we call the background. They know many things we take for granted; for example, they understand the sentence "The apple was tasty" or "I am hungry." However, for a machine to achieve such vast knowledge and be able to talk to normal human beings has been a major challenge for years. Since OpenAI published its LLM 3.5, this problem is no longer an obstacle. The main tools with which people communicate with LLMs are called prompts. In the future, when historians use LLM, they will do it using prompts. In the second part of the book, we try to explain how to write good prompts in the field of computational history. We start with a simple explanation of what the prompt is.

In Chap. 13, we modeled the historian machine. The prompts can be used in two different ways. One is a sensory range input for the LLM entering data and queries, which is the main way we interact with LLM. The prompt is a secondary function that allows the user to program the LLM. For example, we can give an LLM structure in the prompt. This structure is a nutshell program for the LLM to become an expert in specific fields and determine its behavior for the entire conversation.

15.5.1 Precision Versus Innovation

LLMs have a temperature parameter that determines the amount of precision/ creativity they can produce. Usually, when setting the temperature to 0, the LLM is forced to become deterministic and have zero creativity. By setting the temperature to 1 we force the LLM to concentrate on creativity and ignore facts. Setting the temperature above 1 causes the LLM to lose coherence after a few sentences.

This raises the question: to what temperature should we set our LLM? The answer, of course, depends on our motivation.

Each historical event is composed of time, place, and meaning. When trying to compute the time and place of a historical event from the time interval, it is better to set the temperature to 0, since we wish for the time and place of any historical event to be as accurate as possible. We have no interest in the LLM being creative and inventing the location or time of the historical event. When working with meaning at the level of sentences or paragraphs, we wish the temperature to be 0.

Setting the temperature to 0 is also useful when studying historical documents that need to be transformed into a sequence of numbers that will then be analyzed using tools from the realm of exact sciences. In such cases, we cannot be certain of the accuracy of the LLM's translation. Therefore, we recommend to start out by pushing the temperature to 0, which would make it easier to examine the accuracy of the LLM's translation from time interval to numbers.

However, when trying to compute the big picture of a historical narrative composed of many historical events, we use well-known historian models such as Marxism, militarism, feminism, and so on. In this case, we can enjoy the creativity of LLM and set the temperature around 1.

The temperature parameters are also very useful when trying to study fake versus legitimate history. We will explain this in the last chapter of the book Chap. 19.

When precision is needed (for example, when trying to extract data from a historical document), we recommend writing three prompts: one to extract the data, one to ask the LLM whether the data is correct, and one to ask the LLM for proof that the answer is correct. Usually, the LLM provides simple explanations for its answers. However, it is better to break the answers into three different parts, as this makes it easier to manage the sequence of prompts and increases the accuracy of LLM.

15.5.2 Dialogue Versus Monologue

Interactive proofs are famously more powerful and tend to become exponentially shorter than normal proofs. Similarly, dialogues are more effective than monologues. However, preparing dialogues in advance can be challenging, given their inherent exponential structures. Questions must be prepared in advance, and sometimes, the next questions rely on those that preceded them. Some of the historian's theories, such as the dialectic [6] of Hegel, have a simple structure, such as thesis, antithesis, and synthesis, which are easy to program. We will discuss the dialogues in Sect. 17.5

When starting a discussion with LLM, it is recommended to provide some instructions on the framework of the full discussion. For example, we can say that from now on, I would like you to answer only data that was available during the 17th century. This instruction generates a virtual time machine, as discussed in Sect. 4.7.

One standard advice in prompt engineering is to ask LLM to break down any task into several smaller tasks. In general, this may improve the results of the largest tasks if done correctly. Clearly, we can repeat this process forever, and the LLM may lose coherence when broken down into too many parts. This suggests that there is a sweet spot for the number and size of subtasks. Below are several criteria to help decide when to stop this process.

- If a prompt can be answered directly in a satisfactory fashion, it is better not to divide the prompt into subprompts.
- Sometimes, by breaking prompts into subprompts, we simplify the problem by reducing its complexity. Therefore, if the prompts are independent from each other, it is a good time to stop.
- When subprompts cross the conceptual boundary, or when further breakdown is infeasible, it is better to stop before the LLM loses coherence.

15.5.3 Experts

One of the advantages of LLMs is the vast knowledge they were trained on. This means that LLM has expert jargon that covers most of human knowledge. Therefore, we can ask for the opinions of experts on historical documents or events.

To enjoy the expertise of any LLM, we should be able to use the expert jargon. Since we are not experts in most human knowledge, we first need to ask the LLM for a dictionary relative to the expertise we are interested in using; then, we need to use those terms in our prompt so that the result generated by the LLM will be more precise.

This means that when we try to use the expert embedded in the LLM, we will prefer dialogues to monologues. For more on how to use the LLM as an expert system, see Sect. 17.5. In fact, we recommend writing a prompt and then asking the LLM to rewrite it. This can often improve the prompt.

In the next section, we will discuss how to write a good single prompt.

15.6 How to Write a Good Historical Prompt

Writing an effective historical prompt requires *accuracy, analytical depth, and a connection to historical theories.* A well-crafted prompt should not merely ask for a retelling of events; it should encourage critical thinking, interpretation, and argumentation while defining the historical scope as well as the answer to it.

A fundamental principle of historical inquiry is that every historical event has three dimensions: time, place, and meaning. A good prompt must be:

- *Precise in time*: It should specify a clear period rather than an undefined era.
- *Precise in place*: It should define a geographical or political boundary relevant to the event.
- *Scope and size*: It should specify the scope and size of the historical event both in the prompt and in the answer to it.
- *Open to multiple interpretations of meaning*: It should encourage users to analyze different perspectives and historical consequences.

Table 15.1 Evaluation of the prompt

Criteria	Evaluation
Historical accuracy	✓ Strong: The prompt accurately states that Chamberlain has been historically criticized but also acknowledges Britain's military advancements under his leadership
Connection to theories	✓ Strong: The prompt engages with **appeasement theory** and revisionist views on Chamberlain's leadership
Scope and size	✓ Well-defined: The question focuses on **Chamberlain's policies** and their effect on Britain's war preparedness. It is **not too broad** (WWII in general) or **too narrow** (one decision only)
Allows for different interpretations	✓ Strong: The prompt challenges researchers to **reconsider a common historical narrative** and weigh multiple factors. It allows for **both a critical and a positive reassessment of Chamberlain**

15.6.1 Evaluating a Sample Prompt

The next table summarizes the criteria for writing good prompts (Table 15.1).

Two main independent questions are asked during the evaluation process. The first is who is doing the evaluating: human or LLM? The second is what is being evaluated: the prompt itself or the answer to it? Before evaluating the prompts, we must first define a grade scale for each of the different criteria and evaluate the prompt manually by LLM itself.

In the next section, we will provide a simple example of using machine learning technologies to analyze historical documents. Our example will be Julius Caesar's *Commentaries on the Gallic War* [7].

15.7 AI and Two Clocks in the Gallic War

In this section, we analyze the Gallic War using machine learning. First, we use supervised learning to construct a machine that takes time interval and classifies it into one of the categories: Barbarians or Romans. We compute the categories of our supervised learning example by taking the Wiki categories of the Roman people of the Gallic Wars, see [8] , and we tag those paragraphs as Roman for each person. We then take all the Wiki pages in the Wiki category of the Barbarian people of the, Gallic Wars see [9], and categorize those paragraphs as Barbarian. Now, we use these labeled paragraphs to construct a machine classifier that takes the paragraphs and tells us if it is Roman or Barbarian. We can feed all the paragraphs from Julius Caesar's *Commentaries on the Gallic War* [7] to our classifier and define two clocks: one that counts the number of sentences related to the Roman people C_R, and another that counts the number of sentences related to the Barbarians/Gallic people C_B. See Fig. 15.4. Using what we've learned in Chap. 4, we can identify the critical events

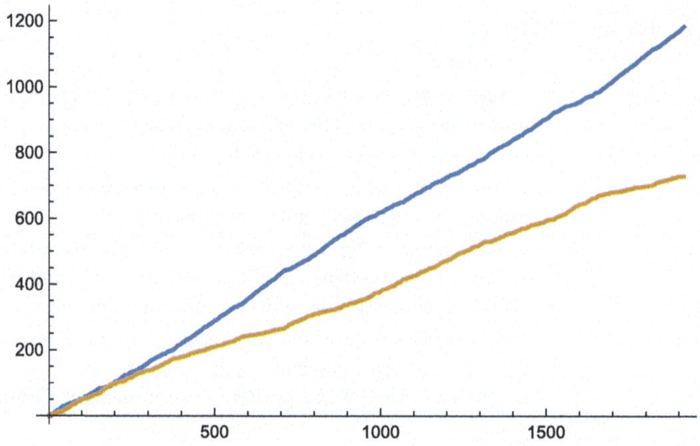

Fig. 15.4 C_R (blue) represents the accumulated number of sentences categorized as Roman, and C_B (orange) represents the accumulated number of sentences categorized as Barbarian

Table 15.2 The critical line in the Commentaries on the Gallic War according to the M-diagram

Colour	Sentence number	Ref	Sentence
Red	954	Book 5 Chap. 52	"From all these things he judges with what danger"
Blue	1658	Book 7 Chap. 88	"Reserving the Aedui and Arverni"

using the M-diagram of the two clocks. The critical points are found in sentences (954, 1658). Looking at Gallic Wars, line 954 describes see (Table 15.2).

Looking at the time interval, the red dot appears near the end of Book 5 in year 54 BC: Caesar led the second expedition to Britain, which diluted the Roman forces in Gaul, while Ambiorix led the revolution in Gaul and successfully destroyed a legion of Roman forces. The second legion was seized and put in danger of total annihilation, but was rescued by Caesar and his reinforcements at the last moment. This was the crucial juncture of the Gallic Wars and the greatest achievement of the Gaelic forces. Had the Gauls managed to destroy the second legion, the history of the war could have had a different outcome.

The blue dot is at the end of Book 7. It shows the moment of the final loss of the Gauls as they were overtaken by the Romans. This is the peak of Roman power during the war (Fig. 15.5).

The choice of prompt type significantly influences the performance of LLMs in different tasks. By understanding and utilizing different prompting strategies, we can optimize AI-generated responses for accuracy, creativity, and relevance across various domains.

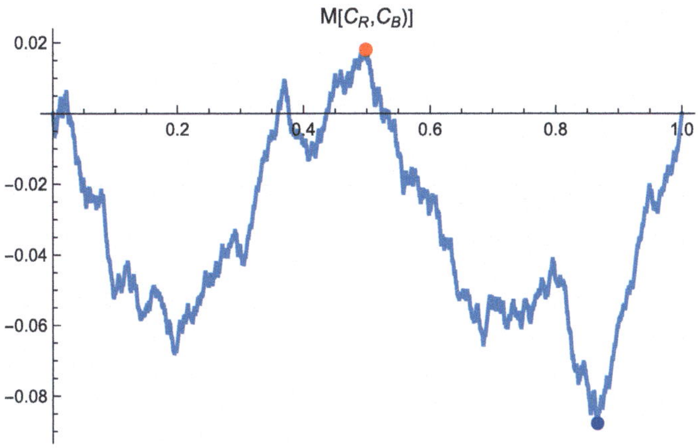

Fig. 15.5 The M-diagram $M(C_R, C_B)$, which allows us to compute a critical sentence in Gallic Wars with respect to the two clocks C_R, C_B

15.8 Conclusion

This chapter provides a general introduction to AI and LLM. We talked about the application of AI and LLM to digital history. Since AI in general and LLM in particular are rapidly evolving spheres, we try to provide a philosophic background to demonstrate how one may use those tools in the world of digital history.

The chapter is divided into two parts. The first part tries to categorize the different tools AI can offer on the subject of digital history. We used a three-dimensional partition: the first is mathematical problems that AI can solve; the second is the information AI needs to solve different problems; and the third is the physics of the data, which has a big impact on the problems one may ask about the data. AI will continue to evolve through these dimensions, preserving the line of thought presented in this chapter.

The second part of the chapter covers another rapidly evolving area: LLM. We try to capture the philosophical nature of LLM, ignoring the technical questions, as we believe that all technical issues will change in the future. We suggested a three-dimensional categorization of LLM prompts: precision versus innovation, expert systems, and dialogue versus monologue. All three dimensions are relevant to the study of digital history and will remain key issues no matter how much technology evolves, since they are in the boundary between humans and machines. All of them incorporate what humans want and what LLM can provide.

15.9 Exercise

1. Analyze each paragraph in Herodotus's *The Persian Wars* by categorization its contents into background, diplomacy, and battle description.
2. Use AI to generate a fake historical document of a Christian refugee girl in the year 310. Use Anne Frank's diary as source material and change the location from Amsterdam to Alexandria. Enter each paragraph into LLM or any other Large Language Models (LLMs) with the instruction to transform Jewish refugees to Christian refugees, German oppression to Roman oppression, Dutch people to Egyptians, and so on.
3. Use Livy's *The History of Rome* as a source to generate a fake historical document describing the First and Second Punic Wars between Rome and Carthage from the point of view of Carthage. The book should present any victory of the Carthaginian army as a positive and any victory of the Roman army as a negative.
4. Use Livy's *The History of Rome* to compute the Syuzhet as a Fabula interval timeline.
5. Use Edward Gibbon's *The Rise and Decline of Roman Empire* as a Rosetta stone to describe the rise and fall of the Persian Empire from the Kourosh Empire in 560 BC to the Muslim conquest of Persia in 633 AD. Use Wikipedia as a source for the relevant names.
6. Use Karl Marx's *The Capital* as a Rosetta stone to generate the philosophy of Spartacus via AI.
7. Discuss how one can distinguish between fake and non-fake historical documents. Do you think that ancient historical documents represent real history or not? Choose your favorite period and source.
8. We discussed the relationship between the background and the central historical narrative. A narrative of any historical conflict has two sides. Consider the war between Greeks and Persians, as described by Herodotus [10]. In this exercise, you are asked to describe how two historical parallel lines, for example, one for the Greeks and the other for the Persians, are mixed into a single historical narrative. Consider two historical narratives, one for each side, and explain how to merge the two caterpillars into a single historical narrative.
9. Write a historical prompt based on the invasion of Poland (1939) using the criteria discussed above. Then, evaluate it using the checklist.

References

1. S.J. Russell, P. Norvig, *Artificial Intelligence: A Modern Approach* (Pearson, 2016)
2. M.P. Deisenroth, *A Aldo Faisal, and Cheng Soon Ong* (Cambridge University Press, Mathematics for Machine Learning, 2020)
3. G. Mizrahi, *Unlocking the secrets of prompt engineering: Master the art of creative language generation to accelerate your journey from novice to pro* (Packt Publishing Ltd, 2024)
4. M. Tabatabaian, *Prompt Engineering Using ChatGPT: Crafting Effective Interactions and Building GPT Apps* (Walter de Gruyter GmbH & Co KG, 2024)

5. Anomaly detection (2023). https://en.wikipedia.org/wiki/Anomaly_detection
6. Wikipedia contributors. wikipedia-dialectic (2024). https://en.wikipedia.org/wiki/Dialectic
7. J. Caesar, *Commentaries on the Gallic War*, 4 Oct. 2022
8. Wiki. Category:roman-people-of-the-gallic-wars (2023). https://en.wikipedia.org/wiki/Category:Roman_people_of_the_Gallic_Wars
9. Wiki. wikisource-category-barbarian-peopleofthegallicwars (2023). https://en.wikipedia.org/wiki/Category:Barbarian_people_of_the_Gallic_Wars
10. Herodotus, A. Denis Godley, *Herodotus: With an English Translation by AD Godley*. (W. Heinemann, 1961)

Chapter 16
Representing Historical Information

In traditional history, the output has always been text. However, this is no longer the case when considering the digital revolution. The output may now be maps, movies, sounds, etc. In this chapter, we develop a methodology for transforming text into video and images.

16.1 Word Clouds

The main idea is to take any text $txt = (w_1, ...w_n)$ composed of n words. We denote the i-th words by w_i. Let $1 = n_1 < n_2, ... < n_k = n$ be an increasing sequence of k-integers. Slice the txt into k chapters $(ch_1, ..ch_k)$, such that chapter i is:

$$ch_i = (w_{n_i}, w_{n_i+1}, ..., w_{n_{i+1}-1}) \tag{16.1}$$

We define the accumulated chapter Ach_i as

$$Ach_i = (w_1, w_2, ..., w_{n_{i+1}-1}) \tag{16.2}$$

To generate a movie out of text, we have to analyze the accumulated chapters. In each chapter, we count the frequency of each word after removing the stop words, see [1], and take p more popular words mp_i. It is recommended to take between 25 to 200 words. Then, we compute the word clouds of these popular words and generate an image im_i.

Compressing historical text by collecting the most common words from the text and breaking down the sentence structure can be likened to a machine's imitation of a stream of consciousness. For more on streams of consciousness, see [2]. The silhouette of the evolving word cloud is an important aspect that serves the purpose of emphasis. The cloud's shape can instantly convey information about the context of

Z. Lotker, *History by Algorithms*,
https://doi.org/10.1007/978-3-031-93627-2_16

Fig. 16.1 Visualization of the relationship between LLM and humans

the words within it. For example, when communicating a stream of consciousness for the book *Man's Search For Meaning* by Viktor E. Frankl, using a silhouette of a man running towards something (Fig. 16.2) communicates the context of the words better than a simpler or less relevant silhouette such as a circle or a flower. Other visual elements, such as the color of the words or the font, can also encode information that complements storytelling.

The advantage of looking at the stream of consciousness instead of reading the complete text is that by watching it, we are left in a fog of narrative, unable to see things clearly. This forces us to imagine what a narrative of such stream would be. Following this process may increase the reader's curiosity and convince the viewer to read the text. Another advantage is that this process is much quicker and can help the viewer decide if they are interested in the text.

16.2 Evolving Graph

Another method of transforming a historical document into a video that can help us understand the narrative is transforming the text into an evolving social network (see Sect. 5.7) according to the following procedures. First, we identify the characters in the text. This can be done manually or mechanically, i.e., algorithmically, with AI that uses the knowledge base. Once we identify the people, we use this list as the node of the Multigraph. We put an edge between two nodes if they appear in the text one after another. We can change the size of the nodes according to how important they are at the current time. An example of such an evolving graph or snapshot can be seen in Fig. 16.3. Here, we see eight frames representing the snapshot of the social network at the end of each of the books about the Gallic Wars by Julius Caesar [3]. Watching the complete video pinpoints the fact that Caesar attacked the Gallic leaders one by one. Another interesting observation is that the size of the node that represents

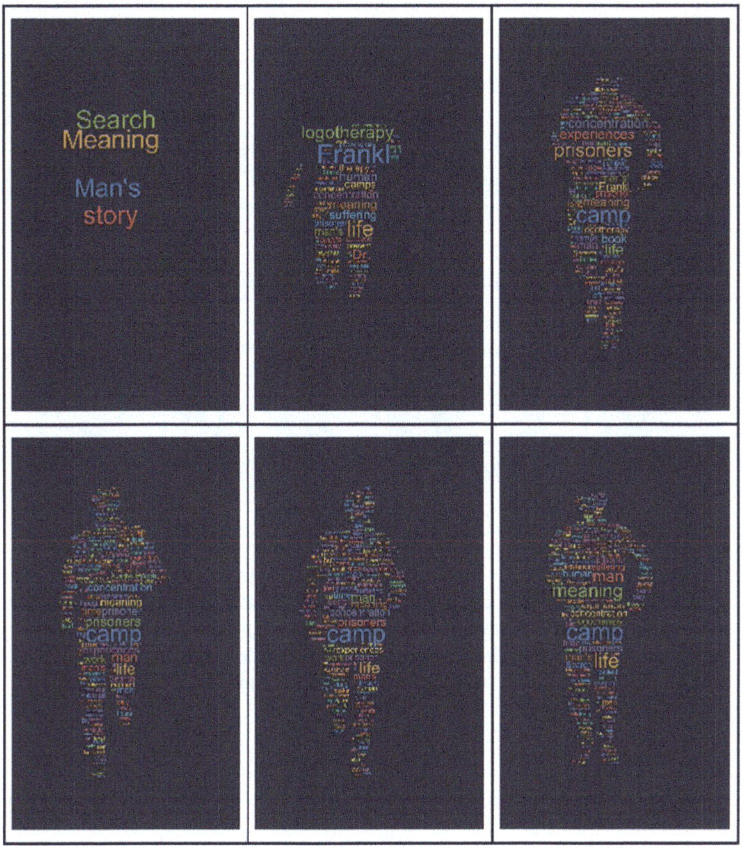

Fig. 16.2 An example of a stream of consciousness generated by a computer about *Man's Search For Meaning* by Victor E. Frankl

Caesar does not change. However, the size of the node that represents the barbarian leader consistently shrinks.

16.3 Generating a Slideshow and Movies of Historical Texts Using AI

LLM tools such as LLM can generate an image from prompts. Moreover, it is possible to take a paragraph from historical text and ask any LLM to devise a prompt that will generate an image out of the paragraph. Once we generate the prompt, we can repeat it several times and get a slideshow or transform it into a video.

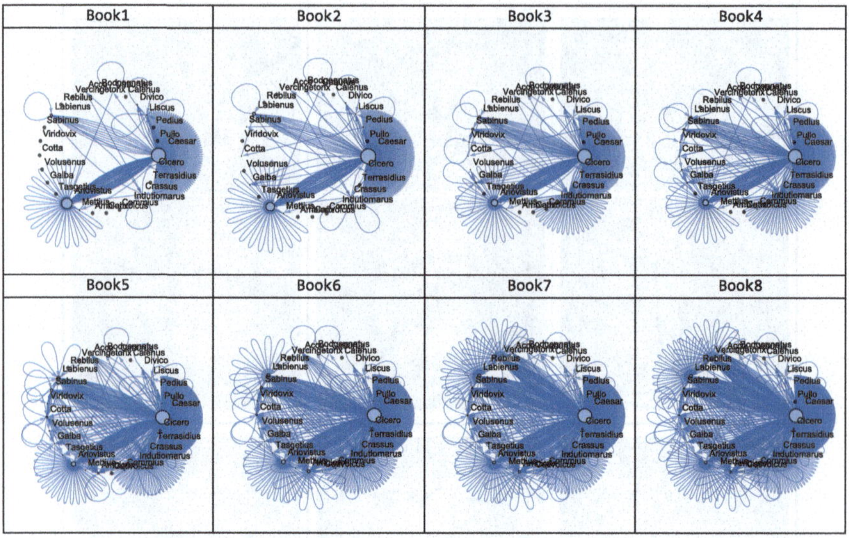

Fig. 16.3 Representation of the 27 characters in all eight books of *Gallic Wars*, each with nodes representing the characters, while the size of these nodes depends on their page rank centrality. The characters appear in the text one after the other, so the eight figures show the evolving social network at the end of each book. The edge (i, j) appears if node i appears after node j in the text and the number of times this edge appears in book k is the same as the number of times node i appears immediately after node j in all books from 1 to k

The drawback of this method is that the image of the i-paragraph is not related to the $i + 1$-paragraph. One way to overcome this problem is to use preexisting, continuous, manmade films or videos.

16.4 Maps

In many cases, historical maps are used to represent historical data. For example, when studying wars, it is common to draw maps explaining the movement of each army. In the past, experts generated maps and students would consume and process this information when studying history.

In general, it is very easy to generate a sequence of maps to describe a historical phenomenon in Wolfram Mathematica, given that historical entities already exist in the Wolfram knowledge base.

For example, Fig. 16.4 shows the expansion and contraction of Nazi Germany during WWII. These maps provide a general narrative of the main battles during WWII. For a complete history of all the battles, generating a full description is a much more complicated task. Perhaps in the future we will be able to generate more

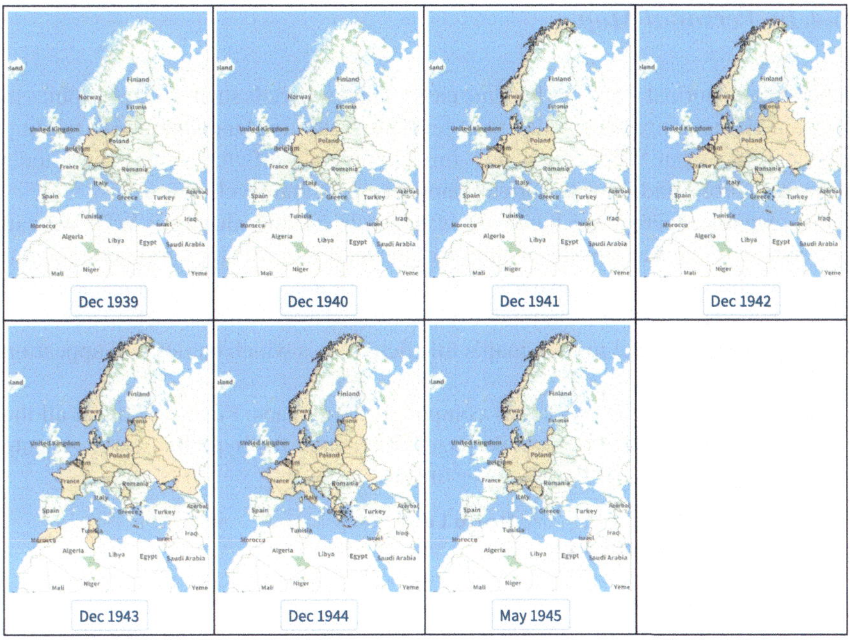

Fig. 16.4 Map representation of the seven main steps of Nazi Germany's border change in WWII

detailed maps. Figure 16.4 shows the seven main steps of the border changes of Nazi Germany during WWII.

1. The beginning of the War
2. The conquest of Poland
3. The French Campaign
4. The peak of the USSR Campaign
5. Invasion of Africa
6. Losing the Africa Campaign
7. Losing the war

For another example, such as the expansion of the Roman Republic, we refer the reader to [4].

It is also very easy to generate maps when our source is a historical timeline from Wikipedia. For example, in Chap. 5 we analyzed the data of the war in Poland and generated a video or a sequence of images, see Fig. 5.5. We encourage the readers to do Exercise 16.6.

In the next subsection, we describe a personal historical map as an example of generating maps by the researcher. We will not be using the historical Wolfram knowledge base, but rather retrieve the data from Wikipedia.

16.4.1 Personal Map

Generating historical maps has become easier thanks to tools such as GeoGraphics in Wolfram Mathematica or any other high-level programming language, and harvesting historical data from Wikipedia is now on a completely different scale. These tools allow us to probe and draw historical maps on a personal level.

For example, there were hundreds of generals on duty during WWII. We can generate maps that describe the evolution of the battles in which each general fought. Then, we can compare the maps and see the differences between those generals. Figure 16.5 represents the accumulated battles waged by Erwin Rommel ("The Desert Fox"). The time interval in each map's title determines which battles will appear on the map.

Next, we describe the process of computing these maps. First, we collect all the hyperlinks from the Desert Fox Wiki page. For each of the hyperlinks, we compute the WikidataID list using the following function:

```
getWikidataIDQ[h_] := Module[{name, h1, s, s1},
  name = StringSplit[h, "/"][[-1]];
  h1 = "https://en.wikipedia.org/w/api.php?action=
      query&prop=pageprops\
&ppprop=wikibase_item&redirects=1&titles=" <> name <>
    "&format=json";
  s = ToString[Import[h1]];
  (*Print[q[[1]]];*)
  If[StringContainsQ[s, "Q" ~~ NumberString],
   s1 = StringTake[s, StringPosition[s, "Q" ~~
       NumberString]][[1]],
   s1 = {};
   ];
  s1
  ]
```

For more on computing the Wiki ID, see [5]. We then import the Wiki knowledge base using the command

```
WikidataData[WikidataData[hLink,"Dataset"], "Dataset"]
```

where *hLink* is a hyperlink in the Roman Wiki page. Next, we check if any of those Wiki data are categorized as battles. For each battle, we double-check if the Desert Fox is mentioned. After double-checking, we get the list of battles in which the Desert Fox played the major role. For each of these battles, we compute the time and place they were fought using the Wiki knowledge base. Once we have this data, we can sort the battles according to their dates and draw accumulated maps using GeoGraphics commands. The outcome of the process can be seen in Fig. 16.5.

In fact, we can compile all those maps into a video describing the movement of the Desert Fox. Here, we have two options: to use time as a proportion for real-time, or just do it uniformly. The advantage of using real-time is that it captures the difficulties of the battles, as long battles can be assumed to be more difficult than short ones. Short battles represent decisive wins. For example, the fact that Germany

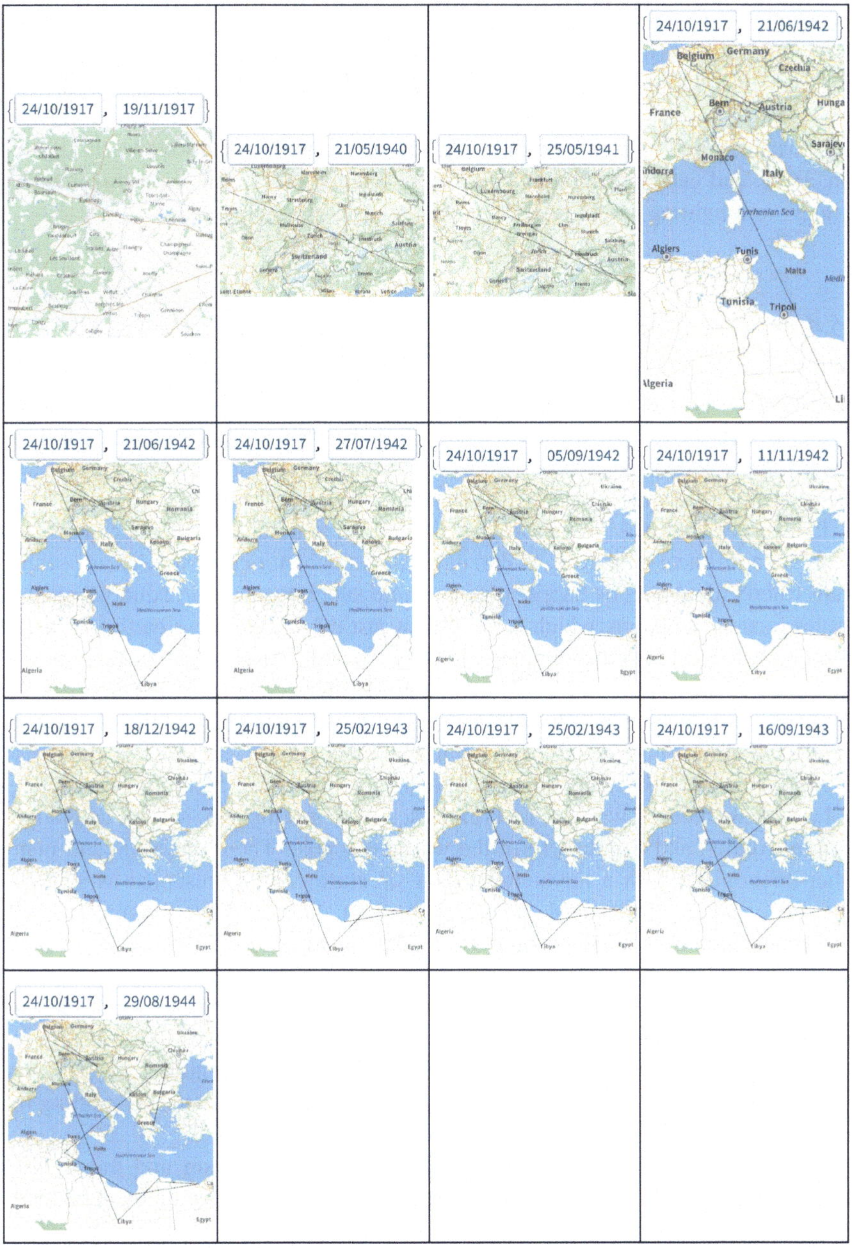

Fig. 16.5 13 maps describing the different battles Erwin Rommel fought during his career. The first battle took place in WWI, and the last at the end of WWII

conquered France in three weeks at the beginning of WWII can be interpreted as a huge difference between the German and French war machines. During the Battle of France, the German army was superior in every parameter, and therefore, the war was decisive.

The campaign for the liberation of France started with Operation Overlord, June 6th, 1944, and ended 11 months later on May 8th, 1945. In this case, two modern armies were fighting and the Allies' advantage was negligible.

16.4.2 Tree Maps

In the previous section, we described how to generate a sequence of maps using Wikipedia and Wiki data knowledge base. We also explain how to transform these images into a video. We succeeded in transforming the sequence of images into a video because the images appear in a chronological, linear order.

In this section, we claim that we can use another structure called a tree map. Tree maps are non-linear and are crucial to understanding historical processes, especially wars where simultaneous happenings generate super events. For example, several battles form a campaign, which is more than just the sum of these battles. You can lose one battle but win the campaign, and vice versa. For example, Napoleon won the Battle of Borodino but lost the campaign. A more complicated example is the Battle of the Bulge, the last almost successful offensive campaign of the Germans in WWII, which was not enough for them to win the war.

Moreover, war, like many other historical narratives, has a hierarchical structure. In large-scale modern warfare, several battles form a campaign, several campaigns form a frontier, and frontiers form the war as a whole. Therefore, wars and many other phenomena with a hierarchical data structure can be presented as a rooted tree. According to Wolfram Mathematica (see [6]), the rooted tree is a tree in which a special ("labeled") node is singled out. This node is called the "root" (or, less commonly, "eve") of the tree.

There are several reasons why the rooted tree is the best structure for describing the history of wars. First, a hierarchical structure is the only possible structure for an army. To see that this is necessary, we can look at the disastrous outcomes for the Roman army in the Battle of Cannae, where Romans had two councils, splitting the line of commands between them on even and odd days.

The second reason is that humans like to use reductionism and hierarchical structures, which allow us to isolate historical events and understand them better.

16.5 Conclusion

In this chapter, we showed three different methods of transforming the text of a historical narrative into video. Other methods are definitely possible. The point of view we wanted to emphasize in this chapter is that the history of the 21st century is not limited to textual media. There are different ways to tell a historical narrative, each medium possibly posing the narrative in a different light. Transforming historical narrative into visual arts can be both enjoyable and educational. As time progresses, more and more historical data will become accessible with video, and tools that extract information from video and images will become more and more important in the study of history.

16.6 Exercise

1. Use the script of *I, Claudius* [7] to generate an evolving social network description of the Imperial Court in Rome at the beginning of the 8th century AD.
2. Use the geographical data of the locations of battles in WWII to generate a map of the German Reich, from the first battle to the last one in Europe. Use the list of battles in [8, 9].
3. Use the timeline in [10] to generate the map of Nazi France from D-day to May 11, 1945.
4. Use the silhouette of a walking man to generate the word cloud that summarizes *The Long March*. You can use the Wiki timeline in [11] to generate the accumulated text in the word cloud.
5. Use the timeline of the Napoleonic era, see [12].
   a. Generate a personal map of Napoleon. For more details on personal map, see Sect. 16.4.1.
   b. Generate a sequence of maps that describe the Napoleonic Empire from 1804 to 1815.
6. In Sect. 16.4.2 we explain how to construct tree maps. In this exercise, you are asked to construct interval graph maps $G(V, E)$, where the nodes $v \in V$ are maps that accumulate historical events over time interval TI_v. Two nodes $v, u \in V$ are connected by an edge $\{v, u\} \in E$ if $TI_v \cap TI_u \neq \emptyset$. Use the attention schema of the seven main battles of WWII that were discussed in Sect. 13.11 to generate an interval graph maps of the seven main arenas.

References

1. Wiki. wikisource-stop-word. https://en.wikipedia.org/wiki/Stop_word, 2025
2. Wiki. wikisource-stream-of-consciousness. https://en.wikipedia.org/wiki/Stream_of_consciousness, 2025
3. Julius Caesar. *Commentaries on the Gallic War*. 4 October 2022
4. animate-the-rise-and-fall. https://www.wolfram.com/language/11/knowledgebase-expansion/animate-the-rise-and-fall-of-historical-countries.html, 2023
5. Wiki. wikisource-wikisource-finding-wikidata-id. https://en.wikipedia.org/wiki/Wikipedia:Finding_a_Wikidata_ID, 2025
6. mathworldrootedtree. https://mathworld.wolfram.com/RootedTree.html, 2023
7. iclaudius. https://www.angelfire.com/tx4/cus/scripts/iclaudius-transcript.html
8. Wariibattleswestern. https://en.wikipedia.org/wiki/List_of_World_War_II_battles#Western_Front
9. Wariibattleswestern. https://en.wikipedia.org/wiki/List_of_World_War_II_battles#Eastern_Front
10. Wiki. wikisource-timeline-liberation-france. https://en.wikipedia.org/wiki/Timeline_of_the_liberation_of_France, 2025
11. Longmarch. https://en.wikipedia.org/wiki/Long_March
12. Wiki. wikisource-timeline-of-the-napoleonic-era. https://en.wikipedia.org/wiki/Timeline_of_the_Napoleonic_era, 2023

Chapter 17
Expert System in History

17.1 Introduction

In this chapter, we discuss how to use an AI expert system when analyzing historical documents. We provide several examples, some of which are easy for AI to tackle, others that are not. In general, tasks which are highly dependent on a place in a sentence are difficult for AI and tasks which are not dependent on a place are much easier. For example, if we study the amount of Violence that appears in some speech, we do not care who "did" the Violence; that is something that a simple counting algorithm can estimate. The idea is to prepare a list of words that describe Violence and count them. The more such words appear in the document, the more violent it is.

Algorithms that try to estimate the psychiatric disorder are much more dependent on the order of words in the sentence. Therefore, a simple word counting algorithm will not suffice. Instead, we have to use LLMs, which work pretty well with tasks such as analyzing disorders of historic leaders (Fig. 17.1).

17.2 Function Diagram for Experts

Using LLM or a simple word counting algorithm as an expert and analyzing a historical text from an expert perspective means taking an archival text and analyzing each paragraph according to d queries where the answer for each query is inside a ring$\mathscr{R}$. The ring is usually $\mathscr{R} = \mathbb{Z}_2$ if it is comprised of yes/no questions[1]. If the algorithm is based on word counting, then the ring is $\mathscr{R} = \mathbb{Z}$ This allows us to transform each historical paragraph into a d-dimensional vector space.

[1] It is possible to use other queries that transfer the text into a sequence of numbers, however, it is always possible to write an equivalent of a set of queries that answers "Yes" or "No." Dealing with "Yes" and "No" questions is much simpler.

Z. Lotker, *History by Algorithms*,
https://doi.org/10.1007/978-3-031-93627-2_17

Fig. 17.1 Cartoon depicting the evolving use of expert systems in historical research and psychiatric diagnosis

$$\begin{array}{ccc} \mathbb{A}rc & \xrightarrow{AI_{exp}} & \mathscr{R} \\ & \searrow^{\pi} & \uparrow{\scriptstyle\gamma} \\ & & \mathbb{T} \end{array} \tag{17.1}$$

Once we transform a text into a vector space, we can use standard analysis tools on those vectors to analyze the historical text. The function π computes the time from the text. It can count the number of paragraphs up to the current paragraph, which is the Syuzhet time in paragraphs, or compute the Fabula time from the paragraph. Again, this could be done using machine or human intelligence. The function γ can be interpreted as the first step in the narrative of the historical document.

17.3 Applying AI as an Expert in Humanities

When applying AI to humanities, we face two main challenges. The first is how to overcome the errors that AI makes, and the second is how to overcome the coherent boundary of the machine, which we discussed in Chap. 14.

There are several methods to overcome the coherence boundary of the machine. First, we break down the text into coherent sections. Then, we can either follow well-prepared methods to fuse the coherent part, or summarize the coherent part by using several clocks and compiling them into a single one. Now, we can analyze the single clock. For more on clocks, see Chap. 4.

In order to summarize a paragraph using clocks, we need to set the temperature to zero. This allows us to achieve the best deterministic answer that AI could provide. It is also good to include a definition and an example of some questions.

To evaluate the accuracy of AI answers, it is recommended to perform some random tests and estimate the accuracy of the questions. We remark that 80% or above is usually good enough for applying clocks to analyze the text.

The main question discussed in Chap. 4 is how to compare two clocks when the clocks start to deviate from each other. This allows us to identify critical points in the narrative of two clocks, which can be used as interpretation tools for the historical text in which the two clocks are defined. In this chapter, we explain how to compute the correlation between two clocks which is defined in the same document. The main problem in computing the correlation between two clocks is that clocks are an increasing monotonic function, which makes the correlation always positive. Therefore, we use the correlation of the M-diagrams of the clocks.

Another application for clocks C when dealing with complex cognitive definitions such as Violence, freedom, psychiatric disorders and so on, is to compare two different documents Doc_1, Doc_2 by computing the C for each of the documents. We can get two different clocks C_1, C_2 which are defined on the Syuzhet time of the documents Doc_1, Doc_2, respectively. Since the clocks C_i capture complicated cognitive definitions, we can use the C_1, C_2 cognitive clock to compare the documents Doc_1, Doc_2 with respect to the cognitive phenomena.

In this chapter, we provide several examples that use use word counting or LLM as an expert in psychology, sociology, and political science.

We start with the task of computing the amount of Violence described in the summary of the UN Summits for 2021–2024. This is a simple task for AI since the phenomenon we study does not depend on the position of the words in the sentence but only on the words themselves.

17.4 Estimating Violence in UN Summaries

Every year around October, the UN holds a summit where each country gets around 20 minutes to speak to the world. The speeches are summarized by the UN institution, and most of them are available on the official UN website [1]. To demonstrate the use, advantages, and disadvantages of LLM, we use these summaries from the years 2021–2024. We excluded speeches that did not appear in all these years consecutively.

Two critical wars took place (and continue to rage on) during this period: the war in Ukraine, and the war in Gaza. In this section, we study the summaries with respect to the amount of Violence in the texts. First, we download each of the summaries of all the countries and ask LLM to rank the amount of Violence in the texts according to the years.

To estimate the performance of the LLM, we ask it to generate a list of 300 words that capture Violence. Then, we count the number of these words in each speech per year and rank them. Figure 17.2 demonstrates the four most violent speeches. Clearly, there is a difference between word counting and LLM estimation. The arrows in the figure point from the most violent to least violent years.

compare

According to a human reader, the summaries of Ukraine's speeches from 2022 are identified as the most violent. This judgment aligns with the results obtained by the word-counting method. As mentioned earlier, the counting algorithm performs

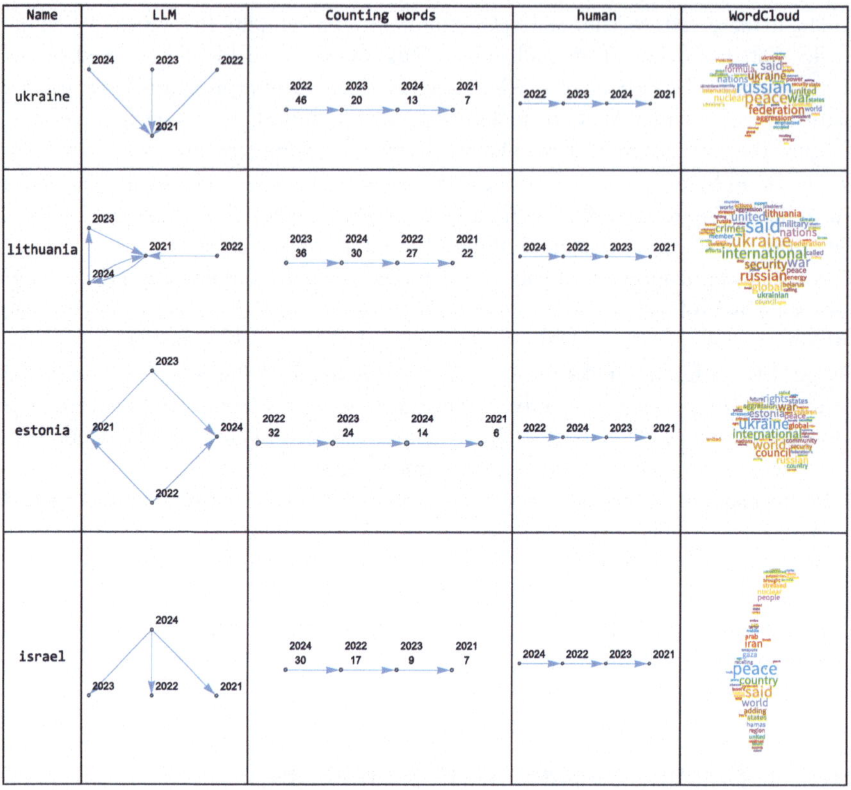

Fig. 17.2 This figure summarizes the amount of Violence in each of the UN summaries. The first column is the name of the country. The second column summarizes the results of the LLM, where we compared all pairs (i, j) of speeches of the same country. We put a direct edge between year i to year j if the speech i of the country was more violent than the speech j according to the LLM. Note that for Lithuania, the edges (2021, 2024) do not make sense. The third column describes the algorithm that counts the number of violent words in each speech. Counting words generates a well-ordered set, where each year can be compared to the others. The fourth column represents the human estimation of the amount of Violence according to the year. Again, it is well-ordered. The last column represents a word cloud summary of all the speeches in the form of the respective country

well mainly because the occurrence of violent words is independent of their position within a sentence.

We note that it is possible to enhance the LLM's performance by incorporating the counting algorithm and providing the model with multiple illustrative examples.

LLM can be used as an expert in almost all disciplines. This means that when studying history using LLM, one can enjoy a discussion on history with those experts. This raises several methodological problems, since the user is not an expert, but may still utilize these experts in the process of conducting historical research.

17.5 Analyzing Historical Documents with AI Expertise

When studying history, one of the main questions concerns the role of the leader in shaping the historical narrative. For example, would WWII have happened if Hitler never existed, and what was Hitler's role in the results of this war? One way to attack such a general question is to analyze the historical figure through different expert systems, trying to understand the leader through the expert system. Such an inquiry of the past demanded expertise in the domain of the expert system and history as well. However, recent advances in AI allowed us to use LLM to pursue such investigations.

In this chapter, we ask AI to analyze historical figures through the texts they left behind. We work in the paragraph scale, i.e., take one paragraph as a single unit for analysis in each historical text. We break the historical text into paragraphs, analyze each paragraph separately, then apply a numerical method to fuse the paragraphs back into a coherent narrative described by numbers instead of text. We call this method AI-numerical history .

We use five steps to generate a psychological estimation of historical texts written by a historical figure. We will first explain the general method, then we demonstrate it by analyzing Hitler's *Mein Kampf* [2] and Churchill's *My Early Life* [3]. Note that both books are autobiographical; therefore, they represent the ideas and perceptions of their authors.

17.5.1 Step One: Diagnosis

The first step is to categorize historical figures p.

17.5.2 Step Two: Processing Historical Text

In this step, we acquire a historical text txt, import it into a computational environment, and structure it for further analysis, which will be done with the help of the LLM.

17.5.3 Step Three: Textual Analyses

In the third step, we need to decide which of the LLM's expertise we wish to use. Denote the set of experts by l_{exp}. Then, for expert $i \in l_{exp}$, construct a list of queries $Qexp_i$ relevant to the expertise.

Next, we construct several LLM queries to analyze the paragraphs separately with respect to the list of queries $Qexp_i$. In this step, we construct several LLM functions which will allow us to question the LLM at the level of a paragraph on all the subjects we wish to examine. Note that when posing a question to the LLM, it usually responds in a natural language, which is not well formatted for extracting the numeric information for further analyses. For example, if we ask the LLM to extract the numbers from the text, we can get the following answers: "1", "one", "One", etc. Moreover, sometimes it becomes creative and inaccurate, which poses another level of difficulty. After overcoming all these issues, we can proceed to the last step.

17.5.4 Step Four: Aggregate

After cleaning the data, we arrive at the step where we can analyze the text through the numbers using mathematical tools. The last step is to aggregate all the paragraphs into a complete, coherent narrative.

In the next few sections we will provide several examples. We will first analyze Adolf Hitler using the *Mein Kampf* [2]. Then, we analyze Winston Churchill using the book *My Early Life* [3]. The advantage of comparing Hitler with Churchill is that they were both leaders who wrote autobiographies. We can use these books as a reflection of their state of mind.

17.5.5 Step Five: Analyses

Instead of computing critical points when comparing two clocks C_1, C_2, we would like to compute the correlations ρ between two different clocks. Since clocks are a monotonic increasing function, it means that the correlation $\rho(C1, C2)$ between any clocks is positive:

$$\rho(C1, C2) \geq 0 \tag{17.2}$$

Therefore, it is recommended to compute the correlation of the M-diagram of the clocks with the Syuzhet time:

$$\rho\left(M\left[C_1, C_s\right], M\left[C_2, C_s)\right]\right) \tag{17.3}$$

There are two main advantages to working with Eq. 17.3. The first is that the correlation is no longer positive. It could be any number between -1 and 1. The second advantage of using Eq. 17.3 instead of

$$\rho(C_1, C_2) \tag{17.4}$$

is that it usually Eq. 17.3 captures the narrative better.

Working with the M in Eq. 17.3 is similar to removing expectation $E[X]$ in the definition of a variance $Var[x] = E[(x = E[x])^2$.

The correlation, according to (17.3), can be interpreted as an indication of a hidden connection in the author's mind. When we analyze autobiographical texts of historical figures, this can reveal their inner world, as we will see in our examples.

17.6 Example

To demonstrate the use of the LLM expert on historical texts, we will use *Mein Kampf* [2] and *My Early Life* [3]. We will analyze each of these historical texts using three experts, listed below.

1. Psychiatric expert
2. Political science expert
3. Emotion analyses expert

After completing the first and second steps of our example, identifying the people we wish to study and finding the expert, which is identifying the text we wish to study. Finally, we move to the third step, which is constructing the query we wish to ask the LLM.

For each expert and subject we wish to query the expert we construct an LLM Yes/No query that is composed out of the definition of the subject, three examples of the Yes instance, and three examples of the No instance. Altogether, we have three experts, 23 queries, 69 Yes examples, and 69 No examples. Note that it is possible to use the LLM to help construct and verify the queries and examples.

We compose a single deterministic function that accepts the input paragraph. It outputs Yes if there is an indication for the specific queries in the input paragraph, and No if there is none.

In this step, we also standardize the LLM's answer using the function *Nearest*. Another problem is that the LLM sometimes does not provide an answer to the question. This can be overcome by asking the same question again and not overloading the LLM—simply waiting one second between any two consecutive questions.

Another problem is that, from time to time, the paragraphs themselves can be written in a language that violates the LLM's censorship rules. There are several methods we can use to overcome this problem. The first is to ignore these paragraphs. Another way is to assume that such paragraphs belong to a specific category. A third option is to break the paragraphs into smaller units of text, such as sentences or even words, and then analyze the levels of those units. It is also possible to overwrite the paragraph in a suitable language. The last option is to cite exactly the source of the paragraph. If the LLM has the source in its memory, it can ease the censorship problem, since it knows that it is dealing with a legitimate source recognized by the historical community.

We note that the method we use is not sensitive to small errors that may occur during the LLM analyses of each paragraph. Since the number of queries is huge, we recommend not to bombard the LLM with queries, but rather send a single query, pause for several seconds, then send the next one. Otherwise, at the current technology, after some time the LLM stops answering the queries correctly.

Since we choose to analyze the texts through three separate expert systems, we need to describe Step Three for each one of them. We start with the Psychiatric expert system, then we move to the Political Science expert system, and we end with the Emotion Analyses expert system.

17.6.1 Step Three: Psychiatric Expert

Before we write the LLM queries, we need to identify the subject we wish to inquire about using the LLM. In the Psychiatric expert system, we wish to inquire about seven psychiatric disorders specified below:

1. Narcissistic Personality Disorder
2. Antisocial Personality Disorder
3. Paranoid Personality Disorder
4. Delusional Disorder
5. Post-Traumatic Disorder
6. Borderline Personality Disorder
7. Psychotic Disorders characterized by symptoms like delusions.

For each of these psychiatric disorders, we construct LLM Yes/No queries that take a paragraph as input and output Yes if the paragraph contains the signs of psychiatric disorders, and No if it does not.

We recommend providing a few examples of Yes instance and a few examples of No instance in the LLM queries. Our next example is the Political Science expert.

We provide a simple example GetAnswerNPD function in Mathematica for the first query. The reader can use this as a Rosetta stone to generate other queries.

```
GetAnswerNPD[paragraph_String, maxAttempts_Integer : 5,
  timeLimit_Integer : 6] :=
  Module[{ans = "TimeOut", attempts = 0, yesExamples, noExamples},
    yesExamples = "Yes Examples:\n1. John often boasted about his
        achievements
claiming that no one in his circle could ever match his talents.
    He demanded constant admiration and lacked empathy, often
    belittling others to make himself feel superior.";

    noExamples = "No Examples:\n1. Emma and her friends spent the
        afternoon \
```

```
discussing their recent trip. They shared stories, laughed, and \
listened to each other's experiences, each person contributing \
equally to the conversation.";

    While[ans === "TimeOut" && attempts < maxAttempts,
      ans = TimeConstrained[
          LLMSynthesize[{LLMPrompt["YesNo"],
            "Narcissistic Personality Disorder is characterized by
                a grandiose sense of self-importance, a need for
               excessive admiration, and a lack of empathy for
               others. Please analyze the following paragraph and
                indicate if it contains themes related to
               Narcissistic Personality Disorder. Answer 'Yes'
               for presence and 'No' for absence of such themes.\
               n\n"<>yesExamples<>"\n\n"<>noExamples<>"\n\n" <>
               paragraph}, LLMEvaluator -> <|"Temperature" ->
               0|>],
          timeLimit, "TimeOut"];
      attempts++];
    ans
  ]
```

17.6.2 Step Three: Political Science Expert

Similar to what we did in the previous subsection, we can analyze political science concepts. The list of subjects we wish to query is detailed below:

1. Dictatorship
2. Democracy
3. Capitalism
4. Socialism
5. Fascism
6. Communism
7. Terrorism
8. Peace

Once we identify the subjects we wish to inquire about, we need to write an LLM query for each. We recommend giving a dictionary definition together with some examples.

17.6.3 Step Three: Emotion Analysis Expert

According to the Wheel of Emotions theory, there are eight basic emotions, listed below:

1. Joy
2. Trust
3. Fear
4. Surprise
5. Sadness
6. Disgust
7. Anger
8. Anticipation

For each of those basic feelings, we write LLM queries that get as an input a paragraph and as output Yes if the paragraph contains the signs of those basic feelings and No otherwise. We refer the interested reader for more information about the wheel of emotions theory to the Wiki page [4].

Our goal is to define for each of those LLM Yes/No queries a clock that measures the number of paragraphs (Syuzhet time) in which this query appears in the text. Once we have defined it, we can use those clocks to analyze the historical text with respect to the connection between the different queries. In the case of autobiographical text, this will indicate the state of a historical figure with respect to the clocks.

In Step Four, we need to run all the queries on the text for each paragraph separately, then organize them in a table, transforming the answers of the LLM in 0/1, where zero is a No and one is a Yes. From time to time, the LLM ignores our instructions to provide Yes/No answers and gives different answers or even errors. We interpret all these answers as No since Yes answers are rather rare and the method can overcome a small number of errors. Other methods for dealing with these difficulties are just as welcome.

Once we have our 0/1 table, we can construct clocks by accumulating those numbers. In the next two sections, we summarize the results for Hitler and then for Churchill.

17.7 Example: Adolf Hitler

17.7.1 Example: Step Two

An English translation of *Mein Kampf* [2] can be found in Project Gutenberg [5]. We can import the book into Wolfram Mathematica or any other high-level language, such as Python, using the Import function, then break it down into paragraphs using the Text Analysis function. There are around 2700 paragraphs, depending on the way you break the document into paragraphs. This highly influences the time and the number of queries LLM needs to complete the text analyses. Since the Yes instance is relatively rare, the number of queries can be reduced by aggregating several paragraphs and then looking for the correct paragraphs. We ignore this possibility. Once we have the source set up as a list of separate paragraphs, we can advance to the next step.

17.7.2 Example: Step Three

Since we are analyzing both Hitler and Churchill with the same experts and the same queries, we already have the expert and subject from Sect. 17.6. Therefore, we summarize the results.

17.7.3 Example: Step Four

After converting the text into a table of numbers, we can apply numerical analysis to it. The simplest analysis is to accumulate the queries into clocks. This is the equivalent of generating a clock for each of the queries. The outcome of those clocks can be found in three figures. Each figure summarizes the experts.

Hitler: Psychiatric Expert

Figure 17.3 looking at the left part of the figure, we see that at the time of writing *Mein Kampf*, the most obvious personality disorder that Hitler presented was ASPD, Antisocial Personality Disorder (253 paragraphs). The second disorder is PPD, Paranoid Personal Disorder (235 paragraphs). This means that when analyzing Hitler's psychology, one should take into account that Hitler was suffering from ASPD and

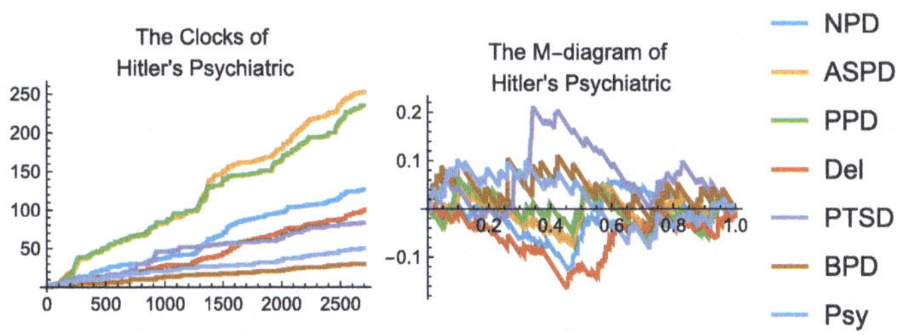

Fig. 17.3 The left part of the figure shows accumulated clocks describing the psychological disorder of Hitler per paragraph. The right part of the figure is the M-diagram of these disorders

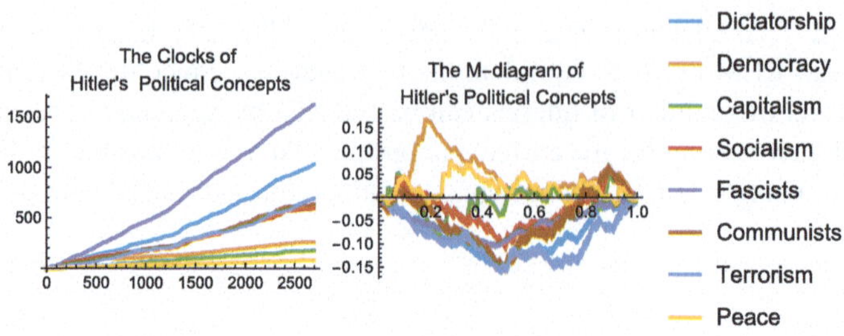

Fig. 17.4 The left part of the figure shows accumulated clocks describing the result of political science experts analyzing the Hitler's book per paragraph. The right part of the figure is the M-diagram of these political science opinions

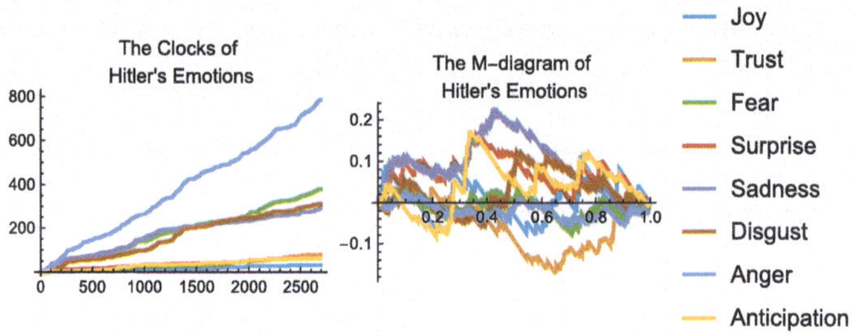

Fig. 17.5 The left part of the figure shows the accumulated clocks describing the result of emotion experts analyzing the Hitler's book per paragraph. The right part of the figure is the M-diagram of the emotion expert

PPD as early as 1925. Note that in the first half of the book, both disorders were almost equal[2]

Hitler: Political Science Expert

We repeat the four main steps for the Political Science expert and the outcome of this process is eight clocks which can be seen in left of Fig. 17.4. Looking at the figure, we see that Hitler is first of all a fascist (1630 paragraphs), then a dictator (1041 paragraphs), and third, a terrorist (698 paragraphs).

Hitler: Emotion Analysis Expert

Again, we summarize the results of the eight subjects List 17.6.3 from which we queried the LLM for each emotion. The result can be seen in Fig. 17.5. The most

[2] Since Mein Kampf is a canonical text in the study of history, it is clear that the LLM knows the text and its many interpretations, so we should be careful in assuming that in 1925 we would get the same analyses.

common emotions are Anger (788 paragraphs), Fear (382 paragraphs), and Disgust (315 paragraphs). The least common emotion is Joy (34 paragraphs).

After computing all the relevant clocks and the M-diagrams, in the next subsection we compute the correlation between the relevant M-diagrams, which sheds some light on Hitler's mind.

17.7.4 Example: Step Five

Since we have three different experts, we compute the correlation between any two and use the correlation to evaluate Hitler's state of mind. We represent each of the three correlations in the table, then and discuss its meaning. We start with the correlation between Political Science and Emotion Analysis.

We scan the Table 17.1 through emotions. In the first column, we see the correlations between the emotion Joy and subjects in the field of political science, the highest correlation (0.46) being with Socialism and Communism. For Trust, the highest correlation is with Terrorism (0.69) and Socialism (0.55). The third column is devoted to Fear, with the highest correlation is with Peace (0.26). The highest negative correlation is with Fascism (−0.27). We move to Surprise and see the highest correlation with Peace (0.57) and the lowest with Fascism (−0.72). The highest Sadness correlation is with Peace (0.46) and the lowest is with Fascism (−0.9). The next emotion, Disgust, is highly correlated with Capitalism (0.43) and least correlated with Democracy (−0.54). For Anger, the highest correlation is with Terrorism (0.44). Anticipation is correlated with Dictatorship (−0.47) and Terrorism (−0.47). One may scan the Table 17.1 according to political science; we leave this as an exercise for the reader, see Exercise 17.11.

Correlations between Psychiatric Disorders and Emotional Analyses can be found in Table 17.2. According to the Psychiatric clocks that were computed in Sect. 17.7.3, the main disorders Hitler suffered from are ASPD and PPD. Looking at the relevant column in Table 17.2 we see that both disorders are highly correlated with Anger

Table 17.1 Political systems and emotional associations

Political/Emotions	Joy	Trust	Fear	Surprise	Sadness	Disgust	Anger	Anticipation
Dictatorship	0.43	0.51	−0.07	−0.54	−0.86	−0.23	0.18	−0.37
Democracy	−0.15	0.45	−0.06	0.38	0.03	−0.54	0.03	−0.34
Capitalism	0.25	−0.23	0.08	−0.23	−0.11	0.43	−0.02	0.25
Socialism	0.46	0.55	−0.12	−0.36	−0.82	−0.28	0.07	−0.35
Fascists	0.44	0.18	−0.27	−0.72	−0.90	0.06	0.01	−0.25
Communism	0.46	0.35	−0.21	−0.49	−0.88	−0.12	0.00	−0.27
Terrorism	0.34	0.69	0.24	−0.36	−0.63	−0.39	0.44	−0.37
Peace	0.00	0.01	0.26	0.57	0.46	−0.21	0.02	0.35

Table 17.2 Psychiatric disorders and emotional associations

Psychiatric/Emotions	Joy	Trust	Fear	Surprise	Sadness	Disgust	Anger	Anticipation
NPD	0.08	−0.37	−0.52	−0.70	−0.51	0.57	0.00	−0.23
ASPD	−0.26	−0.01	0.16	−0.35	−0.29	0.47	0.59	−0.46
PPD	−0.44	0.35	0.50	0.16	0.20	0.06	0.81	−0.49
Del	0.50	0.21	−0.37	−0.61	−0.78	0.00	0.05	−0.29
PTSD	−0.18	−0.31	0.51	0.70	0.72	0.01	−0.15	0.74
BPD	0.11	0.44	0.62	0.67	0.48	−0.41	0.25	0.15
Psy	−0.04	0.76	0.56	0.52	0.30	−0.63	0.51	−0.30

Table 17.3 Psychiatric disorders and political science

Psychiatric/Political	NPD	ASPD	PPD	Del	PTSD	BPD	Psy
Dictatorship	0.35	0.38	0.00	0.78	−0.65	−0.23	−0.04
Democracy	−0.29	−0.11	0.31	−0.16	−0.19	0.12	0.53
Capitalism	0.28	0.19	−0.18	0.17	0.12	0.00	−0.40
Socialism	0.31	0.34	0.02	0.71	−0.62	−0.16	0.02
Fascism	0.59	0.42	−0.16	0.84	−0.64	−0.45	−0.38
Communism	0.47	0.40	−0.08	0.79	−0.63	−0.32	−0.20
Terrorism	0.11	0.38	0.24	0.60	−0.46	0.04	0.27
Peace	−0.41	−0.17	0.18	−0.47	0.50	0.51	0.40

emotions, ASPD with 0.59 and PPD with 0.81. From this we can conclude that anger was the dominant emotion when Hitler wrote *Mein Kampf*.

Table 17.3 shows the correlation between Hitler's Psychiatric Disorders and his Political Science opinion. According to the table, the Psychiatric Disorder that correlated the most with Hitler's Political Science opinion is Delusional.

In the next section we will analyze Winston Churchill with the same method.

17.8 Winston Churchill

To compare Churchill with Hitler, we need to repeat the same process with *My Early Life* [3].

17.8.1 Example: Step Two

We should mention that Churchill's book is roughly 10% the size of t*Mein Kampf*'s three volumes. As mentioned earlier, we analyze the book using three different experts. We summarize the output of each of these experts in two graphs. The left side of the figure represents the different clocks generated by experts, and the right side of the figure represents the computed M-diagram for these clocks.

17.8.2 Example: Step Three

See Sect. 17.6.

17.8.3 Example: Step Four

Since we analyze both Hitler and Churchill with the same experts and the same queries, we have already written the expert and subject in Sect. 17.6. Therefore, we summarize the results.

Churchill: Psychiatric Expert

The result of the process for the psychiatric expert can be seen in Fig. 17.6. Looking at the left side of the figure, we see that the most common disorder Churchill is Post-Traumatic Disorder, with 25 paragraphs related to it. The second is Psychotic Disorders (11 paragraphs) and the third is Paranoid Personality Disorder (10 paragraphs).

We note that the M-diagram functions are much harder to grasp, but computing the correlation between different M-diagrams captures the narrative and the relation between the clocks better. This is because clocks are a monotonic increasing function whereas M-diagrams are not.

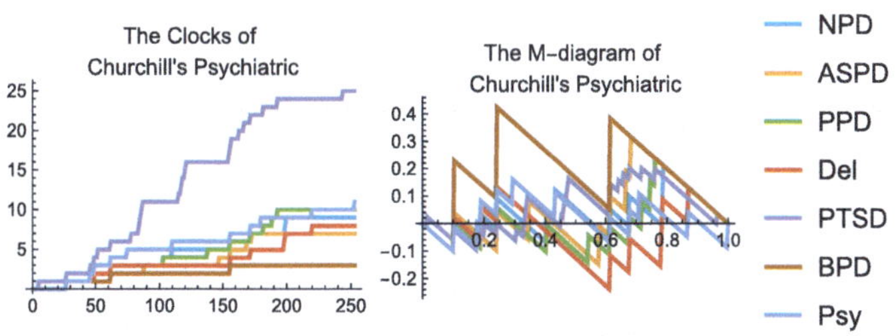

Fig. 17.6 The left side of the figure is accumulated clocks describing the psychological disorder of Churchill per paragraph. The right side of the figure is the M-diagram of these disorders

Churchill: Political Science Expert

The results of the political science experts can be seen in Fig. 17.7. Looking at the figure, we see that democracy and terrorism appear in the same number of paragraphs (21). The third most mentioned political concept is a dictatorship with 15 paragraphs.

Churchill: Emotion Analysis Expert

The last expert is the emotion expert. The outcome of the expertise can be seen in Fig. 17.8. The most common emotion is fear (90 paragraphs), the next is anticipation (57 paragraphs), and the third is sadness (51 paragraphs).

In the next section, we discuss the correlation between the M-diagrams of Churchill's experts similarly as we did with Hitler.

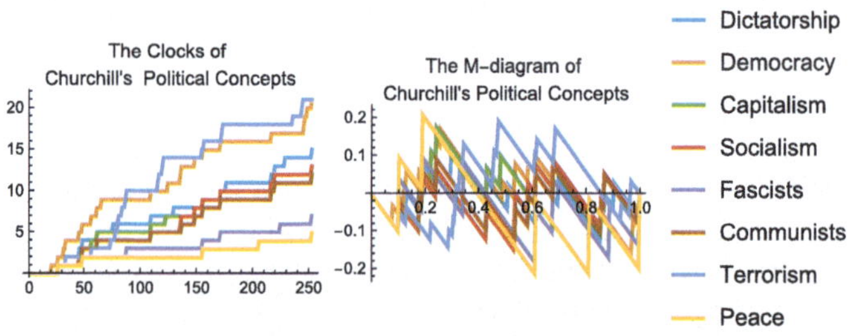

Fig. 17.7 The left side of the figure is accumulated clocks describing the result of political science experts analyzing the Churchill's book per paragraph. The right side of the figure is the M-diagram of these political science opinions

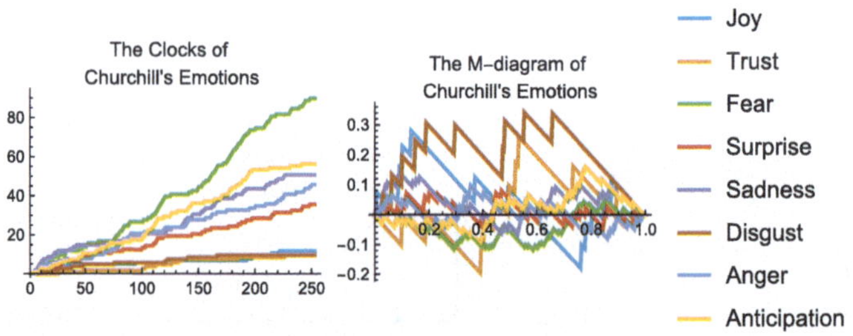

Fig. 17.8 The left side of the figure shows accumulated clocks describing the result of emotion experts analyzing the Churchill's book per paragraph. The right side of the figure is the M-diagram of the emotion expert

17.8.4 Correlation Between Experts

Once we computed the M-diagram of each of the clocks generated by three different experts, we can compute the correlation between two different phenomena. For example, we can ask what is the correlation between the emotions that Churchill had and his psychiatric state. Since we have three experts, we summarize the correlations in three tables. The first table shows the correlation between Political Science and Emotions (Table 17.4).

Looking at the table, it is clear that Churchill gains the most joy from talking about peace, while the least amount of joy comes from talking about terrorism. Looking at the next emotion, Trust, we see that the most negative correlation is with peace, which means he does not believe in peace. The most positive correlation for Trust is terrorism, which can be interpreted as his belief that terrorism is an effective weapon. The next emotion in the table is fear. In this case, all political concepts have negative correlations with fear. The most negative correlation is with democracy, the next is capitalism, and the last is dictatorship. Note that the book was published in 1930 during Churchill's "Wilderness Years" (1929–1939), which may explain his feelings toward democracy. The fact that he was a member of a liberal party in 1904–1924 can explain his feelings toward capitalism. Moving to the emotion of surprise, we can see that the most correlated political concepts are peace (0.5) and democracy (0.35), while the least correlated concept is terrorism (−0.25). For sadness, the most negative correlations are with terrorism (−0.53) and capitalism (−0.44), and the most positive correlation is with socialism (0.32). For disgust, all correlations are positive. The highest numbers are for democracy (0.7) and capitalism (0.49). Anger has the most negative correlation with socialism (−0.37) and the most positive correlation with fascism (0.31).

The next table is the correlation between Emotions and Psychiatric Disorders.

According to the psychiatric clocks, Churchill suffered from PTSD. Looking at Table 17.5, we see that the smallest correlation between any Psychiatric Disorders and Emotional Associations is −0.81, which is the correlation between Joy and PTSD.

The last table is the correlation between psychiatry and political science, see Table 17.6. The highest correlation is between capitalism and psychiatry (0.85).

Table 17.4 Political systems and emotional associations

Political/Emotions	Joy	Trust	Fear	Surprise	Sadness	Disgust	Anger	Anticipation
Dictatorship	0.33	−0.41	−0.47	0.3	−0.25	0.26	0.07	−0.5
Democracy	0.14	−0.13	−0.75	0.35	−0.21	0.7	0.25	−0.46
Capitalism	0.08	−0.13	−0.62	0.31	−0.44	0.49	0.05	−0.31
Socialism	−0.3	0.29	−0.01	−0.18	0.32	0.24	−0.37	0.16
Fascism	0.36	−0.57	−0.26	0.29	0.1	0.15	0.31	−0.6
Communism	−0.08	−0.02	−0.15	0.04	0.21	0.24	−0.26	−0.1
Terrorism	−0.68	0.48	−0.21	−0.28	−0.53	0.39	0.24	0.34
Peace	0.67	−0.65	−0.31	0.5	0.15	0.16	0.13	−0.67

Table 17.5 Psychiatric disorders and emotional associations

Psychiatric/Emotions	Joy	Trust	Fear	Surprise	Sadness	Disgust	Anger	Anticipation
NPD	−0.25	0.29	0.27	−0.13	−0.17	−0.03	−0.60	0.65
ASPD	−0.59	0.24	0.44	−0.52	0.31	0.06	−0.39	0.55
PPD	−0.29	0.17	0.65	−0.31	0.34	−0.22	−0.73	0.68
Del	0.47	−0.57	0.31	0.06	0.24	−0.40	−0.35	−0.14
PTSD	−0.81	0.60	0.16	−0.37	−0.21	0.33	−0.20	0.67
BPD	−0.09	−0.08	−0.47	0.07	−0.18	0.61	0.04	−0.11
Psy	−0.04	−0.26	−0.41	0.17	−0.40	0.43	0.02	−0.16

Table 17.6 Psychiatric disorders and political science

Political/Psychiatric	NPD	ASPD	PPD	Del	PTSD	BPD	Psy
Dictatorship	−0.04	−0.17	−0.17	0.37	−0.28	0.48	0.77
Democracy	−0.22	−0.17	−0.37	−0.16	−0.10	0.66	0.57
Capitalism	0.13	−0.12	−0.20	0.14	0.03	0.62	0.80
Socialism	0.20	0.46	0.32	0.13	0.22	0.36	0.25
Fascism	−0.18	0.02	−0.14	0.32	−0.27	0.40	0.53
Communism	0.18	0.33	0.21	0.31	0.04	0.49	0.50
Terrorism	0.14	0.30	−0.13	−0.63	0.81	0.26	0.16
Peace	−0.16	−0.28	−0.19	0.47	−0.56	0.37	0.41

17.9 Comparing Hitler to Churchill

In the previous sections, we used three expert systems to analyze *Mein Kampf* [2] by Hitler and *My Early Life* [3] by Churchill. In this section, we will show how one may compare those two historical figures mechanically. We follow Cantor's definition of Cardinality in order to define what it means to compare two historical persons[3] According to Cantor, two sets A and B have the same cardinality if there exists a one-to-one correspondence function f between their elements. Formally,

$$f : A \to B \tag{17.5}$$

where f is a bijection from A to B.

Therefore, when we want to compare two historical persons, we need to construct a bijection from person P_1 to person P_2. In our case, we wish to construct a bijection from a property of Hitler to a property of Churchill, or vice versa.

To demonstrate how to do this, we construct three bijections from Hitler to Churchill using our three LLM expert systems. Clearly, this is not the only way to do it, merely a demonstration.

[3] A similar definition can be done to compare two historical entities.

The idea is to construct a correlation matrix $\rho_M = [\rho_{i,j}]$ for each expert LLM. Let $exp = (llm_1, ..., llm_k)$ be the vector query that describes the expert exp. Denote the vector of clocks defined by the expert as

$$\vec{c} = (c_1, ..., c_k) \tag{17.6}$$

The element of the correlation matrix $\rho_M = \left[\rho_{i,j}\right]$ is the correlation between the M-diagrams of clock c_i of each of the clocks we have computed in the previous section. Formally

$$\rho_{i,j} = \rho\left(M\left[c_i, c_s\right], M\left[c_j, c_s)\right]\right) \tag{17.7}$$

where c_s is the Syuzhet time, i.e., the standard clock or an identity function.

When we try to compute the correlation between two clocks c_1, c_2, we face a technical problem as each clock may have a different Syuzhet time. This is exactly what happened in our example, since *Mein Kampf* has roughly 2700 paragraphs, and *My Early Life* has only 253. Therefore, it is difficult to compute the correlation between different vectors. However, when we compute the correlation between the normalized clocks Nc_1, Nc_2, we transform the clocks to be defined on the unit intervals and therefore the domain of the M-diagram

$$M\left[c_i, c_s\right] = N(c_i) - N(c_s) \tag{17.8}$$

is the unit interval. Now we sample both M-diagrams $M\left[c_i, c_s\right]$ at a specific resolution and compute the correlation between the samples. Therefore, transforming the clocks into M-diagrams allows us to compare clocks with different time domains.

Then we transform the correlation matrix ρ into the preferential order on the clocks, use this order to solve the stable marriage problem. For more on the stable marriage problem, see [6].

In the next few paragraphs, we provide numerical details for each LLM expert.

Emotion Correlation

Now we compute the preferential order of Churchill's emotions with respect to Hitler.

Now we compute the preferential order of Hitler's emotions with respect to Churchill.

We now can compute the stable matching of emotion between Churchill's and Hitler's emotions. The outcome can be seen below (Tables 17.7, 17.8, 17.9, 17.10 and 17.11).

We now write the following prompts to any LLM: "Given the following emotional mapping from Adolf Hitler's emotions (Joy, Trust, Fear, Surprise, Sadness, Disgust, Anger, Anticipation) to Winston Churchill's emotions (Fear, Joy, Sadness, Disgust, Anger, Trust, Surprise, Anticipation), write a coherent analytical paragraph that explains the symbolic or strategic significance behind each emotional correspondence, clearly interpreting how each emotion experienced by Hitler conceptually relates or translates to Churchill's emotional response within their historical adversarial context."

Table 17.7 Emotional trait correlation between Churchill and Hitler

Churchill/Hitler	H-Joy	H-Trust	H-Fear	H-Surprise	H-Sadness	H-Disgust	H-Anger	H-Anticipation
C-Joy	−0.21	0.74	0.44	0.35	0.05	−0.48	0.53	−0.53
C-Trust	−0.3	−0.79	−0.27	−0.44	−0.04	0.84	−0.16	0.05
C-Fear	0.52	0.0.17	−0.15	−0.41	−0.74	0	−0.11	−0.08
C-Surprise	−0.32	0.34	0.28	0.23	0.29	−0.16	0.0.38	−0.49
C-Sadness	0.4	0.31	−0.31	−0.31	−0.57	−0.04	0.0.09	−0.38
C-Disgust	−0.4	−0.52	−0.26	0.28	0.5	0.37	−0.12	0.07
C-Anger	−0.08	0.13	0.42	0.42	0.67	−0.13	0.27	0.23
C-Anticipation	0.09	−0.6	−0.28	−0.32	−0.35	0.55	−0.39	0.25

Table 17.8 Hitler's emotional trait preferences for Churchill's emotions

Hitler's emotion	Churchill's emotions (Ranked Preference)
H-Joy	C-Fear > C-Sadness > C-Anticipation > C-Anger > C-Joy > C-Trust > C-Surprise > C-Disgust
H-Trust	C-Joy > C-Surprise > C-Sadness > C-Fear > C-Anger > C-Disgust > C-Anticipation > C-Trust
H-Fear	C-Joy > C-Anger > C-Surprise > C-Fear > C-Disgust > C-Trust > C-Anticipation > C-Sadness
H-Surprise	C-Anger > C-Joy > C-Disgust > C-Surprise > C-Sadness > C-Anticipation > C-Fear > C-Trust
H-Sadness	C-Anger > C-Disgust > C-Surprise > C-Joy > C-Trust > C-Anticipation > C-Sadness > C-Fear
H-Disgust	C-Trust > C-Anticipation > C-Disgust > C-Fear > C-Sadness > C-Anger > C-Surprise > C-Joy
H-Anger	C-Joy > C-Surprise > C-Anger > C-Sadness > C-Fear > C-Disgust > C-Trust > C-Anticipation
H-Anticipation	C-Anticipation > C-Anger > C-Disgust > C-Trust > C-Fear > C-Sadness > C-Surprise > C-Joy

Psychiatry Correlation

For the Psychiatric expert, we provide the correlation matrix and the final comparison function. The correlation matrix of the Psychiatric expert system is given below:

The comparison function between Hitler's and Churchill's psychiatric disorders is given below.

We leave the computing of the preferential psychiatric disorders for Hitler and Churchill as an exercise for the reader, see Exercise 17.11.

After computing the preferential order of Hitler and Churchill's psychiatric disorders, we can compute the stable marriage of these preferences. This gives us a bijection function between Churchill's and Hitler's disorders. The outcome can be seen in Table 17.12.

Table 17.9 Churchill's emotional trait preferences for Hitler's emotions

Churchill's Emotion	Hitler's Emotions (Ranked Preference)
C-Joy	H-Trust > H-Anger > H-Fear > H-Surprise > H-Sadness > H-Joy > H-Disgust > H-Anticipation
C-Trust	H-Disgust > H-Anticipation > H-Sadness > H-Anger > H-Fear > H-Joy > H-Surprise > H-Trust
C-Fear	H-Joy > H-Trust > H-Disgust > H-Anticipation > H-Anger > H-Fear > H-Surprise > H-Sadness
C-Surprise	H-Anger > H-Trust > H-Sadness > H-Fear > H-Surprise > H-Disgust > H-Joy > H-Anticipation
C-Sadness	H-Joy > H-Trust > H-Anger > H-Disgust > H-Surprise > H-Fear > H-Anticipation > H-Sadness
C-Disgust	H-Sadness > H-Disgust > H-Surprise > H-Anticipation > H-Anger > H-Fear > H-Joy > H-Trust
C-Anger	H-Sadness > H-Surprise > H-Fear > H-Anger > H-Anticipation > H-Trust > H-Joy > H-Disgust
C-Anticipation	H-Disgust > H-Anticipation > H-Joy > H-Fear > H-Surprise > H-Sadness > H-Anger > H-Trust

Table 17.10 Direct emotional trait pairings between Hitler and Churchill

H-Joy	C-Fear
H-Trust	C-Joy
H-Fear	C-Sadness
H-Surprise	C-Disgust
H-Sadness	C-Anger
H-Disgust	C-Trust
H-Anger	C-Surprise
H-Anticipation	C-Anticipation

Paired emotional traits of Hitler and Churchill

Table 17.11 Churchill/Hitler and psychiatric disorders

Churchill/Hitler	H-NPD	H-ASPD	H-PPD	H-Del	H-PTSD	H-BPD	H-CPsy
C-NPD	−0.04	−0.13	−0.47	−0.09	0.07	−0.18	−0.40
C-ASPD	0.40	−0.24	−0.62	0.39	−0.17	−0.43	−0.65
C-PPD	0.30	0.01	−0.42	0.44	−0.28	−0.31	−0.41
C-Del	−0.27	−0.12	−0.09	0.22	−0.28	0.15	0.33
C-PTSD	0.30	−0.17	−0.55	−0.06	0.28	−0.39	−0.77
C-BPD	−0.16	−0.41	−0.19	−0.32	0.19	0.00	−0.03
C-Psy	−0.49	−0.63	−0.28	−0.50	0.31	0.16	0.18

Table 17.12 Paired psychiatric traits of Hitler and Churchill

Hitler's Trait	Churchill's corresponding trait
H-NPD	C-ASPD
H-ASPD	C-NPD
H-PPD	C-PTSD
H-Del	C-PPD
H-PTSD	C-Psy
H-BPD	C-BPD
H-Psy	C-Del

Political Science Correlation

We provide the correlation of political science experts, see Table 17.13.

We leave the computing of the preferential political science system for Hitler and Churchill together with the comparison function as an exercise for the reader, see Exercise 17.11.

17.10 Conclusion

There is a famous debate on who controls historical narrative: people or their environment? We believe that AI tools can help history researchers understand the context of the narrative. Most of those who are interested in history lack the theoretical background to analyze historical texts as an expert in psychiatric, psychology, sociology, and political science. Now, with the help of AI tools, researchers can overcome this gap of knowledge and enjoy the expertise of AI in most areas of human knowledge to achieve a better understanding of historical documents. This enriches researchers in both ways: they can achieve a better understanding of a historical document and learn more about the expertise used to analyze the document.

In this chapter, we demonstrate several ways for AI to analyze historical documents as an expert system with respect to psychiatric, political science, and emotion analyses. We show that when we deal with a simple notion such as Violence, the simple word-counting algorithm performs better than complex LLM queries since the existence of Violence is independent from the sentence structure. We then move to more complicated queries, where the phenomena depend on the structure of the sentence. In such cases, the LLM performs much better than the word-counting algorithm.

We then use the results of the LLM to generate clocks and employ M-diagrams to compute their correlation. This allows us to combine different subjects when we analyze a specific historical figure. We end the chapter explaining how to compare two historical figures with the same expert.

Table 17.13 Political Trait correlations between Churchill and Hitler

Churchill/Hitler	H-Dictatorship	H-Democracy	H-Capitalism	H-Socialism	H-Fascism	H-Communism	H-Terrorism	H-Peace
C-Dictatorship	−0.287	0.61	−0.573	−0.199	−0.45	−0.337	−0.195	0.3
C-Democracy	−0.634	0.481	−0.568	−0.563	−0.665	−0.625	−0.517	0.442
C-Capitalism	−0.559	0.375	−0.464	−0.481	−0.588	−0.563	−0.482	0.351
C-Socialism	0.087	−0.39	−0.481	0.079	0.245	0.223	−0.141	−0.159
C-Fascism	−0.169	0.579	−0.451	−0.054	−0.362	−0.215	−0.099	0.122
C-Communism	−0.01	0.323	−0.322	0.079	0.036	0.070	−0.157	0.062
C-Terrorism	−0.606	−0.49	0.224	−0.684	−0.343	−0.546	−0.659	0.058
C-Peace	−0.035	0.742	−0.68	0.049	−0.294	−0.12	0.115	0.235

When repeating these kinds of analyses, one should be careful since AI tends to hallucinate. However, since we force the answer to be Yes or No and deterministic, we minimize such hallucinations. We also recommend a simple counting algorithm to supervise the LLM algorithm and improve its results, as was shown in Sect. 17.4.

17.11 Exercise

1. It is common to assume that the main leaders of WWII were Hitler, Mussolini, Stalin, Churchill, and Roosevelt. Choose two leaders and compare their speeches using the method developed in this chapter.
2. In Sect. 17.7.4 we have scanned the correlation table according to the emotions, which are the columns of the table. Scan Table 17.1 according to the row, not column, i.e., according to Political Science. Does the result make sense?
3. Choose your favorite leader and analyze the evolution of his speech through time.
4. Choose one of your favorite debates and compute both candidates' social locus of control.
5. Use the Wiki page on Abraham Lincoln to analyze Lincoln's psychological state of mind. Repeat the process and compare it with Gandhi and Mandela.
6. Use Gorbachev's book, *Perestroika: New Thinking for Our Country and the World* to analyze the author's psychological state. Using AI tools, try to explain why Perestroika as a strategy failed.
7. Pick any of the presidential debates and analyze the locus of control of each candidate.
8. Use Shakespeare's tragedy *Julius Caesar* to analyze the psychological state of Julius Caesar, Brutus, Cassius, Anthony, and Octavian.
9. Combining the basic word-counting algorithms into the prompt is a straightforward way to improve any prompt. In this exercise, we explain how one may do so. For simplicity's sake, we concentrate on the UN speeches example, see Sect. 17.4. Denote S_i^c the UN summary of the speech of the country c in the year i.

    a. Ask LLM to provide 300 words that describe Violence.
    b. Analyze the summaries of the UN speeches using these 300 words. Denote v_i^c to be the number of Violence words that appear in the summaries of the country c in year i. Find examples where there is a large gap between the amount of Violence in different years of the same country's summaries, i.e.,

$$v_i^c > v_j^c \tag{17.9}$$

Table 17.14 Direct political trait pairings between Hitler and Churchill

Hitler's Political Trait	Churchill's Corresponding Political Trait
H-Dictatorship	C-Dictatorship
H-Democracy	C-Peace
H-Capitalism	C-Terrorism
H-Socialism	C-Communism
H-Fascism	C-Socialism
H-Communism	C-Capitalism
H-Terrorism	C-Fascism
H-Peace	C-Democracy

Use these examples in the prompts as an example for Yes instance, as follows: "For example, speech S_i^c is more violent than speech S_j^c" For No instance, use these examples as follows: "For example, speech S_j^c is not as violent than speech S_i^c." Note that this idea allows us to improve our prompt using programming without human knowledge.

10. In this exercise, we compute prompts that compare Hitler and Churchill using our experts.
    a. Use the correlation matrix in Table 17.11 to compute the preferential psychiatric disorders for Hitler and Churchill. Use these preferential orders to compute the stable matching of Hitler's and Churchill's disorders. Compare your result with Table 17.12.
    b. Write prompts that use stable matching for the psychiatric disorders of both Hitler and Churchill and compare them.
    c. Repeat Exercise 17.11 with the correlation matrix Table 17.13. Compare your result with the stable matching of the political science, which can be seen in Table 17.14.
    d. Write the prompts that use stable matching for the political science opinions of both Hitler and Churchill and compare them.

References

1. Wiki. Un-gadebate (2024). https://gadebate.un.org/en
2. A. Hitler, *Mein Kampf*, vol. 1. (Motilal Banarsidass, 2014)
3. W. Churchill, *My Early Life* (DigiCat, 2022)
4. Wikipedia contributors. wikisource-wheel-of-emotions (2024). https://en.wikipedia.org/wiki/Emotion-classification#Plutchik's-wheel-of-emotions
5. Mein kampf-gutenberg (2023). https://gutenberg.net.au/ebooks02/0200601.txt
6. Wikipedia contributors. wiki-stable-marriage-problem (2024). https://en.wikipedia.org/wiki/Stable-marriage-problem

Chapter 18
Analyzing History with Video

In the previous chapter, we used a Large Language Models (LLMs) as an expert in sociology and applied it to the study of history. In this chapter, we visit the vast field of studying history through Videos rather than texts. History traditionally relies on textual sources. However, at the beginning of the 20th century, following the Edison revolution and the information revolution of the end of the 20th century, videos have become a valid historical resource.

In this chapter, we will use videos of the 2016 US presidential debates as a source for our study of historical documents. Nearly all of the debates can be found on YouTube.

We picked the presidential debates for their significant impact on the final voting results. They are easily accessible for all researchers, in high quality and often accompanied by a transcript, which makes the analyses easier for machines.

In the next sections, we describe some of the capabilities of Wolfram Mathematica to analyze the Video, audio, and transcript of the first presidential debate between Hillary Clinton and Donald Trump.

In general, Wolfram Mathematica can easily transform speech into text. However, we have no need for this function because we already have the transcripts of the presidential debate videos (Fig. 18.1).

18.1 Function Diagram for Analyzing Video

When dealing with Video, the first thing we have to do is to separate the Video into several channels, such as image, audio, and text. The function diagram (18.1) represents the process of splitting the archive, in this case a single Video, into the three channels.

Z. Lotker, *History by Algorithms*,
https://doi.org/10.1007/978-3-031-93627-2_18

Fig. 18.1 A cartoon illustrating how AI can analyze human body movement and how this technology might be used to study historical behaviors, gestures, or reenactments

$$\begin{array}{ccc} \mathbb{A}rc & \xrightarrow{alg} & (Image, Sound, Text) \\ & \searrow^{\pi} & \uparrow{\scriptstyle \gamma} \\ & & \mathbb{T} \end{array} \tag{18.1}$$

The next step will be to analyze each of those channels separately and generate historical meaning out of their signals. The next section concentrates on breaking the videos into the aforementioned information channels.

18.2 Braking the Video Into Several Dimensions

Video is a multi-dimensional experience that incorporates images, sounds, and texts. When analyzing the Video as a historical source, it is important to break it into individual multi-dimensional sensors like images, sounds, and texts. The reason is that it is easier to analyze each channel independently, then combine the results.

18.2.1 Text

The textual portion can be generated in one of two ways. Many canonical historical videos, such as presidential debates, include a transcription that contains the text and timestamp (when it appears in the Video), or scripts that mention the speaker and what he or she said. In our case, all president debate transcripts can be found in [1]. The second method is to use machine learning tools such as the Wolfram Mathematica command SpeechRecognize, which transforms audio into text. Once we have the text, we can analyze the voice of the speaker by using surgery on the

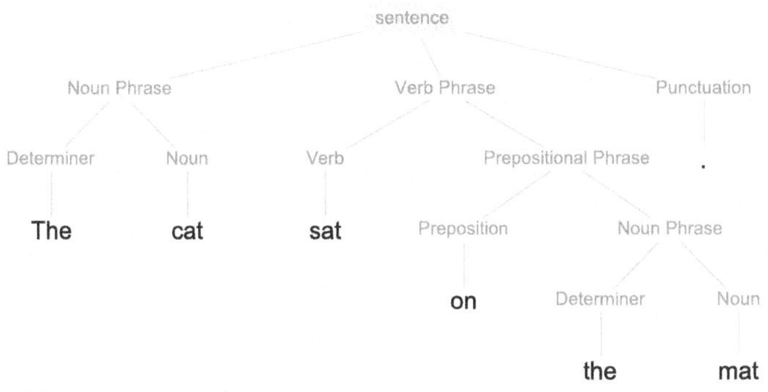

Fig. 18.2 The grammatical structure of the sentence "The cat sat on the mat"

deep learning network that can be found in the net repository. For more information on this subject, see ...

For this subsection, we chose to work with the existing script rather than generate one ourselves. Once we have the text of each candidate, we can start performing grammar analyses on the differences between them. It is well-known that the distribution of the frequency of words is Zip-distribution; see the paper [2]. What is less known is that the distribution of sentence formulas, i.e., grammatical structure, also is Zip-distribution. Consider the following example in the simple sentence: "The dog sat on the mat." The grammatical structure of the sentence can expressed as a tree Fig. 18.2.

When we remove the words from the grammatical tree, we are left with the sentence formula.

The sentence "The dog sat on the mat" has the same formula as "The child ate at the picnic."

When comparing Hillary Clinton's language with Donald Trump's during the first debate, we see that Clinton uses more complicated sentence formulas with fewer words. In contrast, Trump uses more straightforward sentences and repeats himself many times, but his vocabulary is richer.

18.2.2 From Video to Frame

Our first step in analyzing the Video is to upload it into Wolfram Mathematica via a simple Import command. The next step is transforming the Video into a frame. Wolfram Mathematica or many other programs can do this. This can be done by two Wolfram commands: VideoExtractFrames for a single frame and VideoFrameList for a sequence of frames. Denote the set of frames of the first debate:

$$Fram = \{fram_1, fram_2, ..., fram_{t_e}\} \tag{18.2}$$

where $fram_1$ is the first frame, $fram_2$ is the second frame, and $fram_{t_e}$ is the last frame in the Video.

The advantage of working with a frame instead of a video is that there are plenty of machine-learning functions that know how to work with images, but not videos. It is not clear that in the future this step will be necessary; however, with current technologies, it is better to work with frames, not videos.

To demonstrate the ability to analyze historical videos using machine learning, we continue our study of the first presidential debate between Hillary Clinton and Donald Trump.

18.3 Machine Learning Tools

In this section, we study the debates between Donald Trump and Hillary Clinton using machine learning tools that can detect the candidate and compute their skeleton. Once we have identified the skeleton, we can decompose it into two different parts: the head and the hands. Each of those parts can be analyzed using a different neural network to gather information and compare the candidates' body language. For simplicity, we use the complete skeleton, but in the future the researchers will likely use much more detailed information to analyze the body language of politicians. The next subsection deals with recognizing the candidates.

18.3.1 Recognizing the Candidates

At first, we analyzed the total movement of each candidate using the neural network: NetModel["YOLO V8 Pose Trained on MS-COCO Data"]. Before analyzing the body movements of each candidate we need to be able to identify the candidates automatically. This can be done easily with the Wolfram command FaceRecognize. Machine learning is so powerful that it is enough to provide a single image of Lester Holt, the host of the debates, and each candidate: Hillary Clinton and Donald Trump. Machine learning is then able to recognize the face of each figure $MachineRecognisedCandidate$.

The next step is to define the body language and to construct a machine that is able to interpret it. In our case, we define very simple body language, affiliated with onee of two words: "Fascism" and "Democracy." The grammar is also very simple, as each sentence consists of just one word. The next subsection will discuss this process.

Since we can recognize the candidate on the screen, we can compare the time each candidate appears. Figure 18.3 shows the results of our analyses of the first presidential debates.

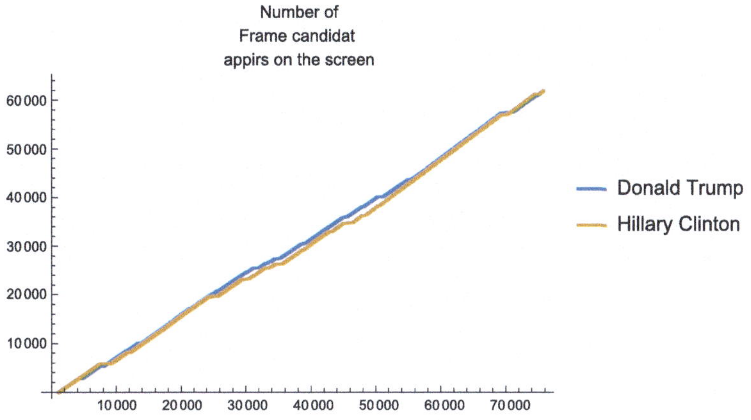

Fig. 18.3 The number of frames each candidate appeared in during the first presidential debate between Hillary Clinton and Donald Trump in 2016. The X-coordinate measures the time by the number of frames, and the Y-coordinate is the number of frames each candidate appears in

18.3.2 Speak, Hands for Me

In this subsection, we follow Shakespeare's quote, "Speak, hands for me," said by Cassius to Caesar before killing him. According to Shakespeare, hands in particular and body movement in general measure the degree of Violence people use. We use the YOLO V8 neural network to generate a person's skeleton from their image, composed of the following 17 points: "Nose", "LeftEye", "RightEye", "LeftEar", "RightEar", "LeftShoulder", "RightShoulder", "LeftElbow", "RightElbow", "LeftWrist", "RightWrist", "LeftHip", "RightHip", "LeftKnee", "RightKnee", "LeftAnkle", "RightAnkle."

In order to construct the body movement language, we need the Rosetta stone that describes the cluster of each of the words in the language. Since our language contains only two words—"Fascism" and "Democracy"—we need to find a Video example of the body language of those two clusters. Choosing these videos significantly influences the outcome of the analyses. In our case, we use the film "The Great Dictator."

Figure 18.4 shows the normalized full skeleton of Charlie Chaplin in two scenes from The Great Dictator. The term "normalized" means that we put the skeleton in the unit square. The first scene shows the dictator speaking to the crowd, where Chaplin imitates Hitler's and Mussolini's mannerisms. We call this scene "Fascism." This is reflected in the left section of Fig. 18.4. On the right is the second scene from the end of the film, where Chaplin speaks to the audience in favor of democracy and against tyranny. We can see that he doesn't use body language during his speech. We call this scene "Democracy."

One feature that YOLO V8 neural network can perform when analyzing an image is to find all rectangles that contain a single person in them. We use this feature to break each frame or image into several sub-images containing Hillary Clinton or

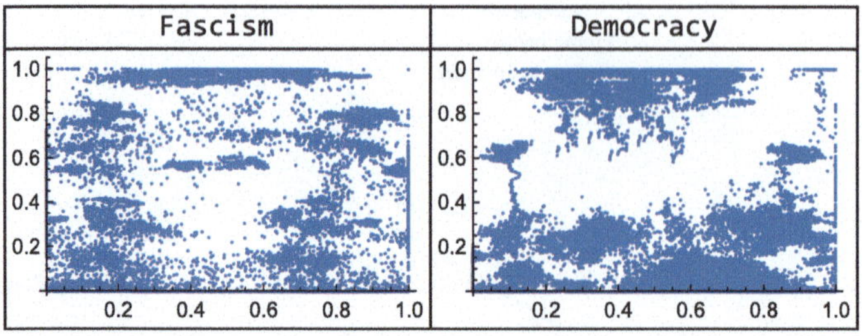

Fig. 18.4 The aggregated skeleton of the scenes from The Great Dictator after taking the skeleton generated by YOLO V8 and putting it in the unit square. The left image shows the scene where the dictator speaks to the crowd, and the right is the last scene of the film where Chaplin speaks in favor of democracy

Donald Trump. We then use the Wolfram Mathematica command FaceRecognize to distinguish between Hillary Clinton, Donald Trump, or the moderator.

Now we have for each frame the skeleton of Hillary Clinton

$$skh_i = hp_i^1, ..., hp_i^{17} \tag{18.3}$$

and Donald Trump

$$skd_i = dp_i^1, ..., dp_i^{17} \tag{18.4}$$

if they appear in the frame. If they do not appear, we set those skeletons to an empty set.

Now, we concentrate on each candidate and repeat this process for both. For simplicity's sake, we describe the process for Hillary Clinton and repeat the same for Donald Trump using the naming convention of replacing the h with d for Donald Trump.

We scan the frames according to the time they appear in the Video, from earliest to latest. We aggregate 200 consecutive frames into a single body sentence approximately 7 s long. Denote the body sentence of Hillary Clinton at time i, BS_i^h to be

$$BS_i^h = \cup_{j=i}^{i+200} skh_j \tag{18.5}$$

The reason we chose 200 consecutive frames is to achieve a vector that contains around 3400 points, the same order of magnitude as a small image, which machine learning tools used to analyze. Note that the set BS_i^h contains 3400 points at most, since Hillary Clinton doesn't appear in all the frames $\{frame_i, frame_{i+1}, frame_{i+200}\}$, and some points appear in several frames.

Table 18.1 presents the results of the machine learning analyses of the body language during the first presidential debates, compared to the body language of Charlie Chaplin in The Great Dictator. According to the machine learning algorithm, Don-

Table 18.1 The computing results of body language in the first presidential debates between Donald Trump and Hillary Clinton

Candidate name	Democracy	Fascism
Donald Trump	181	7376
Hillary Clinton	5858	1473

ald Trump's body language shows 3% Democracy and 97% Fascism, while Hillary Clinton's shows 80% Democracy and only 20% Fascism.

Note that we do not express any moral judgment. We just analyze the body language of the candidates and compare them with the body language in The Great Dictator. For example, in the movie "Silence of the Lambs," Hannibal Lecter, played by Anthony Hopkins, is closer to the "Silent body language" than the "Speaking body language." Indeed, one can claim that Hannibal Lecter is not a populist; he identifies with Nietzsche's concept of the "bermensch."

Next, we would like to use the methodology developed in chapter History Through Clocks Chap. 4 to compute the critical points in the debates with respect to the clocks Fascism versus Democracy and Clinton versus Trump.

18.3.3 From Body Movements to Clocks

In the previous subsection, we developed tools for classifying the body language of each candidate into one of two categories: Fascism or Democracy. In this subsection, we use those two categories to generate the M-diagram of the clocks $M\ [C_D, C_F]$. The clock Democracy clock C_D advances by one each time a candidate's body language is recognized as Democracy, and the Fascism clock C_F advances by one whenever the candidate's body language is recognized as Fascism. With two clocks for each candidate, we can compute the M-diagram of the clocks. Figure 18.5 represents the outcome of those two diagrams.

18.4 Computing Body Language

In the debate a lot of information is encoded in the body language of the candidate. Sometimes, their body language contradicts the message in the text; sometimes, it supports it. Moreover, when one of the candidates is speaking, the other tends to respond through body language. So, trying to understand the debate without analyzing the body language means losing about half the information.

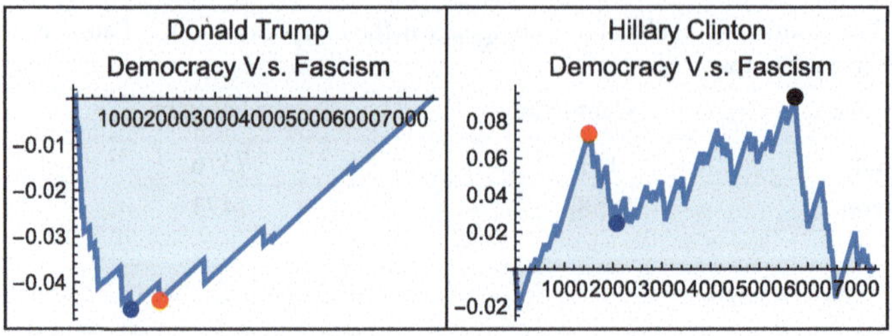

Fig. 18.5 This figure shows the M-diagram of both candidates with respect to the body language clocks. The left is the M-diagram of $M(CD_D, CD_F)$ for Donald Trump with two critical points—Blue and Red, and the right is the M-diagram of $M(CH_D, CH_F)$ for Hillary Clinton with three critical points—Blue, Red, and Black

Again, the debate is a very simple environment and therefore the machine can analyze it much more easily than the general Video. To compute body language, we divided it into several independent areas: the head, the left hand, and the right hand. The head can be split into eyes and mouth, and sometimes the throat; and hands can be split into fingers.

18.5 Conclusion

In this chapter, we demonstrated how historical Video material can be analyzed using machine learning techniques in history studies. Specifically, we analyzed the first presidential debates between Hillary Clinton and Donald Trump. We studied the language complexity of the candidates and shown that while Clinton used a more grammatically complex structure than Trump, Trump used more words than Clinton. In Sect. 18.3, we analyzed the body language of each candidate and compared it to Charlie Chaplin's body language in The Great Dictator. The results are summarized in Table 18.1. They reveal that Trump's body language is much more fascist than that of Hillary Clinton. Later in Sect. 18.3.3 we transformed the body language in the clocks and found the critical points in the first debates using the M-diagram of this clocks.

In this chapter, we demonstrated the use of a machine learning technique to analyze historical Video. In the future, much more complextechniques will most likely be used to analyze historical videos and documents.

The next chapter will explore fake history and the philosophy of science, discussing the problem of fake and true history.

18.6 Exercise

1. In Sect. 18.3, we constructed a machine learning neural network that analyzes the first debate between Hillary Clinton and Donald Trump according to their body language and determined whether it expresses "Fascism" or "Democracy" according to the movement of the candidate's complete skeleton. In this question, you are asked to repeat this process not with the full skeleton but only with the hands of each candidate. Those are the parts in the skeleton named "LeftShoulder", "RightShoulder", "LeftElbow", "RightElbow", "LeftWrist", "RightWrist". Explain the advantages and disadvantages of working with these parts of the body instead of working with the complete skeleton (Fig. 18.6).
2. From Video to Audio: Sometimes it is useful to separate the soundtrack from the Video, particularly when we wish to transform audio into text. This can be done easily using the Wolfram command SpeechRecognize. In this exercise, you are asked to generate the text from a soundtrack of your favorite historical document. Note that when we analyze the audio sources of data, there are several possible formats to work with. Surprisingly, it is not a good idea to work with compressed versions, such as .mp3 or .mp4. Using them comes with the cost of time spent on decompressing each file. It is better to work with a non-compressed format, i.e., .wav instead of .mp3 or .mp4. Separating the audio from the Video can be done with many tools, including Wolfram Mathematica, using the command Import[filename, "Audio"].
3. Facial Expression Analysis: Use the presidential debate between Hillary Clinton and Donald Trump or any other historical Video to analyze the critical points in the Video according to facial expressions.

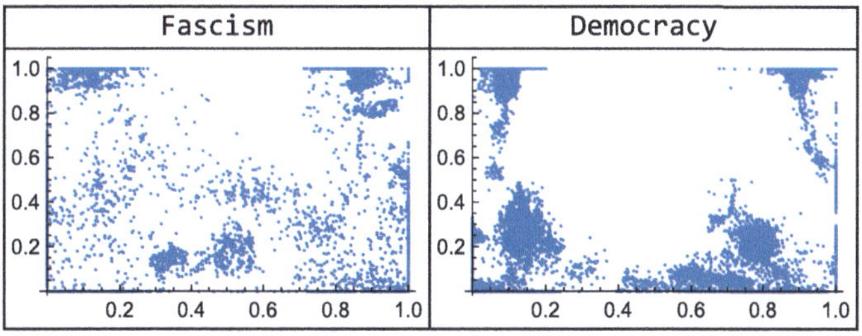

Fig. 18.6 This figure is similar to Fig. 18.4, but it was generated only with 6 points of the skeleton that represent hands

a. Transform the video into frames.
b. For each frame, analyze your subject just as we did in the Sect. 18.3.1.
c. Once you identify your subject, cut out their face from the frame using either YOLO V8 neural network or the Wolfram command FindFaces. Now apply Wolfram command FacialFeatures with the option "Emotion" to compute the subject's emotions in the frame. These computed emotions will be our facial expression language.
d. Repeat the previous section for each frame and compute the clock for each emotion, like we did in Sect. 18.3.3. Use the M-diagram of many clocks as described in Sect. 4.8 to find the critical point of your historical video.

References

1. Debatesorg (2023). https://www.debates.org/voter-education/debate-transcripts/
2. G. Kingsley Zipf, *The Psycho-Biology of Language: An Introduction to Dynamic Philology* (Routledge, 2013)

Chapter 19
Fake History

Since history is a narrative of the past, the question about fake history seems irrelevant. However, our purpose in this chapter is not to reveal fake stories about the past, but instead to generate them. The reason we choose to attack fake history with this method is to demonstrate our vulnerabilities to fake narratives. A good narrative is easy to believe, no matter its truthiness. Therefore, we wish to warn the reader: Be careful! Most of history can be faked, and this chapter shows just how to do that (Fig. 19.1).

To distinguish between fake and true history, we developed several rules of thumb that will allow us to read history narratives with careful scrutiny.

19.1 Function Diagram of Fake History

In Chap. 2, we developed a methodology for categorizing different kinds of history using function diagrams. Since it is complicated to determine if historical facts are true or fake, we try to encourage the reader to engage in the process of generating alternative history in order to enrich the concept of history and develop critical thinking.

We summarize the process of generating fake/alternative history using the function diagram (19.1). We start with a historical archive $\mathbb{A}rc$. Using the intelligent I, which can be either human or artificial, we transform $\mathbb{A}rc$ into a fake/alternative archive. We can accomplish this via several methodologies: inventing events, deleting events, or interpreting events. The time of the events is computed by π and the new narrative is described by the function γ.

$$\begin{array}{ccc} \mathbb{A}rc & \xrightarrow{I} & \overline{\mathbb{A}rc} \\ & \searrow^{\pi} & \uparrow{\scriptstyle\gamma} \\ & & \mathbb{T} \end{array} \tag{19.1}$$

Z. Lotker, *History by Algorithms*,
https://doi.org/10.1007/978-3-031-93627-2_19

Fig. 19.1 A cartoon illustrating fake history

The next sections describe some rules that guide and limit the way one may invent fake/alternative history.

19.2 Copernicus Rule of Thumb for Fake History

According to the Copernican Revolution, the universe does not have one center; Instead, it has many, all of which are equivalent. In the same way, history does not have a single point of view, and different points of view can be exchanged. Upon encountering a sentence heavily biased toward a single point of view, the suspicious reader should immediately generate the opposite point of view and consider both a possibility. Moreover, a narrative that is central according to some point of view can not describe the narrative history in a complete sense. For example, according to Shakespeare, Caesar said that Brutus was an ambitious man. This, of course, is a representation of Caesar's view of Brutus as a danger to Caesar himself - which indeed he was. However, the old adage is true: power corrupts. This is a much more general statement, which can be applied to multiple people and situations. It captures many more points of view than the single statement about Brutus.

To summarize, historical narratives contain events that are independent of the point of view and some that are dependent on the point of view. The independent part belongs to the factual historical information, and the dependent part belongs to the narrative.

Copernicus Rule of Thumb

Historical truth does not depend on the point of view.

For instance the following are irreputable facts: Caesar conquered Gall, andr Germany lost WWII.

19.3 Occam's Razor

Occam's razor prefers a simple explanation over a complex one. The main question when trying to apply Occam's razor to a historical narrative, especially digital history, is "what is simple." Here, we suggest the following interpretation: we split the argument into two types - logic arguments versus fact arguments. Logic arguments do not count toward the complexity of the argument since they are very easy to verify for machines. Historical facts are not so easy to verify. Therefore, we will try to minimize the use of data as much as possible. For us, Fermat's last theorem is much simpler than the fact that Fermat was the first to suggested and prove it, as he claimed.

To summarize, simplicity is not measured in the number of pages. Logic is much easier to verify, even if, for humans, it seems more difficult. In the pursuit of truth, logic does not matter.

Occam's Razor

The correct logical derivation does not matter when measuring historical simplicity. Logic prevails over data.

19.4 Fakes Are Much Easier to Generate Than "Truth"

When dealing with fake history, one should always remember that it is easy to generate small-scale, local fake history. However, when fake history needs to be integrated with many facts, it becomes exponentially harder to generate. Therefore, fake history tends to be local over time and disappear quickly.

One way to generate fake history is to transfer real narrative history into another location. For example, one can take History of the Decline and Fall of the Roman Empire [1] and transform it into the Rise and Fall of China at the end of the 19th century or Japan in the 16th century. In this case, when we transport the narrative from one place to another, it becomes more difficult to find contradictions than just normal small lies that disappear over time. When you notice similar narratives, you should be suspicious as well.

Consistency

Local fake history tends to disappear.

19.5 Conclusion

In this chapter, we discussed the property of fake history. We provided several rules of thumb on how to generate fake history.

The best way to deal with fake history is to be able to see how easy it is to generate fake history. In the exercise below, we encourage the reader to play around with generating fake history to get a feel of how to separate a fake historical narrative from a reasonable one.

19.6 Exercise

1. Is it a fact that Biden won the 2020 election?
2. The main narrative of the 20th century contains two world wars. It is commonly agreed that WWI started in 1914 and ended in 1918, and that WWII started in 1939 and ended in 1945. Based on this statement:
   a. Construct an alternative narrative that there was only one World War, which started with the Boxer Rebellion in 1890 and ended with the Chinese Revolution in 1949.
   b. According to the existing historical narrative, the World Wars were fought between Germany and its allies against England and its allies. Construct an alternative narrative where a single War ancompassing both the First and the Second World Wars, was not fought between the nations, but instead between mankind and machines which subsequently birthed the general purpose computer. In this fake narrative, the machines won the war, while humanity lost.
   c. According to the central narrative, the fascist regime lost to the democratic one. However, at the beginning of the 21st century, fascist movements started reappearing in many countries. Can you change the narrative of the WWII to explain the emergence of fascist movements at the beginning of the 21st century?
3. According to the general narrative, the Middle Ages ended in 1500, at which point the Renaissance began. Can you change the standard narrative so that the Renaissance begins with Saladin's conquest of Jerusalem? In your answer, you can use Fibonacci's revolutionary shift from Roman numerals to decimal numbers and Marco Polo's journey to China as an indication for the new tides in the Christian world.
4. According to Western history, Alexander the Great was the mastermind behind the fall of the Persian Empire.
   a. Can you generate an alternative narrative where the mastermind was Ptolemy rather than Alexander?
   b. Can you describe the outcome of the war between Alexander and Darius III as a tragedy for Western culture and as an enlightenment?

c. According to the standard historical narrative, the war between the Persian Empire and the Greeks started with the Persian attempt to conquer Greece. Can you generate a fake history where none of these battles never happened and instead were invented by Greek authors to justify Alexander's conquest?

5. The next exercise deals with the Roman period of transition from Republic to Empire.

   a. Explain how Julius Caesar succeeded in capturing all his enemies while Augustus's enemies all committed suicide.
   b. One of Augustus's main achievements was reducing the Roman army from about 50 legions to approximately 20. By doing so, he succeeded in eliminating the civil wars. Using this fact, explain why the Battle of the Teutoburg Forest was, in fact, a significant success of Augustus's policy and not a disaster, as the standard narrative suggests.

Reference

1. E. Gibbon, *History of the Decline and Fall of the Roman Empire*, vol. 10 (F. DeFau & Company, 1907)

Appendix A
Mathematical Background for Computational History Using AI Prompts

A.1 Introduction

The book assumes that the reader is familiar with concept in mathematics in the context of exact sciences and engineering at an undergraduate level. Those who lack these fundamentals can overcome the gap by consulting the basic notations and references books proposed below.

Since there are several AI general programs, we will provide relevant prompts to be used as a gateway to the missing material.

We also provide some general introduction to social networks. The basic information can be found in the classic work by Scott [1]. The connection between social networks, economic science and game theory can be found in Easley and Kleinberg's work [2]. Newman's book [3] introduces the use of algorithms in social networks, based on the mathematical and physics approaches. Nevertheless, although these and other books give plenty of information about social networks, we suggest a new approach by using examples from literature for a better understanding of social networks.

A.2 Set Theory

There are several good introductory books to set theory; see [4–6].

Prompt for Set Theory

Definition A.1 The following prompt can help fill in the missing notation in set theory:

1. **Describing a basic set notation:** Explain the notations used to denote a set, its elements, and how to express that a particular object belongs to a

Z. Lotker, *History by Algorithms*,
https://doi.org/10.1007/978-3-031-93627-2

set. For instance, if I have a set containing the numbers 1, 2, and 3, how would I write this? How do I denote that 2 is an element of this set?

2. **What is an empty set in set theory:** Define the concept and significance of an empty set in set theory.
3. **Understanding subsets and proper subsets:** Differentiate between a subset and a proper subset using the appropriate set theory notation. If I have sets A and B, where $A = \{1, 2, 3\}$ and $B = \{1, 2\}$, how would I denote that B is a subset or a proper subset of A?
4. **Power set notation:** Define the power set of a set and the notation used to represent it. If I have a set $C = \{a, b\}$, how would I denote its power set, and what would the elements of that power set be?
5. **Operations on sets—union, intersection, and difference:** Describe the notations used to represent the union, intersection, and difference between two sets. If A and B are two sets, how would you represent the set of elements that belong to either A or B, to both A and B, and to A but not B?
6. **Complement and absolute complement:** Explain the notation used to represent the complement of a set. If we have a universal set U and a set D contained within U, how would we represent the set of all elements in U that are not in D? What distinguishes the absolute complement from the relative complement in notation and meaning?
7. **Cartesian product:** Define the Cartesian product of sets and its importance in set theory.
8. **Define relationship and function in set theory:** Describe the concept of a relationship and a function within the context of set theory.
9. **Overview of the fundamental axioms of set theory:** Provide a concise summary of the main axioms that govern the operations and properties in set theory.
10. **Orderings in set theory:** The concept of orderings is foundational in set theory, providing a structured means to compare and position elements within sets. How would you define and differentiate between the following types of orders?

    a. **Total (or linear) order:** In what way does a total order provide a definitive relationship between every pair of elements within a set?
    b. **Partial order:** Given that not every pair of elements within a partially ordered set is comparable, how does this type of order differ from a total order in terms of reflexivity, antisymmetry, and transitivity?
    c. **Well order:** What attributes distinguish a well-ordered set, especially concerning the least element in every non-empty subset? Can you provide an example of such an order?
    d. **Dense order:** With densely ordered sets having an element between any two distinct elements, how does this order play out among rational numbers?

e. **Lexicographic order:** Often likened to the way words are arranged in a dictionary, how does the lexicographic order work when comparing elements from Cartesian products of sets?
f. **Strict order:** Emphasizing a definitive hierarchy without equivalences, how does a strict order ensure the absolute positioning of elements within a set?

A.2.1 Function

We provide this prompt for a better understanding of a function.

1. **Basic concepts:**
   a. Define a function. What is the domain and codomain of a function?
   b. What is the difference between a function and a relation?
   c. Illustrate with a diagram the concept of a function from set A to set B. Label the domain, codomain, and range.
   d. Explain the vertical line test. Why is it used to determine whether a graph represents a function?
2. **Types of functions:**
   a. What is a one-to-one (injective) function? Provide an example.
   b. Define an onto (surjective) function. How does it differ from an injective function?
   c. Describe a bijective function. Why is it significant to the study of functions?
   d. Provide examples of polynomial, rational, and trigonometric functions. How do they differ from one another?
3. **Function operations and compositions:**
   a. Given two functions f and g, define the sum, difference, product, and quotient of f and g.
   b. Describe the process of composing two functions, f and g, to form the function $f \circ g$.
   c. How is the domain of the composed function $f \circ g$ determined?
   d. Provide an example where the order of function composition affects the resultant function.
4. **Inverse functions:**
   a. Define an inverse function. How is it denoted?
   b. Describe the graphical relationship between a function and its inverse.
   c. Explain the process of finding the inverse of a function algebraically.
   d. Under what conditions does a function have an inverse that is also a function?

A.3 Standard Set of Numbers

Throughout this book, we use standard sets extensively. Below are some prompts that will provide an accessible introduction to the definition of such sets.

1. **N—Natural numbers:** The set of all non-negative integers.

$$\mathbb{N} = \{0, 1, 2, 3, \ldots\} \tag{A.1}$$

2. **Z—Integers:** This set includes all integers that are positive, negative, or zero.

$$\mathbb{Z} = \{\ldots, -2, -1, 0, 1, 2, \ldots\} \tag{A.2}$$

3. **Q—Rational numbers:** These are numbers that can be written as a ratio of two integers.

$$\mathbb{Q} = \left\{ \frac{p}{q} : p, q \in \mathbb{Z}, q \neq 0 \right\} \tag{A.3}$$

4. **R—Real numbers:** This set includes all possible numbers on the number line, including both rational and irrational numbers.

$$\mathbb{R} \tag{A.4}$$

5. **C—Complex numbers:** This set includes numbers that have both a real and an imaginary part.

$$\mathbb{C} = \{a + bi : a, b \in \mathbb{R}\} \tag{A.5}$$

Here are some prompts based on these sets:

1. Prove that for every $n \in \mathbb{N}$, $n^2 + n + 41$ is prime.
2. Show that the sum of any two odd integers in $\mathbb{Z}$ is even.
3. Given that $a, b \in \mathbb{Q}$, prove that their sum and product are also in $\mathbb{Q}$.
4. Prove that $\sqrt{2}$ is not in $\mathbb{Q}$ but is in $\mathbb{R}$.
5. Demonstrate that for all complex numbers $z \in \mathbb{C}$, $|z| = |-z|$.

A.4 Vectors, Matrices, and Their Spaces

The next prompts offer an introduction to the subject of vectors, matrices, and their spaces used repeatedly throughout this book.

1. **Vectors:**

   a. Define a vector in the context of both physics and mathematics. How do the two perspectives relate?

b. Draw a vector in the plane, indicating its magnitude and direction. How can we represent this vector algebraically?
c. Explain the concept of vector addition graphically.
d. What does it mean for two vectors to be linearly independent? Provide a graphical illustration.
e. Describe the difference between a unit vector and a zero vector.

2. **Matrices:**

a. Define a matrix and list some of its primary uses in mathematics and applied sciences.
b. Given two matrices, describe how matrix addition is performed. Provide an example.
c. How does matrix multiplication differ from scalar multiplication and matrix addition?
d. Explain the concept of the identity matrix. Why is it essential in the context of matrix operations?
e. Describe the process of finding the determinant of a matrix. Why is the determinant important?

3. **Vector spaces:**

a. Define a vector space. List the primary properties a set must possess to qualify as a vector space.
b. Describe the subspace of a vector space. Provide an example.
c. Explain the concept of a basis for a vector space.
d. How does the dimension of a vector space relate to its basis?
e. Given a set of vectors, describe the process to determine if they span a particular vector space.

4. **Matrix Spaces:**

a. Describe the space of all $m \times n$ matrices. How is this space related to vectors?
b. Define the row space and column space of a matrix. Why are these concepts important to linear algebra?
c. Explain the Rank-Nullity theorem and its significance.
d. How is the concept of linear independence of vectors extended to the rows and columns of a matrix?
e. Describe an application of matrices in real-world scenarios, emphasizing the significance of the associated matrix spaces.

A.5 Graph Theory

Readers who possess undergraduate knowledge of exact science and engineering at a first and second year level and are familiar with graph theory, elementary set theory, and linear algebra, can skip the Appendix. To deepen their knowledge we recommend these readers to look for the works of Diestel [7], Bollobas [8], and Godsil [9].

Prompt for Graph Theory

Definition A.2 The following prompt can help fill in the missing notation in graph theory

1. **Basic concepts:** Introduce the fundamental concepts of graph theory $G(V, E)$. How would you define a graph, vertex V, edge E, and loop?
2. **Directed versus undirected graphs:** Differentiate between directed graphs (digraphs) and undirected graphs. What are the key characteristics and potential applications of each?
3. **Connectivity in graphs:** Define what it means for a graph to be connected. How do the concepts of paths and cycles relate to the connectivity of a graph?
4. **Trees and forests:** In the context of graph theory, what distinguishes a tree from a forest? How is the concept of a spanning tree significant in network design?
5. **Graphs and their matrices:** Dive deep into the interrelation between graphs and matrices. Particularly, how are graphs represented using the adjacency matrix and the incidence matrix? Furthermore, in this context, what roles do eigenvalues and eigenvectors play in understanding the properties of these matrices and their corresponding graphs?
6. **Weighted graphs and multigraphs:** Explore the concepts of weighted graphs and multigraphs in depth. How do weighted graphs differ from regular graphs in terms of representation and application? Additionally, what distinguishes a multigraph from a simple graph, and in what scenarios might one opt to use a multigraph?
7. **Neighborhood in graphs:** Dive into the concept of the neighborhood of a vertex within a graph. How is the neighborhood of a vertex defined? What significance does it hold in understanding a graph's topology, and how can it be used to derive properties or solve problems in graph theory?
8. **Families of graphs:** Elaborate on the diverse families of graphs that exist within graph theory. How do complete graphs, bipartite graphs, cyclic graphs, and star graphs differ in structure and properties? Additionally, what are some applications or problems uniquely suited to each family of graphs?
9. **Graph algorithms:** Describe some popular algorithms in graph theory, such as Dijkstra's algorithm for shortest paths or Kruskal's algorithm for

minimum spanning trees. Why are these algorithms significant in computer science?

10. **Degrees and the Handshaking Lemma:** Discuss the concept of vertex degrees in a graph. What is the Handshaking Lemma, and how does it provide insights into the structure of a graph?

A.6 Social Networks

Introduction to Social Networks: Prompts

1. **Basics of social networks:**

   a. Define a social network. How do online social networks differ from traditional social networks?
   b. Describe the relationship between social networks and digital history.
   c. Explain the concepts of nodes and edges in the context of social network graphs.
   d. How has the rise of the internet and digital technology influenced the growth and significance of social networks and digital history?

   Next, we provide some prompts on the structure of social networks.

2. **Structure and patterns in social networks:**

   a. Describe the Six Degrees of Separation theory. How does it relate to the concept of network diameter?
   b. Define centrality in a network. Why might centrality be an important metric for influencers or key players in a network?
   c. Explain how Degree Centrality measures a node's importance in a network. Discuss its limitations in directed or weighted networks?
   d. Analyze how the PageRank algorithm ranks nodes based on their influence. How does it differ from other centrality measures in handling directed networks?
   e. Describe the principle behind Eigenvector Centrality and its ability to capture the influence of a node's neighbors. What types of networks benefit most from this measure?
   f. Compare and contrast Degree Centrality, PageRank, and Eigenvector Centrality in identifying influential nodes in a social network. When is each measure most effective?
   g. Explain the concept of a scale-free network. How is it relevant to real-world social networks?

We provide introduction prompts for dynamic evolving social networks.

3. **Dynamics and evolution of social networks:**

   a. Describe how weak ties might be more influential than strong ties in the dissemination of information across a network.
   b. What are some factors that can lead to the growth or decay of a network?
   c. Explain the phenomenon of homophily in social networks.
   d. How can the study of network dynamics provide insights into the spread of behaviors, trends, or even diseases?

4. **Applications and implications:**

   a. How are businesses leveraging social networks for marketing and outreach?
   b. Discuss the role of social networks in shaping public opinion and influencing political dynamics.
   c. What are some privacy concerns associated with online social networks, and how are they being addressed?
   d. Describe how social networks can be harnessed for social good, citing examples of positive social change instigated through these platforms.

Mathematical Definition of Evolving Social Networks: Prompts

1. **Basics of evolving networks:**

   a. Define an evolving social network in the context of graph theory. How does it differ from a static network?
   b. Discuss the significance of nodes (V_i) and edges (E_i) in the representation of a network at time t_i.
   c. Explain the difference between discrete-time and continuous-time evolving networks. How might they be represented differently?
   d. Describe a real-world scenario where capturing the evolution of a social network is crucial for understanding its dynamics.

2. **Models for evolving networks:**

   a. Introduce the concept of temporal networks. How do they emphasize timing and sequencing in network evolution?
   b. Describe the dynamic Erdős–Rényi model. How does it use probabilistic methods to characterize network evolution?

c. Explain agent-based models in the context of evolving social networks. What are the advantages and limitations of such models?

3. **Network dynamics and behavior:**

   a. Discuss the factors that might drive the evolution of a social network, leading to changes in V_i and E_i.
   b. Describe a situation where understanding the dynamics of an evolving network can lead to predictive insights.
   c. How can studying the evolution of a network help with identifying influential nodes or understanding the spread of information?

4. **Applications and implications:**

   a. Consider online social media platforms. How have they provided vast datasets for studying the evolution of social networks?
   b. Discuss potential challenges in modeling and analyzing evolving social networks.
   c. Describe how the mathematical insights from evolving networks might be applied in areas like marketing, epidemiology, or disaster response.

5. **Evolving social network as a function:** How can you represent the dynamic nature of evolving social networks using a functional approach? Can you describe the advantage and disadvantages of this approach compared to the definition of evolving social networks as a sequence of graphs?

A.7 Algorithms and Graphs

Prompts on Algorithms and Graphs

1. **Introduction to graphs and algorithms**:

   - Explain the fundamental concepts of graphs. How are they used to model real-world problems in computer science?
   - Discuss the various types of algorithms commonly used to traverse graphs. How does each algorithm work, and in what scenarios is each most beneficial?

2. **Graph representation**:

   - Describe the differences between an adjacency matrix and an adjacency list. What are the advantages and disadvantages of each representation in terms of space and time complexity?

- How do directed and undirected graphs differ in representation and algorithmic approaches?

3. **Shortest path problems**:
    - Explain Dijkstra's algorithm for finding the shortest path in a graph. How does it differ from the Bellman-Ford algorithm?
    - In what scenarios might the Floyd-Warshall algorithm be preferred over Dijkstra's for computing shortest paths?
4. **Graph traversal**:
    - Contrast Depth-First Search (DFS) and Breadth-First Search (BFS) in terms of mechanism, applications, and efficiency.
    - Describe applications where a DFS might be more appropriate than a BFS and vice versa.
5. **Cycles and connectivity**:
    - Discuss algorithms used to detect cycles in a graph. Why is cycle detection important in certain applications?
    - Describe the concept of graph connectivity. How can algorithms determine if a graph is strongly connected, weakly connected, or disconnected?
6. **Trees and graphs**:
    - How is a tree different from a general graph? Discuss algorithms that determine if a given graph is a tree.
    - Explain the significance of Minimum Spanning Trees (MST). Describe algorithms like Kruskal's and Prim's used to find an MST in a weighted graph.
7. **Flow and matching**:
    - Introduce the concept of network flow. How does the Ford-Fulkerson algorithm work to determine the maximum flow in a flow network?
    - Discuss the concept of bipartite matching in graphs. What algorithms can solve the maximum bipartite matching problem?
8. **Applications of graph algorithms**:
    - Discuss real-world applications where graph algorithms play a crucial role, such as in transportation, social networks, and computer networks.
    - Describe the role of graph algorithms in modern machine learning and data analysis. How are graphs used to model data and find patterns?
9. **Challenges and limitations**:
    - Discuss challenges faced when working with large-scale graphs, such as those in social network analyses.

- What are the limitations of certain graph algorithms, and how can they be addressed or mitigated?

A.8 Queueing Theory

Prompts on Queueing Theory

a. Explain the Poisson distribution, including its probability mass function, key properties, and assumptions. Provide examples of real-world scenarios where it is commonly used and discuss how its parameter $\lambda > 0$ influences its shape and interpretation.
b. Explain the exponential distribution, including its probability density function, key properties, and assumptions. Discuss the role of its parameter $\lambda > 0$ (rate parameter) in shaping the distribution and its relationship with the Poisson distribution. Provide examples of real-world scenarios where it is commonly used to model the time between events.
c. Explain the Birth-Death process in queuing theory, using parameters $\lambda_n > 0$ (birth/arrival rate) and μ_n (death/service rate) to describe transitions. Derive the steady-state probabilities and discuss how these parameters influence system performance. Provide examples of queuing systems that can be modeled using this process, such as call centers or computer networks.
d. Explain the $M/M/1$ queue model, including its assumptions, key components (arrival rate λ, service rate μ, and single server), and the mathematical expressions for performance metrics like average queue length and waiting time. Provide examples of real-world systems where this model is applicable
e. Explain the Jackson network in queuing theory, including its assumptions and components. Use parameters like arrival rates (λ_i), service rates (μ_i), and routing probabilities ($p_{i,j}$) to describe the flow of entities through the network. Discuss the conditions for stability and provide examples of systems that can be modeled using Jackson networks, such as computer systems or transportation networks.
f. Provide a simple, real-world example of a Jackson network, including its nodes, arrival rates (λ_i), service rates (μ_i), and routing probabilities ($p_{i,j}$) to describe the flow of entities through the network Explain how to calculate the total arrival rates, utilization, and key performance metrics such as average queue length and waiting time for each node in the network.

References

1. J. Scott, Social network analysis. Sociology **22**(1), 109–127 (1988)
2. D. Easley, J. Kleinberg, *Networks, Crowds, and Markets*, vol. 6(1) (Cambridge University Press, 2010), pp. 6–1

3. M. Newman, *Networks: An Introduction* (Oxford University Press, 2010)
4. K. Kunen, *Set Theory an Introduction to Independence Proofs*, vol. 102 (Elsevier, 2014)
5. Y.N. Moschovakis, *Descriptive Set Theory* (American Mathematical Society, 2009)
6. P.R. Halmos, *Naive Set Theory. Dover Books on Mathematics* (Dover Publications, 2017)
7. R. Diestel, *Graph Theory*. Electronic Library of Mathematics (Springer, 2006)
8. B. Bollobas, *Graph Theory: An Introductory Course*, vol. 63 (Springer, 2012)
9. C. Godsil, G.F. Royle, *Algebraic Graph Theory*, vol. 207 (Springer, 2013)

Index

Symbols

Z. Lotker, *History by Algorithms*,
https://doi.org/10.1007/978-3-031-93627-2

The manufacturer's authorised representative in the EU is Springer Nature Customer Service Centre GmbH, Europaplatz 3, 69115 Heidelberg, Germany. If you have any concerns regarding our products, please contact ProductSafety@springernature.com

Printed and bound by CPI Group (UK) Ltd, Croydon, CR0 4YY
30/09/2026
02225590-0012